THE EFFECTS OF RELATIVITY IN
ATOMS, MOLECULES, AND THE SOLID STATE

THE EFFECTS OF RELATIVITY IN
ATOMS, MOLECULES, AND THE SOLID STATE

Edited by
S. Wilson
Rutherford Appleton Laboratory
Oxfordshire, United Kingdom

I. P. Grant
University of Oxford
Oxford, United Kingdom

and

B. L. Gyorffy
University of Bristol
Bristol, United Kingdom

PLENUM PRESS • NEW YORK AND LONDON

Library of Congress Cataloging-in-Publication Data

The Effects of relativity in atoms, molecules, and the solid state /
  edited by S. Wilson, I.P. Grant, and B.L. Gyorffy.
      p.   cm.
  "Proceedings of a meeting on the effects of relativity in atoms,
molecules, and the solid state, held March 30-April 1, 1990, in
Abingdon, Oxfordshire, United Kingdom"--T.p. verso.
  Includes bibliographical references and index.
  ISBN 0-306-43888-7
  1. Atomic structure--Congresses.   2. Molecular structure-
-Congresses.   3. Solid state physics--Congresses.   4. Solid state
chemistry--Congresses.   5. Relativity (Physics)--Congresses.
  I. Wilson, S. (Stephen), 1950-   . II. Grant, I. P. III. Gyorffy,
B. L.
  QC172.E34 1991
  539'.1--dc20   91-11449
                                                                 CIP

Proceedings of a meeting on the Effects of Relativity in Atoms,
Molecules, and the Solid State, held March 30–April 1, 1990,
in Abingdon, Oxfordshire, United Kingdom

ISBN 0-306-43888-7

© 1991 Plenum Press, New York
A Division of Plenum Publishing Corporation
233 Spring Street, New York, N.Y. 10013

All rights reserved

No part of this book may be reproduced, stored in a retrieval system, or transmitted
in any form or by any means, electronic, mechanical, photocopying, microfilming,
recording, or otherwise, without written permission from the Publisher

Printed in the United States of America

Preface

Recent years have seen a growing interest in the effects of relativity in atoms, molecules and solids. On the one hand, this can be seen as result of the growing awareness of the importance of relativity in describing the properties of heavy atoms and systems containing them. This has been fueled by the inadequacy of physical models which either neglect relativity or which treat it as a small perturbation. On the other hand, it is dependent upon the technological developments which have resulted in computers powerful enough to make calculations on heavy atoms and on systems containing heavy atoms meaningful. Vector processing and, more recently, parallel processing techniques are playing an increasingly vital role in rendering the algorithms which arise in relativistic studies tractable.

This has been exemplified in atomic structure theory, where the dominant role of the central nuclear charge simplifies the problem enough to permit some prediction to be made with high precision, especially for the highly ionized atoms of importance in plasma physics and in laser confinement studies. Today's sophisticated physical models of the atom derived from quantum electrodynamics would be intractable without recourse to modern computational machinery. Relativistic atomic structure calculations have a history dating from the early attempts of Swirles in the mid 1930's but continue to provide one of the primary test beds of modern theoretical physics.

In non-relativistic molecular structure calculations the use of basis set expansion techniques is almost mandatory. Although it has been known since the work of Kim in the late 1960's that there are problems associated with the use of basis set expansion techniques in relativistic electronic structure calculations, these difficulties have only recently been resolved. By affording a firm foundation for the *ab initio* treatment of molecules containing heavy atoms and the interactions between them these developments open up new horizons for modern theoretical chemistry. Relativistic quantum chemistry has a vast range of potential applications in both the academic and the commercial sectors in fields as diverse as catalysis and pharmacology, molecular electronics and molecular biology. However, relativistic electronic structure calculations for molecules will remain demanding for some time to come both in the requirement for more efficient and reliable numerical methods and algorithms, and for the computer resources which these methods require.

Relativistic solid-state calculations will be just as demanding. In solid-state theory the emphasis has until now been more on deriving physically intuitive approximation schemes which seem tractable. The consensus which has enabled atomic and molecular physicists and theoretical chemists to build general purpose models has, so far, not existed amongst solid-state theorists.

This volume records the proceedings of a meeting on "The Effects of Relativity in Atoms, Molecules and the Solid State" held under the auspices of three of the Science and Engineering Research Council's Collaborative Computational Projects : Project 1 -

Electron correlation in molecules, Project 2 - *Continuum states of atoms and molecules*, and Project 9 - *The electronic structure of solids*. The meeting was held at The Coseners House, Abingdon, Oxfordshire, over the weekend 30th March - 1st April, 1990.

July 1990

S. Wilson
I.P. Grant
B.L. Gyorffy

Contents

Relativistic Effects on Periodic Trends   1
    P. Pyykkö

ATOMS

Relativistic Atomic Structure and Electron-Atom Collisions   17
    I.P. Grant

On the Accuracy of Oscillator Strengths   45
    B.C. Fawcett

Atomic Structure Calculations in Breit-Pauli Approximation   55
    W. Eissner

Relativistic Calculations of Parity Non-Conserving Effects in Atoms   67
    A.C. Hartley and P.G.H. Sandars

High Precision Relativistic Atomic Structure Calculations Using the Finite Basis Set Approximation   83
    H.M. Quiney

Relativistic Calculations of Electron Impact Ionisation Cross-Sections of Highly Charged Ions   125
    D.L. Moores

MOLECULES

Nonsingular Relativistic Perturbation Theory and Relativistic Changes of Molecular Structure   135
    W.H.E. Schwarz, A. Rutkowski and G. Collignon

Basis Set Expansion Dirac-Fock SCF Calculations and MBPT Refinement   149
    Y. Ishikawa

Comments   163

Polyatomic Molecular Dirac-Hartree-Fock Calculations with Gaussian Basis Sets   167
    K.G. Dyall, K. Fægri, Jr. and P.R. Taylor

Open Shell Relativistic Molecular Dirac-Hartree-Fock
    SCF-Program   185
    O. Visser, P.J.C. Aerts and L. Visscher

General Contraction in Four-Component Relativistic
    Hartree-Fock Calculations   197
    L. Visscher, P.J.C. Aerts and O. Visser

Accurate Relativistic Dirac-Fock and MBPT Calculations
    on Argon with Basis Sets of Contracted
    Gaussian Functions   207
    Y. Ishikawa and R.C. Binning, Jr.

Comments   215

Relativistic Many-Body Perturbation Theory of
    Atomic and Molecular Electronic Structure   217
    S. Wilson

SOLID STATE

Relativistic Density-Functional Theory for
    Electrons in Solids   255
    B.L. Gyorffy, J.B. Staunton, H. Ebert, P. Strange
    and B. Ginatempo

Influence of Relativistic Effects on the Magnetic
    Moments and Hyperfine Fields of 5d-Impurity
    Atoms Dissolved in Ferromagnetic Fe   275
    H. Ebert, B. Drittler, R. Zeller and P.H. Dederichs

Relativistic Spin-Polarized Density-Functional Theory:
    Simplified Method for Fully Relativistic
    Calculations   285
    P. Cortona

Theory of Magnetocrystalline Anisotropy   295
    J. Staunton, P. Strange, B.L. Gyorffy, M. Matsumoto, J. Poulter,
    H. Ebert and N.P. Archibald

The Spin Polarized Photoemission from
    Non-Magnetic Metals   319
    B. Ginatempo and B.L. Gyorffy

Theory of Magnetic X-Ray Dichroism   333
    H. Ebert, B. Drittler, P. Strange, R. Zeller and B.L. Gyorffy

Participants   349

Index   351

RELATIVISTIC EFFECTS ON PERIODIC TRENDS

Pekka Pyykkö

Department of Chemistry, University of Helsinki
Et. Hesperiankatu 4, 00100 Helsinki, Finland

1. INTRODUCTION

Whether we work on atoms, molecules or solids, we have in common the Periodic System. At least seven different trends can be found in it, see Figure 1. The last three of them are of relativistic origin. Some of their particular consequences on chemical and physical properties of elements are shown in Figure 2. A few recent examples are quoted below.

2. A REVIEW OF REVIEWS

Grant (1970) reviewed the relativistic calculations on atoms and Pyykkö (1978) the ones on molecules, including a chapter on "Relativity and the Periodic Table". These effects were brought to the attention of many chemists by Pitzer (1979) or Pyykkö and Desclaux (1979), and have now entered the inorganic chemistry textbooks by Cotton and Wilkinson (1988) or Mackay and Mackay (1989).
    The book by Pyykkö (1986) includes 3119 references (and seems to miss about 1% of the relevant material). Some later reviews are mentioned in Table 1.

3. SOME RECENT RESULTS IN CHEMISTRY

The valencies of gold. It was suggested already by Pyykkö (1978) that the relativistic destabilization of the 5d shell would explain the well-known valency increase from Ag(I) to Au(III). Schwerdtfeger (1989) has now compared the calculated stabilities of the (free) halides AuX_2^- and AuX_4^-, and explicitly demonstrated that the preference for the higher oxidation state, III instead of I, indeed is of relativistic origin.

The "gold maximum". The relativistic contraction and stabilization of the ns valence shell (n = 4 to 6) suffers a local maximum at the coinage metals, group 11. Schwarz et al. (1989) analysed the origins of this trend.

The Effects of Relativity on Atoms, Molecules, and the Solid State
Edited by S. Wilson et al., Plenum Press, New York, 1991

"Relativistic compounds". A number of novel organometallic compounds have recently been synthesized and characterized, in which a weak attraction, roughly of 10 kcal/mole or half of an eV, much like in a good hydrogen bond, seems to exist between formally closed-shell Au(I) atoms having a $5d^{10}$ configuration. More generally, bonds between $5d^{10}$, $6s^2$ or the configuration $5d^8$ are found. Without explicit proof, these bonds are presumed to have strong relativistic contributions. In any case, this "relativistic encouragement" has been a source of inspiration for the syntheticians, as evidenced by some of the titles of Balch et al. (1987), Nagle et al. (1988), Raptis et al. (1989), Scherbaum et al. (1988), and Wang et al. (1988).

A particularly striking example is the octahedral, six-coordinate carbon complex $\{(Ph_3PAu)_6C\}^{2+}$ of Scherbaum et al. (1988). The carbon electron octet could then form, in O_h symmetry, an $(a_1)^2(t_{1u})^6$ "8-electron-7-centre bond". In addition, the Au-Au attractions would help. The only available calculation on it is that by Rösch et al. (1989). Both the Au-Au and the C-Au interactions are found to contribute to the bonding.

Perturbative Hartree-Fock-Slater studies on the bonding in the $5d^8$-$6s^2$ molecule $Tl_2Pt(CN)_4$ by Ziegler et al. (1989a) suggest that the interaction between thallium and the tetracyanoplatinate has both strong ionic and covalent components.

Au_2^{2+}. A clean example on the $5d^{10}$-$5d^{10}$ attraction would be the gas-phase Au_2^{2+}, claimed by Saunders (1989). The theoretical situation is summarized in Table 2. The pseudopotential HF or CI calculations and the semiempirical tight-binding calculations give a repulsive ground state. The non-relativistic DVM calculations by Li et al. (1990) are seen to considerably overbind for Au_2^+ and thus lack credibility. Therefore the observation, if true, must come from an excited electronic state (Mukherjee et al. 1990). Both their TB calculations and the present PP-HF ones indeed find such a minimum in the states promoting one $5d\sigma^*$ antibonding electron to the $6s\sigma$ bonding MO.

The ground-state PP-CI curve of Ermler (private communication) can be divided to three domains:
a) $R > 6$ a.u.: A large-R domain with superposed $+1/R$ repulsion and an $-\alpha/R^4$ charge-polarisability-type attraction. The coefficient of the latter gives an Au^+ polarisability, α, of about 14.2 a.u.
b) About $5 < R < 6$: Some extra bonding (about 0.065 eV) but no minimum.
c) $R < 5$: Extra, Pauli repulsion.

Trends along triads. The bond strength trend Cu > Ag < Au was explained quite early as a relativistic effect (see Ziegler et al. (1989b) and, for later literature, Balasubramanian (1989a).) A similar trend is seen in various organometallic compounds of groups 6 (Cr > Mo < W), 8 (Fe > Ru < Os), 9 (Co > Rh < Ir), 10 (Ni > Pd < Pt), and the strengthening has also been shown to be a relativistic effect (Ziegler et al. 1989b). For the same trend among the group 6 dimers, see Ziegler (1987). A recent experimental example on this D_e trend are the diatomic molecules CuIn > AgIn < AuIn (Balducci et al. 1989).

Trends in Periodic System(s)

1. <u>Main vertical trend</u>: First shell of each l anomalously small (1s, 2p, 3d, 4f). Due to "primogenic repulsion", larger $\langle r \rangle$ for larger n.
2. <u>Main horizontal trend</u>: Smaller $\langle r \rangle$ for larger Z.
3. <u>Main periodicity</u>: Filled shells particularly stable. Half-filled non-relativistic shells also.
4. Partial screening effects: d-shell contraction, lanthanoid contraction.

5. Relativistic contraction and stabilization (s, p).
6. Relativistic expansion and destabilization (d, f)
7. Spin-orbit splitting.

No deep group-theoretical principle.

(C) Pekka Pyykkö. March 21, 1990.

Fig. 1.

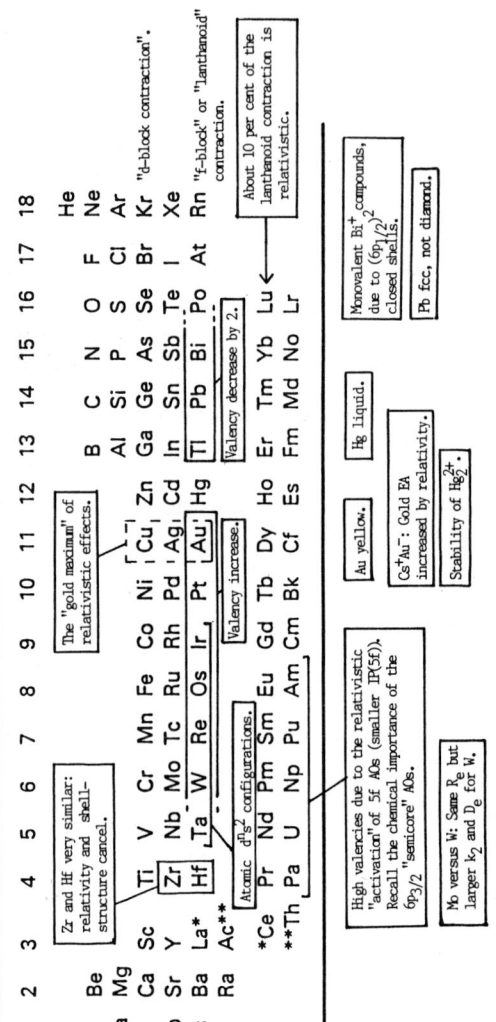

Fig. 2. Chemical consequences of relativistic effects. Reproduced with permission from Pyykkö (1988a).

Table 1. A review of reviews.

Year	Author(s)	Subject
1986	Brooks et al.	Solids.
	Pyykkö	Bibliography.
1987	Balasubramanian and Pitzer	Pseudopotential work on molecules.
	Lindgren (ed.)	Trieste Proceedings (atoms).
1988	Drake and Goldman	Finite basis sets for atoms.
	Ermler et al.	Pseudopotential work on mol.
	Grant and Quiney	Atoms.
	Johnson	Relativistic RPA.
	Langhoff and Bauschlicher	Transition metal systems.
	Malli	AM.
	Pyykkö	Structural chemistry.
	Wilson (ed.)	AM papers by Grant, Gropen, Pyykkö, Quiney, Wilson.
1989	Balasubramanian	Chemical bonding.
	Balasubramanian	p-block diatomics.
	Balasubramanian	Group 11 clusters.
	Crasemann	Atoms.
	Johnson et al.	Santa Barbara Proceedings: Fundamental theory, few-electron atoms.
	Lindgren	MBPT.
	Quiney et al.	MBPT.
	Salahub and Zerner (ed.)	Toronto trans. metal Proc.: Christiansen, Hay, Malli, Morokuma, Ziegler.
1990	Balasubramanian	p-block dimers, trimers.
	Schwarz	Chemical bonding.
	Wilson (ed.)	Abingdon Proceedings.

Table 2. Bond lengths and dissociation energies of Au_2^{n+} (n=0-2); "nb" stands for "not bound". For n=2 the barrier height is given as D_e.

Species	R/pm			D_e/eV			Method
	NR	R	Exp.	NR	R	Exp.	
Au_2	285	254	247.2	1.48	1.85	2.30	CEPA-1[a]
					2.2		PP-CI[b]
		256			1.87		PP-MC[c]
		267					PP-HF[d]
		220			2.28		TB[e]
	260			2.4			DVM[f]
Au_2^+	312	270	–	1.57	1.75		CEPA-1[a]
				–	1.65	1.661(4)	PP-CI[b]
		284					PP-HF[d]
		224			1.93		TB[e]
	270			2.9			DVM[f]
Au_2^{2+} (X $^1\Sigma_g$)		nb			nb		PP-CI[g]
		nb			nb		PP-HF[d]
		nb			nb		TB[e]
	260			0.8			DVM[d]
Au_2^{2+} ($^3\Sigma_u$)		264	–				PP-HF[d]
		236			0.3		TB[e]
Au_2^{2+} ($^1\Sigma_u$)		272	–				PP-HF[d]
		236			0.3		TB[e]

[a] Schwerdtfeger et al. (1989b). The NR values are deduced from their Table II. For earlier R/NR comparisons, see Table IV of Pyykkö (1988a). A typical D_e increase is almost a factor of two.
[b] Balasubramanian and Feng (1989).
[c] Bauschlicher et al. (1989).
[d] Present work using the HF/LANL1DZ option of Gaussian 88.
[e] Mukherjee et al. (1990). Δ = 2.18 eV.
[f] Li et al. (1990).
[g] W.C. Ermler, private communication.

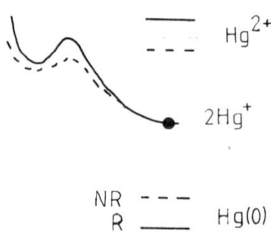

Fig. 3. A schematic representation of relativistic effects on Hg_2^{2+} and its reference states Hg(0), Hg^+ and Hg^{2+}.

For bond lengths it is rather more rare that the 5d element should have a shorter bond than the corresponding 4d element. A notable exception are the hydrides and fluorides of groups 8-12, see Figure 16 of Pyykkö (1988a). New examples of the same type (4d>5d) are the intermetallic diatomics NiPd/NiPt, CuAg/CuAu and AgAg/AgAu (Taylor et al. 1990). Further group 11 diatomic examples are considered by Schwerdtfeger et al. (1989b).

As to the metals, Takeuchi et al. (1989) demonstrate that the cohesive energy trend Ag<Au and the closely similar lattice parameters come from relativistic effects; without them metallic gold would have a clearly larger a_0 and a smaller cohesive energy than silver.

Schwerdtfeger et al. (1989a) discuss the question, why is thallium preferentially Tl(I) in inorganic but Tl(III) in organometallic chemistry. Recall that the lighter elements B-In are preferentially trivalent.

Superheavy elements. Historically, this is the area where relativistic effects were originally thought to have their main relevance in chemistry. Kratz et al. (1989) have now studied the aqueous chemistry of Element 105 and find it to rather resemble that of Nb and Pa, instead of that of Ta. No calculations exist yet.

As to the atomic ground states, Glebov et al. (1989) calculate for Element 104 a $7s^2 6d^1 7p^1$ one, with the $s^2 d^2$, analogous to the lighter homologues Ti-Th, lying 0.5 eV higher. The chemical properties of element 104 are nevertheless predicted to be those of a typical d-element.

Actinoids. For actinoids, Ziegler et al. (1989c) find that the An-X (X=F-I) bonds show only small relativistic contractions but the An-H and An-C(σ) bond energies still are greatly strengthened by relativistic effects.

Liquids. Not much is known about relativistic effects on the electronic and geometric structure or the thermodynamic properties of liquids, such as melted lead. For a recent example, see Jank and Hafner (1989).

The importance of the reference state. Strömberg and Wahlgren (1990) considered relativistic effects on the stability of the mercurous ion, Hg_2^{2+}, on a relatively low 1st-order PT quasirelativistic level and concluded that, relative to the 2 Hg^+ dissociation limit, the relativistically shortened molecule may actually lie higher, i.e. be relativistically destabilized. Near the minimum of the well the kinetic-energy stabilization would still operate, as found by Ziegler et al in 1981 or by Neisler and Pitzer (1987). A further calculation, also yielding a bound well for the free mercurous ion, was published by Durand et al. (1987).

Inspired by this provocative statement, the sketch in Fig. 3 then suggests that, while the mercurous ion is certainly destabilized with respect to the 2 Hg thermodynamic ground state, it is as certainly stabilized with respect to the 2 Hg^{2+} reference state, at least in the gas phase. This would explain why +I is a common oxidation state for Hg, compared to +II.

Trends in one-electron levels. The valence levels of M_2O (M=Al-Tl) show an interesting trend: the metal ns levels go down and the oxygen 2p levels up along the series; the crossing occurs at In. This leads to rather different photoelectron spectra for these molecules (Ruščić et al. 1986).

Another particular example is the influence of the relativistic stabilization of the Hf 6s AO on the energy of the a_1 MOs of the tetrahydroborate (Hohl and Rösch 1986). This trend is quite common.

Electron affinities. Miyoshi et al. (1988) demonstrate explicitly that the (quasi)relativistic contributions diminish the EA of WF_6 by 1.1 eV. It would be tempting to attribute this straightforwardly to the relativistic destabilization of the 5d LUMO of the neutral species.

Colours. Relativistic effects are supposed to strongly influence a number of observable colours, see ch. IV.E of Pyykkö (1988a). The best documented and most conspicuous case is the yellow colour of gold. Preliminary Multiple-Scattering work by El-Issa and Zanati (unpublished) suggests, that the colour difference between the red MoS_4^{2-} and the yellow WS_4^{2-} would mainly come from shell-structure effects. The relativistic destabilization of the LUMO, which is only partially 5d for W, would not be, after all, the main reason.

In the case of pentaphenyl bismuth, Schmuck et al. (1990) suggested, on the basis of a REX/EHT comparison, that the violet colour is caused by the relativistic stabilization of the Lowest Unoccupied Molecular Orbital, which now is an a_1, containing substantial Bi 6s character. This assumption does find support in the MS calculations. Furthermore, all non-relativistic transitions are symmetry-forbidden and only the relativistic one, perpendicular to the C_{4v} axis is allowed, in agreement with experiment (El-Issa, Pyykkö and Zanati, unpublished). For D_{3h} symmetry even all relativistic transitions are forbidden.

Mulliken (1928,1989) pointed out that the red atmospheric oxygen bands come from the $X\ ^3\Sigma_g - b\ ^1\Sigma_g$ transition of O_2. Being spin-forbidden, this transition derives it entire intensity from relativistic effects. The same is true for all phosphorescent effects and, incidentally, for the red line of Aurora Borealis (^1D to ^3P of neutral atomic oxygen at 630.0 nm).

The violet colour of iodine arises from a $5p\pi^* - 5p\sigma^*$ HOMO-LUMO transition, whose 0-0 line lies at 641 nm in the red. As the relativistic stabilization of the σ LUMO is some 1.6 eV (Yang 1976), strong relativistic effects on this transition energy can be suspected.

A further example is the yellow colour of hexachloroplumbate(IV). The corresponding stannate is white (El-Issa, Pyykkö and Zanati, to be published).

References

K. Balasubramanian, 1989a, Relativity and chemical bonding, J. Phys. Chem. 93, 6585 - 6596.

K. Balasubramanian, 1989b, Spectroscopic properties and potential energy curves for heavy p-block diatomic hydrides, halides, and chalconides, Chem. Rev. $\underline{89}$, 1801 - 1840.

K. Balasubramanian, 1989c, Electronic structure of coinage metal clusters, J. Mol.Str. THEOCHEM $\underline{202}$, 291-313.

K. Balasubramanian, 1990, Spectroscopic constants and potential energy curves of heavy p-block dimers and trimers, Chem. Rev. $\underline{90}$, 93-167.

K. Balasubramanian and P.-Y. Feng, 1989, The ionization potentials of Ag_n and Au_n and binding energies of Ag_n, Au_n, Ag_n^+ and Au_n^+ (n=1-4), Chem. Phys. Lett. $\underline{159}$, 452-458.

K. Balasubramanian and K.S. Pitzer, 1987, Relativistic quantum Chemistry, in K.P. Lawley (ed.) "Ab Initio Methods in Quantum Chemistry I", Wiley, New York, pp. 287-319.

A.L. Balch, J.K. Nagle, M.M. Olmstead and P.E. Reedy, Jr., 1987, Formation of luminescent, bent Ir-Tl-Ir and Ir-Pb-Ir chains through the binding of thallium(I) and lead(II) to the iridium metallomacrocycle $Ir_2(CO)_2Cl_2$ $\{\mu-(Ph_2PCH_2)_2AsPh\}_2$, J. Am. Chem. Soc. $\underline{109}$, 4123-4124.

G. Balducci, P.E. DiNunzio, G. Gigli and M. Guido, 1989, Dissociation energies of the intermetallic molecules CuIn, AgIn and AuIn, J. Chem. Phys. $\underline{90}$, 406-412.

C.W. Bauschlicher Jr., S.R. Langhoff and H. Partridge, 1989, Theoretical study of the structures and electron affinities of the dimers and trimers of group IB metals (Cu, Ag, and Au), J. Chem. Phys. $\underline{91}$, 2412-2419.

M.S.S. Brooks, B. Johansson, O. Eriksson and H.L. Skriver, 1986, Relativistic effects in heavy elements, Physica $\underline{144B}$, 1-13.

P.A. Christiansen, 1989, Relativistic effective potentials in quantum Monte Carlo studies, in D.R. Salahub and M.C. Zerner (ed.), "The Challenge of d and f Electrons", ACS Symposium Series No. 394, pp. 309-321.

F.A. Cotton and G. Wilkinson, 1988, "Advanced Inorganic Chemistry", 5th Ed., Wiley, New York. See pp. 209, 776-777, 937, 956, 987.

B. Crasemann, 1989, Relativistic phenomena in atomic and chemical physics: Opportunities for studies with synchrotron radiation, Acta Phys. Hung. $\underline{65}$, 171-181.

G.W.F. Drake and S.P. Goldman, 1988, Relativistic Sturmian and finite basis set methods in atomic physics, Adv. At. Mol. Phys. $\underline{25}$, 393-416.

G. Durand, F. Spiegelmann and A. Bernier, 1987, Electronic structure of doubly charged dimers Mg_2^{2+} and Hg_2^{2+}, J. Phys. B: At. Mol. Phys. $\underline{20}$, 1161-1174.

W.C. Ermler, R.B. Ross and P.A. Christiansen, 1988, Spin-orbit coupling and other relativistic effects in atoms and molecules, Adv. Quantum Chem. 19, 139-182.

V.A. Glebov, L. Kasztura, V.S. Nefedov and B.L. Zhuikov, 1989, Is Element 104 (Kurchatovium) a p-element? II. Relativistic calculation of the electronic atomic structure, Radiochim. Acta 46, 117-121.

I.P. Grant, 1970, Relativistic calculation of atomic structures, Adv. Phys. 19, 747-811.

I.P. Grant and H.M. Quiney, 1988, Foundations of the relativistic theory of atomic and molecular structure, Adv. At. Mol. Phys. 23, 37-86.

D. Hohl and N. Rösch, 1986, On the electronic structure of metal tetrahydroborates: Quasi-relativistic X-alpha SW study of $M(BH_4)_4$ (M = Zr, Hf, Th, U), Inorg. Chem. 25, 2711-2718.

W. Jank and J. Hafner, 1990, Structural and electronic properties of the liquid polyvalent elements: The group-IV elements Si, Ge, Sn, and Pb, Phys. Rev. B 41, 1497-1515.

W.R. Johnson, 1988, Relativistic random-phase approximation, Adv. At. Mol. Phys. 25, 375-391.

W. Johnson, P. Mohr and J. Sucher (ed.), 1989, "Relativistic, Quantum Electrodynamic and Weak Interaction Effects in Atoms", AIP Conf. Proc. 189, AIP, New York, 513 p.

J.V. Kratz, H.P. Zimmermann, U.W. Scherer, M. Schädel, W. Brüchle, K.E. Gregorich, C.M. Gannett, H.L. Hall, R.A. Henderson, D.M. Lee, J.D. Leyba, M.J. Nurmia, D.C. Hoffman, H. Gäggeler, D. Jost, U. Baltensperger, N.-Q. Ya, A. Türler and Ch. Lienert, 1989, Chemical properties of Element 105 in aqueous solution: Halide complex formation and anion exchange into triisoctyl amine, Radiochim. Acta 48, 121-133.

S.R. Langhoff and C.W. Bauschlicher Jr., 1988, Ab initio studies of transition metal systems, Ann. Rev. Phys. Chem. 39, 181-212.

Y. Li, S.N. Khanna and P. Jena, 1990, Comment on 'Charge exchange and metastability of small multiply charged gold clusters', Phys. Rev. Lett. 64, 1188.

I. Lindgren (ed.), 1987, "Relativistic Many-Body Problems", Proc. Adriatico Conf. 1986, Phys. Scripta RS 7, 228 p.

I. Lindgren, 1989, Relativistic and non-relativistic many-body procedure, applied to atomic systems, in D. Mukherjee (ed.), "Aspects of Many-Body Effects in Molecules and Extended Systems", Lecture Notes in Chem. 50, Springer, Berlin, pp. 367-392.

K.M. Mackay and R.A. Mackay, 1989, "Introduction to Modern Inorganic Chemistry", 4th Ed., Blackie, Glasgow and Prentice Hall, Englewood Cliffs, NJ. Ch. 16.10.: Relativistic effects.

G.L. Malli, 1988, Ab-initio relativistic quantum chemistry, in J. Maruani (ed.), Molecules in Physics, Chemistry and Biology, Vol. 2, Kluwer, Dordrecht, pp. 85-144.

G.L. Malli, 1989, Ab initio relativistic quantum chemistry of third-row transition elements and actinides, in D.R. Salahub and M.C. Zerner (ed.), "The Challenge of d and f electrons", ACS Symposium Series No. 394, pp. 291-308.

E. Miyoshi, Y. Sakai, A. Murakami, H. Iwaki, H. Terashima, T. Shoda and T. Kawaguchi, 1988, On the electron affinities of hexafluorides CrF_6, MoF_6, and WF_6, J. Chem. Phys. 89, 4193-4198.

S. Mukherjee, G. Pastor and K.H. Bennemann, 1990, Theoretical study of the metastability of Au_2^{++} clusters, Phys. Rev. B (submitted).

R.S. Mulliken, 1928, Interpretation of the atmospheric bands of oxygen, Phys. Rev. 32, 880-887.

R.S. Mulliken, 1989, "Life of a Scientist", Springer, Berlin, Ch. XI.

J.K. Nagle, A.L. Balch and M.M. Olmstead, $Tl_2Pt(CN)_4$: A non-columnar, luminescent form of $Pt(CN)_4^{2-}$ containing Pt-Tl bonds, J. Am. Chem. Soc. 110, 319-321.

R.P. Neisler and K.S. Pitzer, 1987, The dipositive dimeric ion Hg_2^{2+}: A theoretical study, J. Phys. Chem. 91, 1084-1087.

K.S. Pitzer, 1979, Relativistic effects on chemical properties, Acc. Chem. Res. 12, 271-276.

P. Pyykkö, 1978, Relativistic quantum chemistry, Adv. Quantum Chem. 11, 353-409.

P. Pyykkö, 1986, "Relativistic Theory of Atoms and Molecules. A Bibliography 1916-1985", Lecture Notes in Chem. 41, Springer, Berlin, 389 p.

P. Pyykkö, 1988a, Relativistic effects in structural chemistry, Chem. Rev. 88, 563-594.

P. Pyykkö, 1988b, Semiempirical relativistic molecular structure calculations, in Wilson (1988), pp. 137 - 226.

P. Pyykkö and J.P. Desclaux, 1979, Relativity and the periodic system of elements, Acc. Chem. Res. 12, 276-281.

H.M. Quiney, I.P. Grant, and S. Wilson, 1989, On the relativistic many-body perturbation theory of atomic and molecular electronic structure, in U. Kaldor (ed.), "Many-Body Methods in Quantum Chemistry", Lecture Notes in Chem. 52, Springer, Berlin, pp.307-344.

R.G. Raptis, J.P. Fackler, Jr., H.H. Murray and L.C. Porter, 1989, Structural features suggesting relativistic effects in a dimeric gold complex, Inorg. Chem. 28, 4057-4059.

N. Rösch, A. Görling, D.E. Ellis and H. Schmidbaur, 1989, Aurophilie als konzertierter Effekt: Relativistische MO-Berechnungen für Kohlenstoff-zentrierte Goldcluster, Angew. Chem. 101, 1410-1412.

B. Ruščić, G.L. Goodman and J. Berkowitz, 1986, Photoelectron spectra of Ga_2O, In_2O and Tl_2O, J. El. Sp. Rel. Phen. 41, 357-384.

D.R. Salahub and M.C. Zerner (ed.), 1989, "The Challenge of d and f Electrons", ACS Symp. Ser. 394, ACS, Washington DC, 405 p.

W.A. Saunders, 1989, Charge exchange and metastability of small multiply charged gold clusters, Phys. Rev. Lett. 62, 1037-1040.

F. Scherbaum, A. Grohmann, B. Huber, C. Krüger and H. Schmidbaur, 1988, "Aurophilie" als Konsequenz relativistischer Effekte: Das Hexakis(triphenylphosphanaurio)methan-Dikation $\{(Ph_3PAu)_6C\}^{2+}$, Angew. Chem. 100, 1602.

A. Schmuck, P. Pyykkö and K. Seppelt, 1990, Struktur und Farbe von substituiertem Pentaphenylbismut, Angew. Chem. 102, 211-213.

W.H.E. Schwarz, 1990, in Z. Maksić (ed.), "Theoretical Models of Chemical Bonding", Springer, Berlin (in press).

W.H.E. Schwarz, E.M. van Wezenbeek, E.J. Baerends and J.G. Snijders, 1989, The origin of relativistic effects of atomic orbitals, J. Phys. B: At. Mol. Opt. Phys. 22, 1515-1530.

P. Schwerdtfeger, 1989, Relativistic effects in gold chemistry. 2. The stability of complex halides of gold(III), J. Am. Chem. Soc. 111, 7261-7262.

P. Schwerdtfeger, P.D.W. Boyd, G.A. Bowmaker, H.G. Mack and H. Oberhammer, 1989a, Theoretical studies on the stability of Tl-C σ bonds in aliphatic organothallium compounds, J. Am. Chem. Soc. 111, 15-23.

P. Schwerdtfeger, M. Dolg, W.H.E. Schwarz, G.A. Bowmaker and P.D.W. Boyd, 1989b, Relativistic effects in gold chemistry. I. Diatomic gold compounds, J. Chem. Phys. 91, 1762-1774.

D. Strömberg and U. Wahlgren, 1990, First-order relativistic calculations on Au_2 and Hg_2^{2+}, Chem. Phys. Lett.(in press).

N. Takeuchi, C.T. Chan and K.M. Ho, 1989, First-principles calculations of equilibrium ground-state properties of Au and Ag, Phys. Rev. B 40, 1565-1570.

S. Taylor, E.M. Spain and M.D. Morse, 1990, Spectroscopy and electronic structure of jet-cooled NiPd and PdPt, J. Chem. Phys. 92, 2710-2720.

S.-N. Wang, J.P. Fackler, Jr., C. King and J.C. Wang, 1988, Luminescent organometallic compounds with relativistic metal-

metal bonds. Synthesis and characterization of AuTl{Ph$_2$P(CH$_2$)S}$_2$, a one-dimensional metal-metal bonded polymer in the solid state, J. Am. Chem. Soc. 110, 3308-3310.

S. Wilson (ed.), 1988, "Methods in Computational Chemistry. 2. Relativistic Effects in Atoms and Molecules", Plenum, New York, 291 p.

C.Y. Yang, 1976, Relativistic X-alpha scattered-wave calculations for C$_2$ and I$_2$, Chem. Phys. Lett. 41, 588-592.

T. Ziegler, 1987, On the relation between relativity and periodic trends within a triad of transition metals, in J. Avery et al. (ed.), "Understanding Molecular Properties", Reidel, Dordrecht, pp. 521-532.

T. Ziegler, J.K. Nagle, J.G. Snijders and E.J. Baerends, 1989a, Theoretical study of the electronic structures and absorption spectra of Pt(CN)$_4^{2-}$ and Tl$_2$Pt(CN)$_4$ based on density functional theory including relativistic effects, J. Am. Chem. Soc. 111, 5631-5635.

T. Ziegler, J.G. Snijders and E.J. Baerends, 1989b, Relativistic effects on compounds containing heavy elements. The influence of kinetic energy on chemical bonds, in Salahub and Zerner (1989), pp. 322-338.

T. Ziegler, V. Tschinke, E.J. Baerends, J.G. Snijders and W. Ravenek, 1989c, Calculation of bond energies in compounds of heavy elements by a quasi-relativistic approach, J. Phys. Chem. 93, 3050-3056.

ATOMS

RELATIVISTIC ATOMIC STRUCTURE AND ELECTRON–ATOM COLLISIONS

I.P. Grant

Department of Theoretical Chemistry
5 South Parks Road
Oxford, OX1 3UB, U.K.

RELATIVISTIC ATOMIC STRUCTURE CODES USING FINITE DIFFERENCE METHODS

Codes for the relativistic calculation of atomic structures and properties have continued to develop in response to the demands of physicists for models with greater physical realism. The MCDF/BENA packages and the modules associated with them (Grant et al. 1980, McKenzie et al. 1980) have been very widely used, and have proved particularly effective in making predictions of energy levels and wave–functions for highly ionized atoms. These wave–functions are the starting point for calculations of many other properties, for example line strengths or collision cross–sections.

The 1980 computer codes allow the user to construct models with quite realistic physics, limited primarily by the power of one's computer and the cost of computation. Efficiency and economy are therefore as important as physical realism in code design. These codes construct the full Dirac–Coulomb Hamiltonian in a basis of open–shell configurational states (CSF) of more or less arbitrary character. The radial wave–functions can be generated either as the solution of some model potential problem or as the solution of some self–consistent field calculation. In either case, the nucleus can be modelled as a point charge or can be given a finite size. The fully retarded transverse photon interaction between electrons in Coulomb gauge can be added as a perturbation to the Hamiltonian matrix in the CSF basis. Polarization of the vacuum due to the nuclear charge can be computed from the Uehling potential, and the self–energy of the electrons is approximated by a screened hydrogenic model.

However, the code has its weaknesses. For example, excited virtual orbitals are often difficult to compute, and there is no module for calculating radiative transitions. The need to remove these and other deficiencies led to a major code revision, resulting in the General–purpose Relativistic Atomic Structure Package, GRASP, (Dyall et al. 1989). Much of the original FORTRAN IV code has been rewritten in FORTRAN 77 to standardize the code and improve its portability, and all known errors have been corrected. There have been many technical improvements: for example, the original package NJSYM (Burke 1970, 1971) for calculating angular recoupling coefficients has been replaced by the much more efficient NJGRAF package (Bar–Shalom and Klapisch 1988), and the TRANSFORM module (Dyall 1986), which enables results to be output in non–standard angular coupling schemes, has been added. The OSCL module permits the calculation of line strengths. Many algorithmic improvements enable the package to solve previously intractable problems, and the data handling has been modified to accommodate the growing trend towards more elaborate CSF descriptions of the wave-function. The code is now supplied in a form which requires the use of a pre-processor module to adapt the code to the target computer and to set dimensions and other basic parameters before compilation. GRASP has now been implemented on a very wide range of computers: enhanced PCs, SUN workstations, VAX computers, CONVEX and CRAY supercomputers amongst others. A User's Manual is provided to give complete specification of the input options available, a description of the program structure, and some help with the set-up and use of the package. A review of the underlying theory of relativistic atomic structure calculations updating that of Grant (1970) with a detailed account of the use of diagrammatic methods of matrix element calculation in a basis of open shell CSFs has been published recently (Grant 1988).

Since the design of GRASP was frozen some 2 years ago, we have made further improvements which are embodied in a completely new version, GRASP^2, (Parpia, Grant and Fischer, to be published). The code has been completely rewritten to be fully compatible with FORTRAN 77 and to make it more compact, efficient and easier to use. More accurate numerical algorithms are employed, and a new iterative scheme adapted from the methods of Fischer (1986) has allowed us to solve many problems on which earlier versions failed. It is now possible to include one electron excitations to orbitals of the same symmetry ($nlj \longrightarrow n'lj$) in orbital wave-function calculation, as well as including the corresponding off-diagonal one-electron matrix elements in the Hamiltonian matrix, which is all that was possible in earlier versions. Other new options are the evaluation of volume isotope shift parameters, and the use of a relativistic local-density approximation for exchange and correlation effects.

Problems treated with GRASP^2 include the following:

High precision calculations of the levels of helium-like systems

We have studied the low-lying states of helium-like systems using the MCDF-OL procedure for $1 \leq Z \leq 36$. The method of calculation, described in a recent paper (Parpia and Grant 1990), uses a systematic enlargement of the CSF basis set, combined with an extrapolation algorithm, to give a limiting value for the relativistic correlation correction to Dirac-Fock energies in the "no-pair" approximation. Dirac-Coulomb energies can be converged to 1 part in 10^7 by this method; the rate of convergence when the transverse electron-electron interaction is added is rather slower, and we have only achieved a precision of 1 part in 10^6. We have not calculated the remaining contributions in this study, of which the radiative and nuclear mass corrections are the most important. The correlation contributions are in excellent agreement with the results of Accad et al (1971) for the lowest members of the sequence.

As far as we know, this is the first time that an MCDF-OL study of correlation energies has been successfully carried out. Previous versions of GRASP (or the older MCDF) were very difficult to use in OL problems, and relativistic correlation studies would have been severely inhibited by convergence failures.

Many-body effects in the 4p spectrum of Sr I

The 4p hole spectrum of Sr I presents a much more complex physical problem. The subvalence inner-shell spectra of the alkaline earths Ca, Sr and Ba show a remarkable increase in complexity as the nuclear charge increases (Baig et al 1983) which cannot be explained within an independent-electron scheme. In the case of the 4p spectrum of Sr I, recent high resolution data by Connerade, Baig and

Sweeney (1990) reveal supernumerary Rydberg series, Figure 1, whose presence is attributable to d-orbital contraction and attendant 's–d' mixing in the parent ion. We have attempted to model this situation with GRASP2 using the MCDF–AL method. For some symmetries, up to 748 interacting levels must be taken into account. We have calculated both energy levels and line strengths for selected transitions, and have been able to confirm a multi-channel quantum defect analysis of the five Rydberg series shown in Figure 1. Our assignments are shown in Table 1. Some progress has also been made in the analysis of low-lying members of the spectrum. A full account of this work will be reported elsewhere (Parpia, Grant and Connerade, to be published).

ELECTRON–ATOM COLLISIONS USING THE RELATIVISTIC R–MATRIX METHOD

Our first calculations using a relativistic generalization of the well-known R–matrix method were published by Norrington and Grant (1981) for Ne II, and further work on Fe XXIII and Fe VII appeared in Norrington and Grant (1987). Progress has been hindered by a number of factors, of which one of the most important was the difficulty of constructing high quality wave-functions for excited states of the target atom or ion using GRASP. This problem has now been completely overcome with GRASP2. The R–matrix method (Burke and Robb 1975) depends upon the division of configuration space into two regions: an inner one centred on the nucleus of the target atom of radius a, large enough to completely contain the target charge distribution, and an external or "asymptotic" region. The full interaction between the scattering electron and the N electrons of the target is treated by solving for the energy levels and wave functions of the $(N+1)$–electron system when it is confined to the region $r \leq a$. The additional orbitals required to describe the excited states of the scattering electron in this region are obtained by solving the Dirac equation using the frozen core potential of the target. Orthogonality of these basis orbitals to the target orbitals of the same symmetry is enforced by Lagrange multipliers. The usual boundary conditions are employed at $r = 0$, and the small component $q_i(r)$ and the large component $p_i(r)$ are related on the boundary $r = a$ by the formula (Norrington and Grant 1981, 1987)

$$2ac\, q_i(a)/p_i(a) = b + \kappa_i$$

where c is the velocity of light (≈ 137 in atomic units), and κ_i is the Dirac angular symmetry quantum number. In the Schrödinger limit $c \to \infty$, we have

$$p_i(a) \to P_i(a) + O(1/c^2), \quad 2ac\, q_i(a) \to aP'_i(a) + \kappa_i P_i(a) + O(1/c^2),$$

so that our choice of boundary condition ensures that the Schrödinger amplitude $P_i(a)$ satisfies the usual R–matrix boundary condition

$$aP'_i(a)/P_i(a) = b$$

in the nonrelativistic limit. The wave function for the $(N+1)$–electron system is assumed to have a CI structure and has components of scattering type as well as bound components that produce resonances in the cross-section. Exchange is neglected for $r > a$, and the scattering electron is assumed to move in the effective potential of the target. It is the matching of the inner and outer trial solutions on the boundary at $r = a$ that allows us to deduce the scattering matrix as a function of the energy of the incident electron.

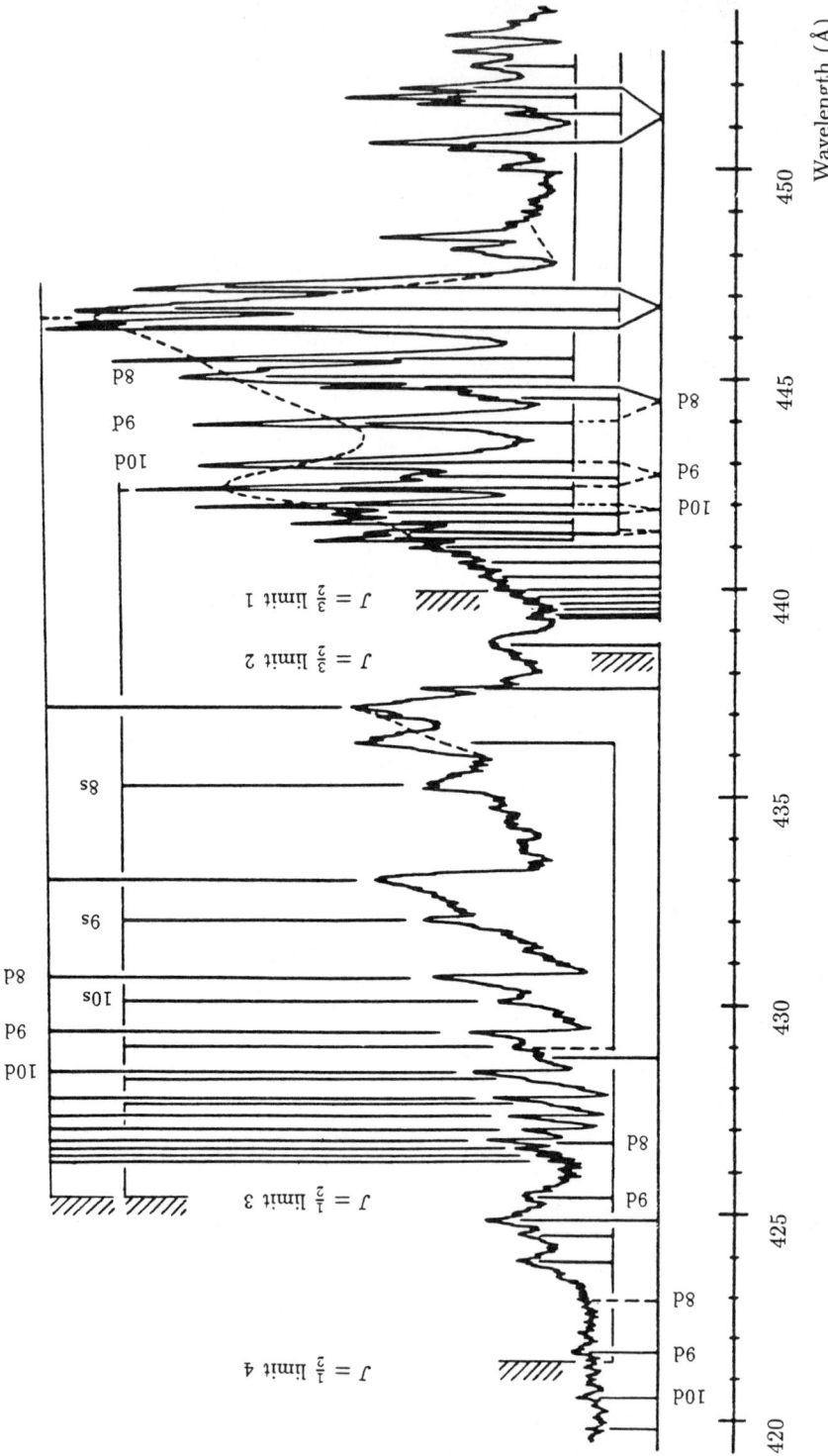

Figure 1. Photometric trace of the absorption spectrum of Sr I between 420 and 450 Å, showing the three upper and two lower limits, and the series converging on them. The most prominent autoionising profiles have been extrapolated down in energy, showing how the interchannel interaction persists as an intensity perturbation below the $^2P_{3/2}$ threshold.

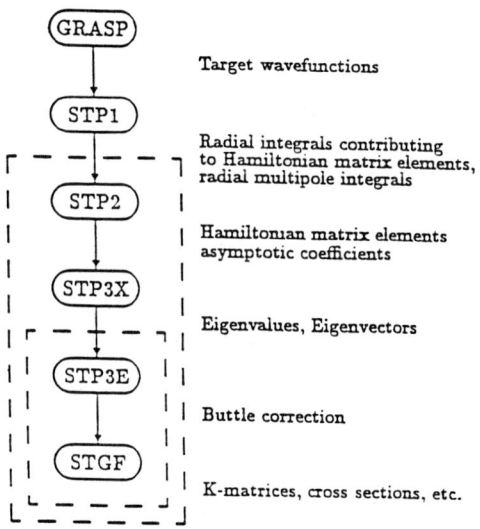

Figure 2. Relativistic R–matrix modules showing the sequence of calculations. The outer loop runs over partial waves defined by quantum numbers J^π. The inner loop runs over energies of the incident electron the $(N+1)$–electron eigenvalues and eigenvectors. As only a finite number of basis orbitals are taken into account in the inner region, it is necessary to correct for the remainder by the method of Buttle (1967); STP3E computes the necessary corrections. Finally, STGF computes the K–matrices, collision strengths and cross–sections required. STP1 is a completely new module; the others are largely taken over or adapted with little change from those used in current versions of the Belfast nonrelativistic R–matrix code (Berrington et al 1978).

Table 1. GRASP[2] predictions of levels of Sr II with appreciable contributions (> 10%) from the configurational states $4p^55s^2$ $J^\pi = 1/2^-, 3/2^-$.

Limit[a]	J^π	Weight of dominant CSF[b]	Intervals (cm^{-1})[c] Calc'd.	Measured
1	$3/2^-$	32.68% [4p]		
			6017	767
2	$3/2^-$	51.62% [4p]		
			7994	7042
3	$1/2^-$	34.37% [4p̄]		
			1365	2196
4	$1/2^-$	20.30% [4p̄]		
			393	2177
5	$1/2^-$	33.33% [4p̄]		

[a]We interpret the levels of Sr⁺ as the limits of the spectral series $4p^65s^2$ $J^\pi = 0^+$ ⟶ $4p^55s^2n\ell$ $J^\pi = 1^-$ resulting from the dipole photo-excitation of an electron from the 4p valence subshell.

[b]The notation indicates the weight of the primary photo-induced subshell vacancy. The computed levels are all very mixed, with a composition consisting of a large number of contributing CSFs of low weight.

[c]Although the computed separations do not agree well with those measured by Connerade et al. (1990), the sequence of limits is in accordance with experiment. The quantitative discrepancies are due to the strongly mixed composition of the eigenstates which makes the results very sensitive to small perturbations. The details will be fully discussed in a paper by Parpia, Grant and Connerade (in preparation). The total energy of the lowest state was computed to be −701 731 067 cm^{-1}.

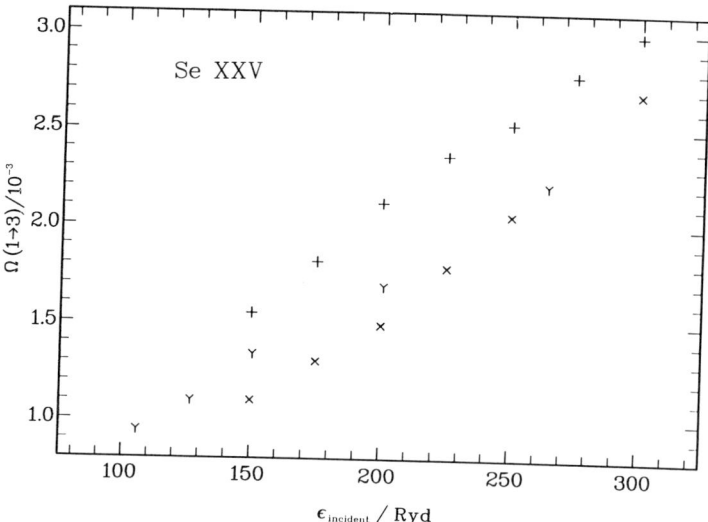

Figure 3. Calculated collisions strengths, Ω, for transition $1-3$ ($2\bar{p}^22p^4\ 0^+ - 2\bar{p}^22p^33s\ 1^-$) in Se XXV. Data points as follows: +, present calculation; Y, Coulomb–Born exchange calculation of Zhang et al., 1987; ×, Breit–Pauli R–matrix calculation of Gupta et al., 1989.

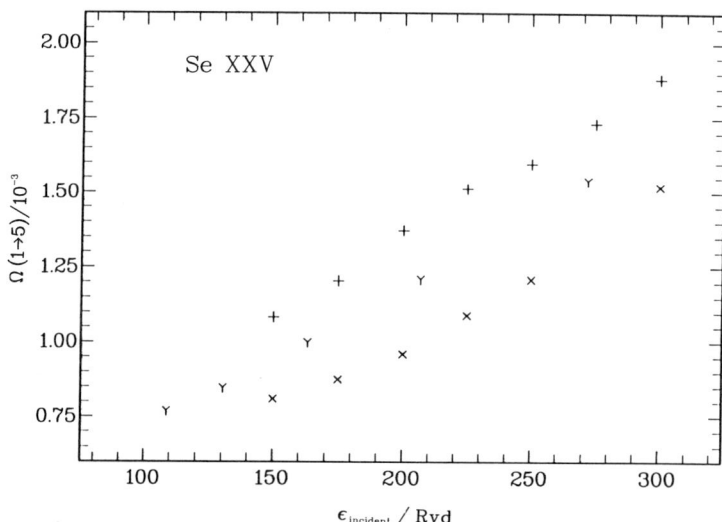

Figure 4. Calculated collisions strengths, Ω, for transition $1-5$ ($2\bar{p}^22p^4\ 0^+ - 2\bar{p}2p^43s\ 1^-$) in Se XXV. Data points as follows: +, present calculation; Y, Coulomb–Born exchange calculation of Zhang et al., 1987; ×, Breit–Pauli R–matrix calculation of Gupta et al., 1989.

The block structure of the new code, JJMTRX, is illustrated in Figure 2. STP1 generates the additional orbitals in $r \leq a$ and computes all the radial integrals required. This part of the calculation is effectively the same as a bound state calculation, and is done with modified modules from GRASP². STP2 assembles the Hamiltonian matrix and the coefficients needed to construct the potentials in $r > a$, and STP3X diagonalizes it to give the $(N+1)$-electron eigenvalues and eigenvectors. As only a finite number of basis orbitals are taken into account in the inner region, it is necessary to correct for the remainder by the method of Buttle (1967); STP3E computes the necessary corrections. Finally STGF computes the K-matrices, collision strengths and cross-sections required. STP1 is a completely new module; the others are taken over or adapted with little change from those used in current versions of the Belfast nonrelativistic R-matrix code (Berrington et al 1978).

A first application of JJMTRX, giving a practical demonstration that we now have a satisfactory state-of-the-art production code, has been to study the collision strengths between low-lying states of neon-like Se XXV, an ion which is of some interest as a candidate for an X-ray laser system. Previous studies have used the nonrelativistic Coulomb-Born method (Zhang et al 1987) and nonrelativistic and relativistic distorted wave calculations were carried out by Reed et al (1985). Gupta et al (1989) have recently carried out R-matrix calculations using the Belfast Breit-Pauli code in an attempt to see if the discrepancies between predicted and observed laser amplification in some transitions was due to neglect of relativistic effects. Our calculation (Wijesundera and Grant, to be published) used 27 j-j coupled target states arising from the $1s^2 2s^2 2p^6$ and $1s^2 2s^2 2p^5 3l$, $l = s, p, d$ (nonrelativistic) configurations. Table 2 shows the energies of the 10 lowest states along with the predictions from other recent calculations. Sixteen orbital basis functions were constructed, using the STP1 module, for each of the symmetries $\kappa = \pm 1, \pm 2, \ldots, \pm 15$, taking the R-matrix boundary at $a = 2.8$ a.u. The dimension of the largest Hamiltonian matrix was 1750 for the total angular momentum/parity combination $J^\pi = 9/2^+$. J values up to 23/2 were required to ensure convergence of the collision strengths to about 1%. The whole calculation took some 2½ hours on the CRAY X-MP/28 at ULCC. Results for selected transitions are shown in Figures 3 and 4; agreement with the Breit-Pauli results of Gupta et al (1989) is much closer for the other transitions of their study, showing that it is primarily in the description of scattering involving excitation of the s electrons in which the Breit-Pauli calculation is most deficient.

RELATIVISTIC ATOMIC STRUCTURE USING SPINOR BASIS SETS

The radial wave-functions constructed in both GRASP and JJMTRX have been generated by traditional finite difference numerical integration methods. Despite their many virtues, this procedure has limitations. In particular, the generation of highly excited states and of continuum functions can prove very laborious, simply because each one-electron orbital has to be constructed by numerical integration, either independently, or as part of the solution of some coupled system of equations. The calculation of processes such as photoionization or electron scattering in which one or more electrons is ejected is therefore expensive, and the construction of sufficiently complete sets of functions for CI or MBPT calculations is difficult. In particular, the integration over continuum states required in MBPT proves messy and time consuming to do accurately (Kelly 1964, 1966).

Theoretical chemists are familiar with an alternative method in which the wave-functions are expanded in a set of suitable functions, ideally functions which can form a complete set in the target Hilbert space. Such ideas have been explored in relation to relativistic atomic and molecular structure; many problems

Table 2. Comparison of the lowest GRASP² target levels (Column (a)) for the calculation of electron scattering from the Se^{+24} ion (neon–like) by the relativistic R–matrix method with results (Column (b)) from the nonrelativistic CIV3 code (Hibbert 1975) including Breit–Pauli corrections (Glass and Hibbert 1978) used in the Breit–Pauli R–matrix calculation of Gupta et al. (1989).

Level	J^π	Dominant CSF	Level energies (rydbergs)	
			(a)	(b)
1	0^+	$2\bar{p}^2 2p^4$	0.0	0.0
2	2^-	$2\bar{p}^2 2p^3 3s$	105.430	105.950
3	1^-	$2\bar{p}^2 2p^3 3s$	105.640	106.155
4	0^-	$2\bar{p}\, 2p^4 3s$	108.622	108.990
5	1^-	$2\bar{p}\, 2p^4 3s$	108.736	109.096
6	1^+	$2\bar{p}^2 2p^3 3\bar{p}$	108.931	109.463
7	2^+	$2\bar{p}^2 2p^3 3\bar{p}$	109.115	109.652
8	3^+	$2\bar{p}^2 2p^3 3p$	109.759	110.208
9	1^+	$2\bar{p}^2 2p^3 3p$	109.813	110.274
10	2^+	$2\bar{p}^2 2p^3 3p$	110.065	110.518

have been encountered, as related, for example, by Kutzelnigg(1987). Spinor basis functions completely overcome these problems in atoms, and it is reasonable to expect that spinor basis sets will be applied in molecular structure calculations before too long. Once a choice of basis set has been made, the numerical approximation of orbitals involves the construction of a finite matrix representation of the Hamiltonian followed by its diagonalization. This process generates high quality representations of the orbitals representing the lower bound states of the effective potential and also square integrable representations, or wave packets, of the higher excited states and the free states in both components of the Dirac continuum. This is much more economical than finite differences if it can be made sufficiently accurate. To be acceptable, the computed values of observable quantities must converge with sufficient rapidity to the correct limit as the basis set is enlarged systematically. We have developed efficient computational machinery which has sufficient mathematical underpinning to make us confident that our results will be physically meaningful.

As a demonstration of the performance of our procedures, we present results for the Dirac–Fock ground–state of the mercury atom using both the Dirac–Coulomb and Dirac–Coulomb–Breit models with a point charge nucleus. The S–spinor basis functions used will be defined presently; we took a basis set, Table 3, of 18s/16p/14d/12f even–tempered S–spinors, giving a total of 36+32+32+28+28+24+24 = 204 orbital wave–functions. Table 4(a) shows that this basis set leads to a ground state total energy that agrees with the one obtained from GRASP to 7 significant figures at the Dirac–Coulomb level and also with the energy obtained by taking the Breit interaction into account as a first order perturbation. The

Table 3. Dirac–Fock calculation for the ground state of mercury. Exponent parameters, equation (14), for even-tempered S-spinor basis.

	s	p	d	f
κ	-1	$+1,-2$	$+2,-3$	$+3,-4$
M	18	16	14	12
α_M	0.378	0.800	0.724	0.808
β_M	1.423	1.415	1.452	1.615

agreement with orbital eigenvalues and mean radii is of similar precision so that the wave functions are also of high quality. We could probably increase the agreement by working to higher precision in GRASP and by slightly adjusting the parameters of the basis set. Table 5 shows the most interesting feature of this calculation, the noticeable change in the properties of the innermost orbitals when the Breit interaction is treated self-consistently. The effect on the total energy is barely significant at this level of precision. This calculation, which was to our knowledge the first basis set calculation for such a heavy atom at this level of precision, took just 27s on the CRAY X-MP/48

Mann (1969) was the first to include the Gaunt interaction (the unretarded form of the Breit interaction) self-consistently in a Dirac–Fock calculation of the mercury ground state. The large number of integrals required, typically 5–10 times as many as the Coulomb interaction integrals, has discouraged further attempts of this sort until Lindroth et al (1989) did it again recently. They used a GRASP-style calculation based on numerical integration, and it is encouraging to note that the shifts, $\delta\epsilon$, in the orbital eigenvalues due to the self-consistent treatment of the Breit interaction are in excellent agreement with our work, Table 5. Lindroth et al (1989) give an extensive analysis of the use of the Breit interaction in Dirac–Fock calculations to which we refer the reader. The persistence of orbital modifications even in outer shells is remarkable and quite different from the perturbative corrections due to the Breit interaction, even showing sign differences for $4f$ and $5d$ orbitals.

Although this calculation took about the same time as GRASP, it yielded far more useful information. The method of approximation generates a complete set of states in a finite dimensional subspace of the target Hilbert space which can then be used, for example, as intermediate states in a many-body calculation (Quiney, 1990). The discrete representation of the basis fits naturally into the Furry (1951) picture of QED, so that one has immediately a natural framework for relativistic perturbation calculations of electron correlation and radiative corrections. The pseudo-states in the continuum form natural quadrature points, reducing the integration over continuum energies to matrix operations, ideal for vector-processing computers (Quiney, Grant and Wilson 1985). Some mathematical considerations which support this approach have been discussed by Grant (1989).

We have previously published results for other systems, in particular for

Table 4. S–spinor calculation of the ground state of Mercury.

(a) Total energies. Breit interaction included as a first order perturbation.

	Dirac–Coulomb	Breit (pert)	Total
SWIRLES	−19653.65587	22.68331	−19630.97256
GRASP	−19653.65042	22.68312	−19630.96730

Differences between results from GRASP and SWIRLES are consistent with estimated errors of about $5 \cdot 10^{-8}$ in both calculations.

(b) Orbital energies and mean radii (atomic units). S — results form SWIRLES; G— results from GRASP^2.

		ϵ	$\langle 1/r \rangle$	$\langle r \rangle$	$\langle r^2 \rangle$
1s	S	−3076.157	97.67030	0.165794(−1)	0.380456(−3)
	G	−3076.158	97.67130	0.165794(−1)	0.380439(−3)
2s	S	−550.5407	24.05544	0.691889(−1)	0.573119(−2)
	G	−550.5411	24.05584	0.691888(−1)	0.573071(−2)
2p̄	S	−526.8621	23.94112	0.569984(−1)	0.406429(−2)
	G	−526.8623	23.94112	0.569982(−1)	0.406422(−2)
2p	S	−455.1448	19.40088	0.656327(−1)	0.524658(−2)
	G	−455.1452	19.40089	0.656326(−1)	0.524657(−2)
3s	S	−133.1793	8.837376	0.179737	0.371083(−1)
	G	−133.1796	8.837219	0.179735	0.371018(−1)
3p̄	S	−122.6431	8.729097	0.170398	0.339115(−1)
	G	−122.6406	8.728806	0.170400	0.339083(−1)
3p	S	−106.5416	7.550064	0.186118	0.402456(−1)
	G	−106.5418	7.550079	0.186117	0.402455(−1)
3d̄	S	−89.43349	7.429123	0.162252	0.309167(−1)
	G	−89.43372	7.429129	0.162252	0.309165(−1)
3d	S	−86.01696	7.162074	0.167047	0.326428(−1)
	G	−86.01720	7.162083	0.167046	0.326426(−1)
4s	S	−30.66489	3.722426	0.398932	0.179671
	G	−30.66501	3.722400	0.398933	0.179672

(continued)

Table 4(b) (Continued)

		ϵ	$\langle 1/r \rangle$	$\langle r \rangle$	$\langle r^2 \rangle$
$4\bar{p}$	S	−26.12446	3.608480	0.401551	0.183491
	G	−26.12429	3.608446	0.401554	0.183493
$4p$	S	−22.18739	3.218961	0.433983	0.214069
	G	−22.18751	3.218956	0.433982	0.214068
$4\bar{d}$	S	−14.79572	3.013852	0.441584	0.225328
	G	−14.79581	3.013844	0.441584	0.225327
$4d$	S	−14.05159	2.920722	0.452489	0.236367
	G	−14.05169	2.920716	0.452489	0.236365
$4\bar{f}$	S	−4.472243	2.557048	0.476739	0.273326
	G	−4.472276	2.557049	0.476737	0.273319
$4f$	S	−4.311054	2.521213	0.483169	0.280713
	G	−4.311087	2.521214	0.483166	0.280705
$5s$	S	−5.106213	1.488458	0.914917	0.939100
	G	−5.106222	1.488448	0.914912	0.939073
$5\bar{p}$	S	−3.537875	1.364281	0.987108	1.101820
	G	−3.537754	1.364251	0.987121	1.101850
$5p$	S	−2.841597	1.223317	1.079173	1.318555
	G	−2.841583	1.223309	1.079177	1.318562
$5\bar{d}$	S	−0.649795	0.920068	1.431269	2.419592
	G	−0.649796	0.920059	1.431277	2.419603
$5d$	S	−0.574389	0.877052	1.498860	2.664936
	G	−0.574392	0.877045	1.498865	2.664927
$6s$	S	−0.328294	0.450909	2.841955	9.343602
	G	−0.328302	0.450914	2.841928	9.343240

Table 5. S–spinor calculation of the ground state of Mercury.
Effect of self–consistent treatment of the Breit interaction.

Orbital energies and mean radii (atomic units).

DC – SWIRLES results for the Dirac–Coulomb Hamiltonian (Table 4).
DCB – SWIRLES results for the Dirac–Coulomb–Breit Hamiltonian
$\delta\epsilon$ is the shift in the eigenvalue due to the self–consistent
inclusion of the Breit interaction.

The total energy is shifted by –0.032104 a.u. relative to the
total energy –19630.97256 a.u. (which includes the Breit inter–
action only as a first order perturbation) reported in Table 4(a).

		ϵ	$\delta\epsilon$	$\langle 1/r \rangle$	$\langle r \rangle$	$\langle r^2 \rangle$
1s	DC	−3076.157		97.6703	0.01658	0.38046(−3)
	DCB	−3065.178	10.9792	97.4914	0.01661	0.38198(−3)
2s	DC	−550.5407		24.0554	0.06919	0.57312(−2)
	DCB	−549.3095	1.2311	24.0331	0.06928	0.57476(−2)
2$\bar{p}$	DC	−526.8621		23.9411	0.05700	0.40643(−2)
	DCB	−524.7947	2.06739	23.8609	0.05716	0.40870(−2)
2p	DC	−455.1448		19.4009	0.06563	0.52466(−2)
	DCB	−453.8400	1.30476	19.3589	0.06576	0.52671(−2)
3s	DC	−133.1793		8.8374	0.17974	0.37108(−1)
	DCB	−132.9513	0.22791	8.8291	0.17994	0.37192(−1)
3$\bar{p}$	DC	−122.6431		8.7291	0.17040	0.33911(−1)
	DCB	−122.2497	0.39335	8.7077	0.17075	0.34051(−1)
3p	DC	−106.5416		7.5501	0.18612	0.40246(−1)
	DCB	−106.3070	0.23466	7.5388	0.18638	0.40359(−1)
3$\bar{d}$	DC	−89.43349		7.4291	0.16225	0.30917(−1)
	DCB	−89.26268	0.17081	7.4179	0.16248	0.30999(−1)
3d	DC	−86.01696		7.1621	0.16705	0.32642(−1)
	DCB	−85.90710	0.10987	7.1546	0.16721	0.32704(−1)
4s	DC	−30.66489		3.7224	0.39893	0.17967
	DCB	−30.61813	0.04676	3.7185	0.39934	0.18003
4$\bar{p}$	DC	−26.12446		3.6085	0.40155	0.18349
	DCB	−26.04109	0.08337	3.6006	0.40228	0.18415

(continued)

Table 5. (Continued)

		ϵ	$\delta\epsilon$	$\langle 1/r \rangle$	$\langle r \rangle$	$\langle r^2 \rangle$
4p	DC	−22.18739		3.2190	0.43398	0.21407
	DCB	−22.14380	0.04359	3.2148	0.43450	0.21458
4d̄	DC	−14.79572		3.0139	0.44158	0.22532
	DCB	−14.77274	0.02297	3.0103	0.44204	0.22578
4d	DC	−14.05159		2.9207	0.45249	0.236367
	DCB	−14.04203	0.00956	2.9187	0.45278	0.236663
4f̄	DC	−4.472243		2.5570	0.47638	0.273326
	DCB	−4.472246	0.00581	2.5562	0.47684	0.273408
4f	DC	−4.311054		2.5212	0.48317	0.280713
	DCB	−4.322537	0.00115	2.5214	0.48307	0.280553
5s	DC	−5.106213		1.4885	0.91492	0.939100
	DCB	−5.099166	0.00705	1.4870	0.91576	0.940818
5p̄	DC	−3.537875		1.3643	0.98711	1.101820
	DCB	−3.525751	0.01213	1.3615	0.98884	1.105670
5p	DC	−2.841597		1.2233	1.07917	1.318555
	DCB	−2.841583	0.00482	1.2219	1.08029	1.321248
5d̄	DC	−0.649795		0.9201	1.43127	2.419592
	DCB	−0.649556	0.00024	0.9194	1.43218	2.422531
5d	DC	−0.574389		0.8771	1.49886	2.664936
	DCB	−0.575481	−0.00109	0.8771	1.49865	2.663792
6s	DC	−0.328294		0.4509	2.84196	9.343602
	DCB	−0.327835	0.00046	0.4505	2.84462	9.361141

argon (Quiney, Grant and Wilson 1989) in which the Breit interaction was treated self-consistently. Although this is extremely expensive when finite difference methods are used, the computational overheads are negligible for this basis set method because the most time-consuming elements, the two-electron integrals, have already been computed for the Coulomb part of the interaction. (The finite wavelength contribution in the full transverse electron-electron interaction, which is essential for integrals in which both positive and negative energy states occur, needs additional terms which require further computing. However, the effect of these terms is rather small.)

The use of basis set methods in atomic structure may therefore significantly extend the range of effects that can be calculated economically on today's computers. Already, we can see advantages for self-consistent fields and for the study of electron correlation effects using many-body methods. There seems little difficulty in calculating vacuum polarization corrections, and we hope to be able to find a way to include the leading terms in the electron self-energy within the scheme. If successful, this will remove one of the greatest uncertainties in present day atomic structure calculations due to the need to rely on unsatisfactory screening arguments and hydrogenic results (Mohr 1974, 1975, 1982; Johnson and Soff 1985) to estimate the self-energy. Finally, the same ideas have implications for relativistic molecular electronic structure calculations, as will be apparent from the papers by Dyall (1990) and Ishikawa (1990) in this volume (see also Clementi 1989).

Criteria for the selection of basis sets for relativistic atomic structure

The foundations of relativistic atomic structure theory were discussed by Grant and Quiney (1988), and that article should be consulted for detailed explanations. For simplicity, we shall only discuss the Dirac equation with a central potential; the extension to Dirac-Fock equations is obvious and is treated by Grant and Quiney (1988) and in the talk by Quiney (1990) in this volume.

The radial Dirac equation can be written (using atomic units)

$$h_D u(r) = \epsilon u(r), \qquad (1)$$

where $\epsilon = E - mc^2$ is the energy relative to the bottom of the electron continuum,

$$u(r) = \begin{bmatrix} P(r) \\ Q(r) \end{bmatrix}, \text{ and } h_D = \begin{bmatrix} c^2 + V(r) & c\left[-\dfrac{d}{dr} + \dfrac{\kappa}{r}\right] \\ c\left[\dfrac{d}{dr} + \dfrac{\kappa}{r}\right] & -c^2 + V(r) \end{bmatrix}.$$

The radial density in $(r, r+dr)$ is

$$D(r) = P^2(r) + Q^2(r)$$

and we require that both

$$\int_0^\infty D(r)\, dr \text{ and } \int_0^\infty D(r) V(r)\, dr$$

are finite. These conditions have important implications for the behaviour of $P(r)$ and $Q(r)$ at the end-points, especially at the singularity $r=0$. We approximate $P(r)$ and $Q(r)$ by finite expansions of the form

$$P_M(r) = \sum_{j=1}^{M} \pi_j(r) \, p_j$$

$$Q_M(r) = \sum_{j=1}^{M} \rho_j(r) \, q_j$$

where $\pi_j(r)$ and $\rho_j(r)$ are regarded as the radial components of a spinor $s_j(r)$ such that

$$s_j(r) = \begin{bmatrix} \pi_j(r) \\ \rho_j(r) \end{bmatrix}.$$

We refer to the set $\{s_j(r), j = 1, 2, \ldots, M\}$ as a "spinor" basis set of dimension N.

We next construct the Rayleigh quotient

$$W[u_M] = \langle u_M | h_\kappa | u_M \rangle \,/\, \langle u_M | u_M \rangle,$$

where u_M denotes our M-dimensional approximation to $u(r)$, and make this stationary with respect to independent variations in the expansion coefficients $\{p_j\}$ and $\{q_j\}$, giving the matrix eigenvalue problem

$$\begin{bmatrix} V^L & c\,\Pi^{LS} \\ c\,\Pi^{SL} & -2c^2 S^S + V^S \end{bmatrix} \begin{bmatrix} p \\ q \end{bmatrix} = \epsilon \begin{bmatrix} S^L & 0 \\ 0 & S^S \end{bmatrix} \begin{bmatrix} p \\ q \end{bmatrix}, \qquad (2)$$

where the elements of the various $M \times M$ matrices are given by

$$S^L_{ij} = \int_0^\infty \pi_i^\dagger(r)\, \pi_j(r)\, dr, \qquad S^S_{ij} = \int_0^\infty \rho_i^\dagger(r)\, \rho_j(r)\, dr;$$

$$V^L_{ij} = \int_0^\infty \pi_i^\dagger(r)\, V(r)\, \pi_j(r)\, dr, \qquad V^S_{ij} = \int_0^\infty \rho_i^\dagger(r)\, V(r)\, \rho_j(r)\, dr;$$

and

$$\Pi^{LS}_{ij} = \int_0^\infty \pi_i^\dagger(r) \left[-\frac{d}{dr} + \frac{\kappa}{r} \right] \rho_j(r)\, dr,$$

$$\Pi^{SL}_{ij} = \int_0^\infty \rho_i^\dagger(r) \left[\frac{d}{dr} + \frac{\kappa}{r} \right] \pi_j(r)\, dr.$$

Integration by parts gives $\Pi^{LS} = \Pi^{SL\dagger}$ whenever the basis functions satisfy bound state boundary conditions.

We should like equations (2) to reduce correctly to the matrix Schrödinger equation as $c \to \infty$ and also that both matrix equations should have solutions approximating the exact *bound state* solutions as close as we please as $M \to \infty$. The first condition essentially means that the operator which represents the nonrelativistic radial kinetic energy,

$$t^L = \tfrac{1}{2}\Pi^{LS}(S^S)^{-1}\Pi^{SL}, \tag{3}$$

must approach the matrix T of the operator

$$T = -\frac{1}{2}\frac{d^2}{dr^2} + \frac{\ell(\ell+1)}{2r^2}$$

as $c \to \infty$. (For this identification of t^L by elimination of q from equations (2) see Grant 1982, 1986.) This is easily secured by requiring that

$$\rho_i(r) \longrightarrow \left[\frac{d}{dr} + \frac{\kappa}{r}\right]\pi_i(r), \quad i = 1, 2, \ldots, M \tag{4a}$$

as $c \to \infty$ (Nonrelativistic Limit Theorem for Basis Spinors). Notice that both $\rho_i(r)$ and $\pi_i(r)$ may depend upon c, and that *the theorem only requires (4) to hold in the limit*. In order to ensure a balanced treatment of the negative energy (or positron) states, it is highly desirable to require also that

$$\pi_i(r) \longrightarrow \left[-\frac{d}{dr} + \frac{\kappa}{r}\right]\rho_i(r), \quad i = 1, 2, \ldots, M. \tag{4b}$$

These limiting conditions are related to the "Kinetic balance" condition which has been advocated by a number of investigators. In its original form (Lee and McLean 1982, Stanton and Havriliak 1984, Aerts and Nieuwpoort 1986) "kinetic balance" meant adding derivative terms to the small component basis set in order that constructions of the type of (4a) could be accomodated. Others (Ishikawa, Binning and Sando 1984, Ishikawa, Baretty and Binning 1986) assumed that (4a) should be replaced by

$$\rho_i(r) = \left[\frac{d}{dr} + \frac{\kappa}{r}\right]\pi_i(r), \quad i = 1, 2, \ldots, M, \tag{5}$$

so *imposing* Pauli's approximation for the nonrelativistic limit of the small components (Bethe and Salpeter (1957), eq. (12.5)) on the relativistic basis set. The success of this form of kinetic balance in a number of studies (see, for example, the chapter by Mohanty and Clementi in Clementi 1989) is therefore not surprising. However, we emphasize that (5) is unnecessarily restrictive and that (4a,b) need only hold in the nonrelativistic limit.

A second theorem that has played an important role in the discovery of relativistic spinor basis sets is the Dirac Matrix Separation Theorem (Grant 1982, 1986). This states that if we define

$$V_{min} = \inf_{p} p^\dagger V^L p / p^\dagger S^L p = \inf_{q} q^\dagger V^S q / q^\dagger S^S q$$

and suppose that (4a) and (4b) hold along with $V_{min} > -2mc^2$, then the $2M$ eigenvalues of the matrix Dirac equation (2) separate into two disjoint sets:

$$S_- = \{\epsilon_i < -2mc^2, \, i = 1, 2, \ldots, M\}$$

and

$$S_+ = \{\epsilon_i > V_{min} > -2mc^2, \, i = M+1, \ldots, 2M\}.$$

It is important to realize that V_{min} always exists in practice, even if the potential V is unbounded. Any acceptable trial function must yield a finite expectation value for V. The proof of this result amounts to a demonstration that switching on a sub-critical potential V does not depress the energy of a just free electron enough for it to enter the negative energy continuum. Since bound states can only exist in the interval $-2mc^2 < \epsilon \le 0$, this ensures that there is a *fixed lower bound* to the eigenvalue set S_+, so that the familiar nonrelativistic Rayleigh–Ritz variational theorems continue to work for the Dirac Hamiltonian. Statements that there is nothing to stop Dirac eigenvalues from "collapsing" into the lower continuum are therefore wrong, as the successful applications described above demonstrate.

The L–spinor basis functions for point nucleus models

The most satisfactory method we have devised to meet all the criteria given so far uses the L–spinor basis functions, which can be viewed as relativistic analogues of the nonrelativistic Coulomb Sturmian functions. The latter are defined by Rotenberg (1962, 1970) as the non-trivial normalised solutions of

$$\left[-\frac{d^2}{dx^2} + \frac{\ell(\ell+1)}{x^2} + \lambda^2 - \frac{2\lambda n}{x} \right] S_{nl}(x) = 0$$

vanishing at $x = 0$ and $x = \infty$, where $n = 1, 2, 3, \ldots$ and $\ell = 0, 1, \ldots, n-1$, and λ is an arbitrary positive constant. The solutions coincide with the hydrogenic eigenfunction corresponding to the values of n and ℓ whenever $\lambda = Z/n$. The Sturmian functions are given by the formula

$$S_{nl}(x) = N'_{nl} e^{-\lambda x}(2\lambda x)^{\ell+1} L_{n-\ell-1}^{(2\ell+1)}(2\lambda x),$$

where $L_{n-\ell-1}^{(2\ell+1)}(2\lambda x)$ is an associated Laguerre polynomial and N'_{nl} is a normalizing constant. It is convenient in many applications to choose the normalization

$$\int_0^\infty S_{nl}(x) \frac{2}{x} S_{n'l}(x)\, dx = \delta_{nn'}, \tag{6}$$

In this case the Gram matrix of this system for each angular symmetry ℓ takes the very simple form

$$g_{nl,n'l} = \begin{cases} n/2\lambda, & n' = n \\ -[(n+\ell+1)(n-\ell)]^{\frac{1}{2}}/4\lambda, & n' = n+1 \\ 0 & \text{otherwise} \end{cases}$$

The polynomials oscillate, the number of nodes in $S_{nl}(x)$ being $n-\ell-1$, the last peak appearing when x is of order $(n-\ell-1)/\lambda$, after which the exponential factor cuts $S_{nl}(x)$ off rapidly.

The L–spinors are the relativistic versions; their radial components are given by

$$f_{n\kappa}^T(r) = N_{n\kappa} r^\gamma e^{-\lambda r}\left\{ -(1-\delta_{n,0}) L_{n-1}^{(2\gamma)}(2\lambda r) \right.$$
$$\left. \pm \frac{N_n - \kappa}{n + 2\gamma} L_n^{(2\gamma)}(2\lambda r) \right\} \tag{7}$$

where the superscript T takes the values L or S to denote respectively the "large" and "small" components; the "+" sign goes with L, the "−" sign with S. The angular symmetry is denoted by κ, the 'cusp' exponent $\gamma := [\kappa^2 - \alpha^2 Z^2]^{\frac{1}{2}}$, the 'apparent principal quantum number' is $N_n := (n^2 + 2n\gamma + \kappa^2)^{\frac{1}{2}}$, and λ is, as before, an arbitrary constant. In this case n must be interpreted as the "inner" quantum number, the "principal" quantum number being $n+|\kappa|$. Expressing $f_{n\kappa}^L$ and $f_{n\kappa}^S$ as functions of $x = 2\lambda r$, we find that the basis functions satisfy the Dirac–like equation

$$\begin{bmatrix} \left[\frac{1}{2} - \frac{2\zeta_n^L}{x}\right] & \left[-\frac{d}{dx} + \frac{\kappa}{x}\right] \\ \left[\frac{d}{dx} + \frac{\kappa}{x}\right] & \left[-\frac{1}{2} - \frac{2\zeta_n^S}{x}\right] \end{bmatrix} \begin{bmatrix} f_{n\kappa}^L(x) \\ f_{n\kappa}^S(x) \end{bmatrix} = 0. \qquad (8)$$

In terms of an eigenvalue parameter μ_n, the two parameters ζ_n^L and ζ_n^S are given by

$$\zeta_n^L = \frac{Z\mu_n^2}{2c}, \quad \zeta_n^S = \frac{Z}{2c\mu_n^2} \qquad (9)$$

so that

$$\zeta_n^L + \zeta_n^S = N_n, \qquad \zeta_n^L - \zeta_n^S = N_n(1 - Z^2/N^2 c^2)^{\frac{1}{2}}, \qquad (10)$$
$$\zeta_n^L \zeta_n^S = Z^2/4c^2, \qquad \zeta_n^L / \zeta_n^S = \mu_n^4.$$

As before, we recover the Dirac–Coulomb functions when $\lambda = \lambda_n := Z/N_n$, so that

$$P_{n\kappa}(r) = \mu_n f_{n\kappa}^L(2\lambda_n r), \qquad Q_{n\kappa}(r) = \mu_n^{-1} f_{n\kappa}^S(2\lambda_n r)$$

apart from an overall normalization.

It is straightforward, though lengthy, to demonstrate that the L–spinors satisfy the conditions (4a) and (4b) required in the Nonrelativistic Limit Theorem. The nonrelativistic orthogonality relation is replaced by

$$(\zeta_n^L - \zeta_{n'}^L)\langle f_{n'}^L| 1/x |f_n^L\rangle + (\zeta_n^S - \zeta_{n'}^S)\langle f_{n'}^S| 1/x |f_n^S\rangle = 0 \qquad (11)$$

This orthogonality relationship does not seem easy to exploit directly for computational purposes although it has a formal mathematical importance. This is most easily seen by passing to the nonrelativistic limit using the results $\zeta_n^L \rightarrow n + |\kappa|$, $\zeta_n^S \rightarrow 0$, as $c \rightarrow \infty$, so that only the first term survives in the form

$$(n - n')\langle f_{n'}^L| 1/x |f_n^L\rangle = 0,$$

leading to the nonrelativistic orthogonality relation (6) when $n \neq n'$. The Gram matrices of the large and small component systems are tridiagonal, with non–zero elements given by

35

$$S^T_{nn'} = \begin{cases} 1, & n = n', \\ \mp \frac{1}{2}\left[\left(1 - \frac{\kappa}{N_n}\right)\left(1 + \frac{\kappa}{N_{n+1}}\right)\right]^{\frac{1}{2}}, & |n - n'| = 1, \\ 0 & \text{otherwise,} \end{cases} \qquad (12)$$

where, as before, the label T is either L or S, the minus sign is taken with the former and the plus sign with the latter. Thus the two Gram matrices have the same eigenvalue spectrum, and therefore the same linear independence properties. The completeness and linear independence of the L–spinors can be analysed in the manner of Klahn (1981).

The construction of the remaining submatrices of the Dirac operator (2) is relatively straightforward. For a point nucleus Coulomb potential, the V matrices are diagonal in the non–relativistic limit, and the off–diagonal terms are $O(1/c^2)$. The Π matrices have a simple algebraic structure permitting recursive evaluation of the various contributions. The construction of matrices of two–electron interaction operators is more complicated, and will not be discussed here.

In practice, one chooses values of n up to some value $n_{max}(\kappa)$; the completeness and linear independence properties of the L–spinors ensure convergence to any desired accuracy by picking $n_{max}(\kappa)$ sufficiently large. Usually, one fixes λ at around $Z/(\ell+1)$, to give about $n_{max}(\kappa)/4$ bound states, with the remaining $3n_{max}(\kappa)/4$ states representing the upper continuum and $n_{max}(\kappa)$ states with energies in the lower continuum. The eigenvalues are stationary with respect to variations in λ, so that one does not need to choose it precisely. However, the distribution of excited states does change with λ, and this parameter can therefore be used to "tune" the basis for particular purposes. This is in fact very easy to do as all the matrices scale as a simple multiple of λ in the case of a Coulomb potential:

$$\begin{bmatrix} c^2 S^L_\kappa - \lambda Z \bar{V}^L_\kappa & c\lambda \bar{\Pi}^{LS}_\kappa \\ c\lambda \bar{\Pi}^{SL}_\kappa & -c^2 S^S_\kappa - \lambda Z \bar{V}^S_\kappa \end{bmatrix} \begin{bmatrix} p \\ q \end{bmatrix} = E \begin{bmatrix} S^L_\kappa & 0 \\ 0 & S^S_\kappa \end{bmatrix} \begin{bmatrix} p \\ q \end{bmatrix}.$$

where the subscript κ has been inserted as a reminder that we are considering a single angular symmetry and the bars indicate that the matrices have been evaluated with $\lambda=1$, $Z=1$. See the contribution by Quiney (p.   ) for further details.

The S–spinor basis functions for point nucleus models

Although L–spinors have many attractive features, they do have practical drawbacks when it comes to calculating matrices of the electron–electron interaction in many–electron calculations makes it advisable to examine alternatives. One such alternative is the S–spinors which we have used in all the many–electron calculations we have published so far. S–spinors can be viewed as the relativistic analogue of the Slater type functions (functions having the general form $r^n \exp(-\lambda r)$) whose simplicity has ensured their popularity in nonrelativistic atomic structure calculations (see, for example, Hibbert 1975, Wilson 1987). S–spinors are most readily defined as the nodeless, or nearly nodeless, functions obtained from L–spinors, equation (7), with minimum index n. Ignoring common normalizing factors, we take $n = 0$ for *negative* κ,

$$f_{m\kappa}^{T}(r) = r^{\gamma}\exp(-\lambda_m r) \qquad \kappa < 0, \qquad (13a)$$

and $n = 1$ for *positive* κ,

$$f_{m\kappa}^{T}(r) = (A^T + \lambda_m r) r^{\gamma}\exp(-\lambda_m r) \qquad \kappa > 0, \qquad (13b)$$

where, as usual, $\gamma = [\kappa^2 - Z^2/c^2]^{\frac{1}{2}}$,

$$A^L = \frac{(\kappa + 1 - N)(2\gamma + 1)}{2(N - \kappa)}, \qquad (13c)$$

$$A^S = \frac{(\kappa - 1 - N)(2\gamma + 1)}{2(N - \kappa)}, \qquad (13d)$$

and

$$N^2 = 1 + 2\gamma + \kappa^2.$$

It is immediately obvious that these functions, regarded as individual L–spinors, satisfy the requirements of the Dirac Matrix Separation Theorem. For negative κ, the same basis set can be used for both large and small components, (13a), but the algebraic form inherited from the L–spinors for positive κ, (59b), is more complicated. The reason for this becomes obvious when we compare the two symmetry types, $\kappa = +1$ ($\bar{p}$ symmetry) and $\kappa = -1$ (s symmetry) which have the common γ value, $\gamma = [1 - Z^2/c^2]^{\frac{1}{2}}$. As $c \rightarrow \infty$, both s-type components go over smoothly into the same s-type Slater function

$$f_{ms}^{T}(r) = r \exp(-\lambda_m r).$$

However, it is easy to see that for $\kappa = +1$,

$$A^L \rightarrow 0(Z^2/c^2), \quad A^S \rightarrow -3 + 0(Z^2/c^2),$$

so that the large component goes over smoothly into a p-type function,

$$f_{mp}^{L}(r) \rightarrow r^2 \exp(-\lambda_m r).$$

as $c \rightarrow \infty$, but

$$f_{mp}^{S}(r) \rightarrow (-3r + \lambda_m r^2)\exp(-\lambda_m r),$$

which is a linear combination of s– and p-type Slater functions satisfying the Pauli relation, (5), apart from an unimportant phase. Clearly, there is no way in which the naive choice of a monomial to describe both components as in (13a) can achieve this result. The behaviour of S–spinors and L–spinors is (by design) the same as that of true Coulomb eigenfunctions near $r=0$. Although the eigenfunctions for distributed charge nuclei have different cusp structure, the same qualitative behaviour — vanishing of the s-type contribution to the large component of a p-type function in the nonrelativistic limit — has to occur, and this must not be forgotten when constructing basis functions for such models. So far, strict kinetic balance, (5), is the only way that has been found to do this approximately.

The selection of a suitable family of exponents, $\{\lambda_m\}$, is crucial to the efficiency of the S-spinor algorithm. We use the members of a geometric sequence

$$\lambda_m = \alpha \beta^{m-1}, \quad m = 1, 2, \ldots \tag{14}$$

where α and β are positive constants. This *even-tempered* prescription was first suggested by Reeves (1963) on empirical grounds, and exploited by Raffenetti (1973), Feller and Ruedenberg (1979), Schmidt and Ruedenberg (1979) and in other papers by Ruedenberg and his collaborators.. One important attraction of this choice is the ease with which highly efficient algorithms can be constructed for evaluating the various matrices needed to high accuracy (Quiney 1990). There are some mathematical complications to be taken into account (Klahn 1981). In particular, the sequence (14) with fixed α and β gives an S-spinor set which is formally linearly independent for any finite β but is *incomplete* in the Hilbert space $L^2(\mathbb{R}_+)$. A way out of this difficulty is to use a sequence of finite sets of even-tempered functions of the form

$$\{\exp(-\lambda_m r) \mid \lambda_m = \alpha_M \beta_M^{m-1}, \quad m = 1, 2, \ldots M\} \tag{15}$$

where

$$\alpha_M \to 0, \quad \beta_M \to 1$$

from above as $M \to \infty$. The prescription suggested by Schmidt and Ruedenberg (1979) and Feller and Ruedenberg (1979),

$$\alpha_M = \left[\frac{\beta_M - 1}{\beta_{M-1} - 1}\right]^{\frac{1}{2}} \alpha_{M-1}, \qquad \ln \beta_M = \left[\frac{M}{M-1}\right]^{-\frac{1}{2}} \ln \beta_{M-1} \tag{16}$$

seems quite effective. In practice, values of β_M in the range 1.3 to 1.5 give good accuracy, and α_M must be chosen so that the largest exponent, λ_M, is of order Z. A near optimal set can be constructed by doing a *nonrelativistic* self-consistent field calculation and searching variationally for the 'best' values of α_M and β_M for some relatively small dimension $M=8$, say, using (16) to obtain values for some larger value of M for the final relativistic runs. The mercury calculation reported in Tables 3–5 was done in this way.

All the one electron matrices are based on simple primary functions $v_{\kappa ij}(2\gamma+p)$, depending on the components $\{\lambda_{\kappa i} = \alpha_\kappa \beta_\kappa^{i-1}, i=1, 2, \ldots, M_\kappa\}$ for each symmetry κ, which satisfy the recurrence relations

$$v_{\kappa ij}(2\gamma+p+1) = (2\gamma+p+1) \, v_{\kappa ij}(2\gamma+p)/\lambda_{\kappa ij}, \tag{17}$$

where $\lambda_{\kappa ij} = \lambda_{\kappa i} + \lambda_{\kappa j}$, which is valid for all positive and negative integer values of

p for which this makes sense. The seed value is

$$v_{\kappa i j}(2\gamma) = \left[\tfrac{1}{2}(\beta_\kappa^{\frac{1}{2}(i-j)} + \beta_\kappa^{\frac{1}{2}(j-i)})\right]^{-2\gamma-1},$$

which depends only on β_κ. Thus the Gram matrix for symmetry κ is the same for both large and small components; it has matrix elements

$$S_{\kappa i j} = v_{\kappa i j}(2\gamma) \qquad\qquad \kappa < 0, \qquad (18)$$
$$\phantom{S_{\kappa i j}} = v_{\kappa i j}(2\gamma) C_{\kappa i j}(2\gamma) \qquad\qquad \kappa > 0,$$

where

$$C_{\kappa i j}(2\gamma) = 1 - \frac{(2\gamma+2)(N-\kappa)^2 \xi_{\kappa i j}^2}{(2\gamma+1) + (N-\kappa)^2} \quad\text{and}\quad \xi_{\kappa i j}^2 = \left[\frac{\lambda_{\kappa i} - \lambda_{\kappa j}}{\lambda_{\kappa i} + \lambda_{\kappa j}}\right]^2.$$

Thus S_κ is a full matrix rather than a banded one. Similarly the Coulomb potential matrices are given by

$$V_{\kappa i j}^T = v_{\kappa i j}(2\gamma - 1), \qquad\qquad \kappa < 0,$$

$$\phantom{V_{\kappa i j}^T} = v_{\kappa i j}(2\gamma-1) \frac{(A_\kappa^T)^2 + 2\gamma A_\kappa^T + \tfrac{1}{4}\gamma(\gamma+\tfrac{1}{2})(1 - \xi_{\kappa i j}^2)}{(A_\kappa^T)^2 + (2\gamma+1)A_\kappa^T + \tfrac{1}{4}\gamma(\gamma+\tfrac{1}{2})},$$
$$\kappa > 0, \qquad (19)$$

and the Π matrices have a similar form which is too lengthy to be worth including here.

The calculation of electron–electron interaction matrices is the major component of any many–electron electronic structure calculation, and we have found that this can be done with great accuracy and high efficiency in S–spinor calculations. Each matrix element can be expressed as a linear combination of radial (Slater) integrals over the one–electron orbitals, as described in detail by Grant (1988). These integrals can, in their turn, be expressed as linear combinations of two–electron integrals over S–spinor components, so that in the case of the Coulomb interaction, we ultimately have to evaluate combinations of primitive integrals of the form

$$I_{mnpq}^{\nu, ijkl} = \int_0^\infty \int_0^\infty dr\, ds\, g_m^i(r)\, g_n^j(r)\, U_\nu(r,s)\, g_p^k(r)\, g_q^l(r) \qquad (20)$$

where

$$g_m^i(r) = r^{\gamma_m + i} e^{-\lambda_m r}, \quad \gamma_m = [\kappa_m^2 - (Z/c)^2]^{\frac{1}{2}}$$

and

$$U_\nu(r,s) = r^\nu/s^{\nu+1} \text{ if } r < s, \text{ or } s^\nu/r^{\nu+1} \text{ if } r > s.$$

The I-integrals can be written in terms of well-known transcendental functions, namely

$$I^{\nu,ijkl}_{mnpq} = \Gamma(a)\left\{\lambda_{mn}^{-a}\lambda_{mn}^{-b}B(a,b;z) + \lambda_{mn}^{-a'}\lambda_{mn}^{-b'}B(a',b';z')\right\} \tag{21}$$

where $\Gamma(a)$ is a gamma function, $B(a,b;z)$ is an un-normalized incomplete beta function and

$$\lambda_{mn} = \lambda_m + \lambda_n,$$
$$a = \gamma_m + \gamma_n + i + j + \nu + 1,$$
$$a' = \gamma_m + \gamma_n + i + j - \nu,$$
$$z = \lambda_{mn}/(\lambda_{mn} + \lambda_{pq})$$

$$\alpha = \gamma_m + \gamma_n + \gamma_p + \gamma_q + i + j + k + l + 1,$$
$$b = \gamma_p + \gamma_q + k + l - \nu,$$
$$b' = \gamma_p + \gamma_q + k + l + \nu + 1$$
$$z' = \lambda_{pq}/(\lambda_{mn} + \lambda_{pq}).$$

Most program libraries include procedures for evaluating the gamma function. The incomplete beta function is more troublesome, and demands special treatment (Quiney 1990). Isolated values can be readily evaluated by using continued fraction representations which can be made to work to machine precision. However, (21) shows that we need to evaluate a large number of values of $B(a,b;z)$ for fixed z in which the parameters a and b differ by integers, and it turns out that stable recurrence relations exist that enable this to be done very efficiently and accurately for a family of vectors with components $B(a,b;z_i)$, $i = 1, 2, \ldots$ The details are given by Quiney at p.   of this volume.

One bonus of this approach is that the same primitive integrals (20), occur in the evaluation of the low-frequency Breit interaction

$$B(1,2) = -\frac{1}{2}\left\{\frac{\alpha(1)\cdot\alpha(2)}{R} + \frac{\alpha(1)\cdot R\ \alpha(2)\cdot R}{R^3}\right\} \tag{22}$$

where R is the relative vector separation of the two electrons as in the evaluation of the instantaneous Coulomb interaction. This means that the Breit interaction integrals cost virtually nothing extra to compute, a striking contrast to the situation in conventional numerical codes such as GRASP. The magnetic and retardation parts of the electron- electron interaction can therefore be treated together in the determination of the orbital basis, so minimizing gauge-dependent effects. Treating the relativistic electron-electron interaction as a single, rather than as a composite, operator in which each part is handled as a separate perturbation, simplifies perturbation calculations dramatically. Of course, (22) is only valid for matrix elements connecting orbitals whose energies are small compared with mc^2, but it is possible to make the necessary finite wavelength corrections in integrals connecting positive and negative energy states without too much difficulty.

A major question which we do not have the space to treat in detail relates to the convergence of S-spinor or L-spinor approximations. This was discussed by Grant (1989) in terms of ideas of Klahn and Morgan (1984). The main conclusions are that Rayleigh-Ritz approximations for bound state eigenvalues and orbitals in Dirac one-electron and Dirac -Fock atomic problems are likely to converge in the mathematical sense as they appear to do in practice. Moreover, the expectation values and matrix elements of most of the operators that are likely to be encountered in atomic physics will also converge mathematically. In particular, because the radiative multipole operators are all bounded (provided the full frequency dependence is retained), their matrix elements can be relied upon to converge when evaluated with S–spinor or L–spinor wave–functions. The computation of such matrix elements is central to the calculation of higher order terms in quantum electrodynamics, and this finding is of vital importance for future developments employing these techniques.

CONCLUSIONS

The technical advances in GRASP will enable us to study more complex configurations of atoms and ions with less cost and greater physical realism, as illustrated by a number of state–of–the–art calculations. The JJMTRX code is now able to produce cross–sections for electron scattering from atoms and ions both in the continuum and in the resonance region. It will now be developed to generate further observable scattering features and, in particular, to compute photo–ionization cross–sections for heavy elements for which there is a good deal of experimental information awaiting interpretation.

The SWIRLES code is likely to be of real importance in the study of highly ionized atoms in the future. S–spinors provide a cheap and highly accurate method of generating relativistic wave–functions for bound state problems, and also of carrying out calculations of higher order corrections using many–body perturbation theory. The negative energy states can be handled quite straighforwardly within this method, and it seems likely that SWIRLES will be used to compute QED corrections in the near future. A combination of S–spinor and L–spinor techniques seems likely to prove very powerful in this context. High–Z atoms and ions can also be modelled successfully using basis set techniques, and an S–spinor version of GRASP looks an attractive proposition.

Basis sets are essential for the treatment of the electronic structure of molecules. We have made some preliminary investigations of relativistic electronic structure using methods based on those described in this paper. The method seems promising, although it is still too early to present any results.

ACKNOWLEDGMENTS

I am very grateful to my colleagues Farid Parpia, Harry Quiney and Wasantha Wijesundera for generously allowing me to make use of hitherto unpublished material and for much useful discussion.

REFERENCES

Accad, Y., Pekeris, C. and Schiff, B., 1971, Phys. Rev. A 4:516.
Aerts, P.J.C., and Nieuwpoort, W.C., 1986, Chem Phys. Lett. 125:83.
Baig, M.A., Connerade, J.P., Mayhew, C. and Sommer, K., 1983 J. Phys. B. 17:371.
Bar–Shalom, A. and Klapisch, M. 1988, Computer Phys. Commun. 50:375.

Berrington, K.A., Burke, P.G., Le Dourneuf, M., Robb, W.D., Taylor, K.T., and Vo Ky Lan, 1978, Computer Phys. Commun. 14:367.
Bethe, H.A. and Salpeter, E.E., 1957, "Quantum Mechanics of One- and Two-electron Atoms", Springer-Verlag, Berlin.
Burke, P.G., 1970, Computer Phys. Commun. 1:241.
Burke, P.G., 1971, Computer Phys. Commun. 2:173.
Burke, P.G., and Robb, M.A., 1975, in, "Advances in Atomic and Molecular Physics, Vol. 11", D.R.Bates and B.Bedersen, eds., Academic Press, New York.
Buttle, P.J.A., 1967, Phys. Rev. 160:719.
Connerade, J.P., Baig, M.A. and Sweeney, M., 1990, J. Phys. B. 23:713
Clementi, E., 1989, "Modern Techniques in Computational Chemistry: MOTECC-89" Escom Science Publishers, Leiden.
Dyall, K.G., 1986, Computer Phys. Commun. 39:141.
Dyall, K.G., Grant, I.P., Plummer, E.P., Johnson, C.T., and Parpia, F.A., 1989, Computer Phys. Commun. 55:425.
Dyall, K.G., 1990, this volume, p.
Feller, D.F., and Ruedenberg, K., 1979, Theor. Chim. Acta 52:231.
Fischer, C.F. 1986, Computer Phys. Rep. 3:273.
Furry, W.H., 1951, Phys. Rev. 81:115.
Grant, I.P., 1970, Adv. Phys. 19:747.
Grant, I.P. 1982, Phys. Rev. A 25:1230.
Grant, I.P., 1986, J. Phys. B. 19:3187.
Grant, I.P., 1988, Relativistic atomic structure calculations, in: "Methods in Computational Chemistry, Vol. 2.", S. Wilson, ed., Plenum Press, New York.
Grant, I.P., 1989, Notes on basis sets for relativistic atomic structure and QED, in, "Relativistic,Quantum Electrodynamic and Weak Interaction Effects in Atoms" (A.I.P. Conference Proceedings 189), W.R.Johnson, P.J.Mohr and J.Sucher, eds., American Institute of Physics, New York.
Grant, I.P., McKenzie, B.J., Norrington, P.H., Mayers, D.F. and Pyper N.C., 1980, Computer Phys. Commun. 21:207.
Grant I.P. and Quiney, H.M. 1988, Foundations of the Relativistic Theory of Atomic and Molecular Structure, in, "Advances in Atomic and Molecular Physics, Vol. 23", D.R.Bates and B.Bedersen, eds., Academic Press, New York.
Gupta, G.P., Berrington, K.A., and Kingston, A.E., 1989, J. Phys. B. 22:3289.
Hibbert, A. 1975, Computer Phys. Commun. 9:141
Ishikawa, Y., Baretty and Binning, R.C., 1986, Chem. Phys. Lett. 121:130.
Ishikawa, Y., Binning, R.C., and Sando, K.M., 1984, Chem. Phys. Lett. 101:111.
Ishikawa, Y., 1990, this volume p.
Johnson, W.R. and Soff, G., 1985, At. Data Nucl. Data Tables 33:405.
Kelly, H.P., 1964, Phys. Rev. 136:896.
Kelly, H.P., 1966, Phys. Rev. 144:39.
Klahn, B., 1981, Review of the Linear Independence Properties of Infinite Sets of Functions used in Quantum Chemistry, in, "Advances in Quantum Chemistry, Vol. 13" P.-O. Löwdin, ed., Academic Press, New York.
Kutzelnigg, W., 1987, Int. J. Quant. Chem. 25:107.
Lee, Y.S., and McLean, A.D., 1982, J. Chem. Phys. 76:735.
Lindroth, E., Mårtensson-Pendrill, A.-M., Ynnerman, A., and Öster, P., 1989, J. Phys. B 22:2447.
Mann, J.B., 1969, private communication.
McKenzie, B.J., Grant, I.P., and Norrington, P.H., 1980, Computer Phys. Commun. 21:233.
Mohr, P.J., 1974, Ann. Phys. (NY) 88:26.
Mohr, P.J., 1975, Phys. Rev. Lett. 34:1050.
Mohr, P.J., 1982, Phys. Rev. A 26:2338.
Norrington, P.H. and Grant, I.P. 1981, J. Phys. B. 14:L261.
Norrington, P.H. and Grant, I.P. 1987, J. Phys. B. 20:L735.
Parpia, F.A. and Grant, I.P. 1990, J. Phys. B. 23:211.
Quiney, H.M., 1990, this volume, p.

Quiney, H.M., Grant, I.P. and Wilson, S. 1990, J. Phys. B. 23: L271.
Raffenetti, R.C., 1973, J. Chem. Phys. 59:5936.
Reed, K.J., 1985, Internal Memo., Lawrence Livermore National Laboratory.
Reeves, C.M., 1963, J. Chem. Phys. 39:1.
Rotenberg, M., 1962, Ann. Phys. 19:262.
Rotenberg, M., 1970, Theory and Application of Sturmian Functions in, "Advances in Atomic and Molecular Physics, Vol. 6, p.233", D.R.Bates and B.Bedersen, eds., Academic Press, New York.
Schmidt, M.W., and Ruedenberg, K., 1979, J. Chem. Phys. 71:3951
Stanton, R. and Havriliak, S., 1984, J. Chem. Phys. 81:1910
Wilson, S., 1987, Basis Sets, in, "Ab Initio Methods in Quantum Chemistry —I", Lawley, K.P., ed., Wiley, New York.
Zhang, H., Sampson, D.H., Clark, R.E.H., and Mann, J.B., 1987, At. Data Nucl. Data Tables 37:17.

ON THE ACCURACY OF OSCILLATOR STRENGTHS

B. C. Fawcett

Rutherford Appleton Laboratory
Science and Engineering Research Council
Chilton, Didcot, Oxfordshire, OX11 0QX, England

ABSTRACT

It is important to investigate whether non—relativistic errors in calculated oscillator strengths and related atomic data are large enough to obscure corrections applied in relativistic calculations. The magnitude such errors is however difficult to estimate. This is just one reason why the accuracy of the atomic data needs careful scrutiny. Application to the analysis of line intensities in laboratory and astropyhsical plasmas is a more general one. This review therefore focuses on errors that can affect the accuracy of calculated oscillator strengths. Precautions that can be taken to minimise errors arising from different causes are surveyed. Special attention is given to their reduction through the improvement of eigenvectors with Slater parameter optimisation methods. These semi—empirical methods for improving the accuracy of oscillator strengths can be important in fully relativistic calculations for a range of ions.

1. INTRODUCTION

When relativistic corrections are applied to atomic—structure calculations of oscillator strengths and energy levels their true effects may be masked completely by a variety of sources of errors resulting from many different physical causes. A detailed understanding of these sources of error and how to reduce them is therefore of vital importance when making relativistic calculations. Such information is also of crucial importance for plasma diagnostics as accurate oscillator strengths are required for the interpretation of spectral line intensities to estimate densities, temperatures and absorption in astrophysical and laboratory plasmas. Moreover some of the factors affecting the accuracy of oscillator strengths also affect calculated collision strengths which are often required to enable the analysis of spectral line intensities.

Although many of the numerous factors which contribute to oscillator strength errors are described in the literature, it is evident unfortunately that some are frequently overlooked, resulting in the publication of numerous unreliable data. This makes it very difficult for users to discriminate between valid and unsatisfactory material. The situation is made more problematic by the shortage of accurate measured data. It is therefore necessary to examine the problem in detail. This review therefore gathers together descriptions of the many physical sources of error and emphasises the serious difficulties confronting those seeking to reduce and determine errors in oscillator strength calculations. Some examples of the techniques for reducing errors on oscillator strengths are cited, mainly selected from work conducted at the Rutherford Appleton Laboratory and the reference list is extended(42–53) to complete the bibliography of this work.

There are various contributions to oscillator strength errors which can be catalogued under different headings each requiring a different technique for minimisation. There are for instance errors arising from the quality of the eigenvectors applied in the computations. There can be errors in the radial dipole matrix elements. Cancellation, if not correctly modelled, can create large errors. Configuration interaction has to be adequately included or errors will arise. Other possible sources of perturbation must be considered. The influence of the inner shells or of levels lying in the continuum will vary from level to level. The quality of the wavefunctions is another fundamental consideration. These as well as other atomic quantities can depend on the method of calculation.

A variety of theoretical methods can be applied. There is the RCN/RCN2/RCG code (1–4), which incorporates a pseudo relativistic Hartree–Fock (HFR) stage and was developed by Dr. R. D. Cowan of Los Alamos Scientific Laboratory; (1–2) or the Multiconfiguration–Hartree–Fock (6,7) (MCHF) code of Prof. C. F. Fischer of Vanderbilt University. Sophisticated fully relativistic codes are the Multiconfiguration–Dirac–Fock (8) (MCDF) and advanced versions called Generalised Relativistic Atomic–Structure Programs (GRASP (9) and GRASP2 (10)), written at Oxford University by Dr. I. P. Grant and co–workers or other relativistic codes written by Dr. J. P. Desclaux (11). Many more methods are available such as the parametric potential method pioneered by Klapisch (12), the Hibbert CIV3 code (13), polarised potential methods and those using scaled potentials such as SUPERSTRUCTURE (14) (see Ref. 1 pp 190–199 for a more detailed survey and description of methods).

Sub–sets of the aforementioned atomic–structure codes may include length, velocity or acceleration oscillator strength formulations, relativistic approximations or quantum electrodynamical (QED) corrections (1,8–14). The appropriate choice of theoretical method can depend critically on the complexity of the ion under investigation. In simple ions for instance the adoption of sophisticated wavefunctions and comprehensive treatment of configuration interaction can make the calculations sufficiently accurate that quite small relativistic corrections can be sensibly added and detailed QED corrections included. At the other extreme, for complicated atomic systems, these types of errors are usually dwarfed by others arising from configuration interaction effects. They result partly from the limitation in the number of configurations that can be included in feasible computations. Although relativistic effects become dominant with heavier ions it must be emphasised that the aforemenioned sources of error can still be very serious.

2. RELATIVISTIC EFFECTS

For heavy elements and high degrees of ionisation fully relativistic Multi–Configuration–Dirac–Fock codes (8–11) are necessary. At intermediate atomic numbers up to Z=40 pseudo–relativistic approximations (see Ref 1, pp 200) may be adequate while non–relativistic calculations break down at much lower atomic numbers. A study of the application of the fully relativistic Multi–Configuration–Dirac–Fock Grant code compared with the pseudo–relativistivic Hartree–Fock Cowan code was made for Fe XV (15). It was shown that the pseudo–relativistic calculations were in close agreement with the fully relativistic ones for this ion up to Z=30 and that MCDF was necessary for higher Z. As the Dirac–Fock equations involve multi–component wavefunctions, array sizes in the computations are much larger than for Hartree–Fock. It can therefore be very important to use these two types of code in conjunction as the one gives the advantage of fully relativistic calculations and the other gives the ability to include more configurations in the model due to the much smaller arrays inherent in the matrices. The more comprehensive inclusion of configuration interaction effects can be essential for many complex ions. In this context it must be emphasised that there is a range of ions for which fully relativistic calculations are absolutely essential and the relativistic codes are unable to include an adequate number of configurations. Therefore semi–empirical modelling (see the detailed discussion in sections 4 and 5) can be of crucial importance for such cases. The GRASP2 code allows semi–empirical eigenvector optimisation, but this has yet to be exploited for relativistic cases.

3. CANCELLATION EFFECTS

The calculation of line strengths and oscillator strengths involves a double summation over amplitudes whose correct relative phases must be known. Two different cancellation effects (1,16,17) are involved. One is due to the signs of the eigenvectors and the other the signs of the radial dipole integrals. If either are incorrect gross errors can occur on individual values. An explanation of cancellation (1,16) is as follows:

In terms of the line strength, S, the weighted oscillator strength gf is given by

$$gf = 8\pi^2 mca_0^2 \sigma S / 3\hbar$$

where a_0, σ, and $\hbar$ are the Bohr unit of length, transition wavenumber, and Planck's constant over 2π, in cgs units. If the level energy eigenfunction $|\gamma J\rangle$ is expanded in terms of eigenvector components and (LS–coupled) basis functions $|\beta J\rangle$ according to the equation

$$|\gamma J\rangle = \sum_\beta y^\gamma_{\beta J} |\beta J\rangle$$

and the qth component of the classical dipole moment of the atom in units of $-ea_0$ is taken as

$$P^1_q = \sum_{i=1} r^1_q(i)$$

Then the square root of the line strength is given by the double summation

$$2\sqrt{S_{\gamma\gamma'}} = \sum_\beta \sum_{\beta'} y^\gamma_{\beta J} \langle \beta J \| P^1 \| \beta' J \rangle y^{\gamma'}_{\beta' J'}$$

The summation is one of amplitudes. Its magnitude may be considerably diminished as a result of destructive interference effects, especially when strong basis function mixing is present due to intermediate coupling and/or configuration interaction perturbations. The cancellation factor

$$CF = \left[\left| \sum\sum y^\gamma_{\beta J} \langle \beta J \| P^1 \| \beta' J \rangle y^{\gamma'}_{\beta' J'} \right| / \sum\sum \left| y^\gamma_{\beta J} \langle \beta J \| P^1 \| \beta' J \rangle y^{\gamma'}_{\beta' J'} \right| \right]^2$$

will then be quite small. When this is the case, there could arise random errors in the resulting gf values. Provisions are made in the atomic–structure codes written by Dr. R.D. Cowan to print cancellation factors on individual transitions. Small cancellation factors therefore indicate which oscillator strengths are subject to errors from eigenvector phase summations. It should also be noted that there is also some measure of cancellation in the calculation of different dipole intergrals: with some being very sensitive to wavefunction details.

4. CONFIGURATION INTERACTION

It is well known that the first step in setting up any model for atomic–structure calculations is to select all the configurations necessary to provide a valid approximation of the configuration interactions which affect the transition arrays under investigation. It is advisable to choose a computer code which allows the inclusion of all the necessary configurations. A large computer memory may be needed to store huge matrices that deal with the many interacting configurations and fast processing power may be required to complete the runs in a reasonable time. Often the problem is not soluble in the most desirable manner as matrices become too large for the computer to handle or else because many of the outer configurations added are themselves so perturbed, by excluded levels, that their addition to the model has adverse effects. Forced approximations for modelling configuration interaction therefore may have to be made and these usually affect the quality of eigenvectors. Steps that can be taken to improve eigenvector representation in such cases will therefore be discussed in the next section.

5. EIGENVECTOR REPRESENTATION

The quality of eigenvectors involved in atomic–structure calculations affect many atomic data including Landé g–values, oscillator strengths and collision strengths. Eigenvectors computed *ab initio* are unfortunately seriously inaccurate for many complex atomic structures because of a multitude of configuration interactions and perturbations. They can however be improved with semi–empirical techniques. The method for the computation of empirical eigenvectors is well established and was developed in the first place as an aid in the classification of atomic spectra and to study the integrals inherent in the calculations. The well known method is to adjust Slater parameters by least–squares optimisation to obtain agreement between observed and calculated energy level differences. The optimised Slater parameters are then used in the Hamiltonian H–matrix to compute eigenvalues and eigenvectors for possible wavelength, oscillator strength or collision strengths calculations. Classical spectroscopy has applied the optimisation technique with great success to establish energy level designations but the demands of level classification on the method are far less stringent than for obtaining eigenvectors for oscillator strength calculations. Therefore its application to the calculation of oscillator strengths is open to criticism if not properly applied. The reason for this is that for level designation the user may be satisfied to learn the main LS, or other coupling scheme, basis–state contribution to the level. An approximate eigenvector may serve to label the level, but this information is not usually adequate to obtain oscillator strengths. This is because for oscillator strengths it is essential to know **all** the basis state contributions to the level, through the amplitudes of the eigenvectors, with a few per cent accuracy together with the phase of the eigenvector for both upper and lower configurations. Without these conditions being met the errors would be unacceptable. To achieve this end great care has to be taken in the parameter optimisation process as the eigenvectors will become meaningless if the Slater parameters are all allowed to drift to physically unacceptable values. Least squares fits are not unique: for if enough adjustable parameters are available all manner of solutions can be obtained, but only one will be correct in nature. Valid solutions can however be achieved if certain ground rules are observed.

To apply the parameter optimisation method it is first necessary to calculate an *ab initio* set of Slater parameters with a reliable method ensuring that that all the main strong or medium interactions are taken into account in the calculations by including all the necessary configurations. The parameters concerned are the centers–of–gravity energy (E_{av}), the single–configuration direct (F^k) and exchange (G^k) coulomb–interaction radial integrals, the spin–orbit parameters (ζ) and the configuration interaction (R^k) integrals. These are then adjusted using least squares optimisation codes so that calculated energy levels agree with measured ones. Eav generally presents no problem but the ζ values are often small and, if left free, can absorb perturbations in a physically unacceptable manner. A sensible solution is to adopt ζ values computed with the Blume–Watson method (18). Likewise small values of F^k and exchange G^k could be seriously affected by unrelated interactions and therefore may need restricting. The optimisation codes usually allow specified parameters to be tied together in such a way that their fitted ratios are allowed to vary in line with their *ab initio* ratios. Fixed *ab initio* ratios could be maintained for all F^k and or all G^k or else for F^k and G^k values between the same orbitals with considerable physical justification. Other prudent choices are either to vary all R^k integrals together in line with their *ab initio* ratios or else to fix them somewhere between 85% and 100% of their *ab initio* values. Usually, to be physically reasonable, the adopted F^k and G^k values should lie between 90% and 100% of their *ab initio* values but there can be exceptions to this criteria. For satisfactory results clearly there must be more measured energy levels than free parameters. Moreover the free parameters should also be sensitive to the available measured levels. This sensitivity can be checked from the matrix of partial derivatives of eigenvalues with respect to the parameters which is printed in some level–fitting codes (1). Optimisation is more likely to be feasible for complex configurations in which many energy levels are governed by a few parameters. For very complex configurations there may be too few parameters to adjust to obtain a satisfactory solution. The effects of far–off interactions may have to be modelled by the introduction of effective operators (see Ref.1 pp 478 for a detailed description). Based on theoretical considerations these are added to the matrices to mimic systematic trends. For instance Trees (see Ref.1 pp 478) introduced a parameter with a $L(L+1)$ dependence. Such parameters can be applied with care to reduce residual errors which would otherwise wrongly predict mixing for sensitively interacting levels.

Further insight into the problems can be gained by considering the specific examples arising in different groups of ions. Multiply–ionised atoms with complex structures such as d^q nl or f^q nl with q>2 (eg: (19) Fe III to Fe VI $3d^q$ nl) are ideal for optimisation because the configurations consist of many levels which are gathered together into well defined regions and are situated well below the continuum. This is not the case for many neutral atoms such as Fe I (16,17) where single configurations extend from near the ground level right up into the continuum. The resulting perturbations, with unclassified levels lying in or near the continuum and from inner shell interactions, are a mystery. It is therefore not surprising that attempts to date to calculate oscillator strengths for Fe I are unsatisfactory. The problems for Fe I are probably not confined to eigenvector inadequacies but are also compounded through uncertainties in radial dipole integrals. Other less complex neutral ions involve perturbation problems which are also difficult to resolve conclusively. These result from the overlapping of complexes of different principal quantum number. In several ions levels of the lowest principal quantum number can reside in the continuum. It is dangerously easy to run a computer code and generate a set of numbers for such ions, but difficult to generate reliable data. Fortunately observational data may be more readily available for neutral ions. In Fe I for instance measured rather than calculated data should be trusted. On the other extreme for hydrogen, helium or lithium like ions, the simpler structures would not generally be studied with the aforementioned fitting techniques. The over–riding considerations are entirely different. For these simpler systems the degree of sophistication of the *ab initio* codes can be of paramount importance along with the quality of wavefunctions and added relativistic and QED (8–11) corrections. Such finesse however has little relevance to ions of intermediate complexity such as Fe X to Fe XII (20–26) as the best *ab initio* codes fail to predict energy levels with sufficient accuracy due to the dominance of configuration interaction problems. Furthermore each configuration in an ion may be subject to different influences which require different solutions. Those with high principal quantum number lying close to the continuum can be problematic since they can be subject to unknown perturbations from unidentified and unmeasured energy levels. This highlights the need for the future identification of these high energy levels, not because they necessarily contribute directly to observed spectra of interest to astrophysicists or other users, but because knowledge of their position allows the determination of their perturbation influence on levels of primary concern.

With optimisation completed and eigenvectors and Slater parameters adopted, independent checks as to their validity are needed. Measured oscillator strengths are available for many ions (27—29) and these can be compared with values calculated using the optimised eigenvectors. The usefulness of the comparison is however limited by the fact that the measured data itself is often subject to large error-bars and uncertainties and the eigenvectors are only one of the atomic quantities affecting the accuracy of the calculated data. Another check is to use the eigenvectors to calculate Landé g–values and compare these with measured values where they are available. In intermediate coupling Landé-g values are critically dependent on the eigenvector alues. The eigenvector representation in the adopted model will therefore not adversely affect the accuracy of atomic data for transitions between levels where the computed and measured Landé g–values are in agreement. For example, our recent calculations in the rare earth elements Ce II (30) and Yb II (31) used the comparison of calculated with measured Landé g–values to aid the least squares optimisation while in Fe II (32) similar comparisons were applied to check earlier results (33,34).

Although the method of Slater parameter optimisation is more usually essential for the more complex configurations there are examples where it may be needed to correct data for some levels in quite simple structures. For instance, oscillator strengths were calculated for beryllium–like ions (35,36) . Although their simpler structure meant that *ab initio* calculations could be made with acceptable accuracy for most levels, the 2s3p ^{1}P and 2s3p ^{3}P levels were found to be hypersensitive to mixing so that the precise eigenvectors obtained through optimisation were necessary.

6. SMALL OSCILLATOR STRENGTHS

Small oscillator strengths can be subject to gross errors if one of the levels is mixed with a level associated with a large oscillator strength. Imperfections in computed eigenvectors give rise to small predicted differences in the level's composition and therefore greatly overestimate the smaller oscillator strength. On the other hand large oscillator strengths are relatively insensitive.

7. RADIAL DIPOLE INTEGRALS

The precision of the calculated radial dipole integrals will also affect the accuracy of oscillator strengths. These depend on the computed one–electron radial wavefunctions and not on the eigenvectors and so results from the least–squares fitting are not applicable to them. One radial integral can be computed for each transition array and contributions are summed for each transition. If these are in error then some general scaling of the oscillator strengths might be expected. Any resulting inaccuracies in the oscillator strength values tend to be systematic for an entire transition array and may vary with transition array. It is therefore worth critically examining measured oscillator strengths or lifetimes to see if they are sufficiently reliable to obtain scale factors for the adjustment of calculated values. There are a number of recent accurate lifetime measurements which can be utilised. For instance laser–excited fluorescence lifetime measurements for Fe II were used to infer an implied scaling down by 0.73 for Fe II oscillator strengths and slightly different scale factors were obtained for Fe I to Fe VIII (16,17,19,33,37). It is not however clear whether the necessary scaling is due to radial dipole integrals or interaction effects. Nevertheless this method of correction would seem to be a reasonable remedy for *ab initio* radial dipole integral errors. Semi–empirical methods such as Coulomb and scaled Thomas–Fermi are available to estimate values for these radial integrals. These estimates are often surprisingly good.

8. LENGTH, VELOCITY AND ACCELERATION FORMULATION

When calculating electric dipole matrix elements for oscillator strength calculations length, velocity or acceleration forms can be used. Given perfect wavefunctions they should be identical but, with the exception of simple atoms, wavefunctions are usually approximate and the consequent results are not in agreement. The quality of the wavefunctions is therefore a fundamental consideration. The accuracy of wavefunctions as well as other atomic quantities can depend on the method of calculation. In some cases variational wavefunctions can have sufficient accuracy to justify velocity–type oscillator strength calculations (which are sensitive to errors at small radius) but more frequently approximate wavefunctions like Hartree–Fock are chosen on the grounds that they can have the correct asymptotic form and hence be useful in length formulation oscillator strength calculations (see Ref. 1 pp 401). Length–type calculations have in fact been compared extensively in the literature with measured data and reasonable correlation has meant that these are often adopted. The acceleration values are unstable and seldom relied on.

9. APPLICATIONS TO COLLISION STRENGTH CALCULATIONS

It was also demonstrated, with collision strength calculations for Fe XIII (38,39), that the procedure involving the adjustment of Slater parameters to improve energy levels and eigenvectors can be applied to improve collision strengths as well since the collision target is described by the same eigenvectors. To carry out this project it was necessary to modify the DISTORTED WAVE (40) collision code and SUPERSTRUCTURE (14) atomic–structure code which mutually interface via a transformation code JAJOM (41) which processes their output. These codes were written at University College London. Similar adaptations were made to the Belfast CIV3 (13) atomic–structure code for possible interface with the Belfast close coupling R–matrix codes (42). Other collision strengths (43) were calculated for Fe XIX.

10. DISCUSSION

It is worth noting here that for line classification purposes only, many classic papers have been securely based on single configuration approximations, and that the mere adoption of such a single configuration model will usually not suffice for oscillator strength calculation. Furthermore before the advent of modern fast and large–memory computers restricted configuration sets were applied to oscillator strength calculations which should now be superseded. Errors can now be reduced through application of methods discussed in this paper while carefully considering all contributing factors. This can yield data which is useful especially when well resolved spectra are available and where the study of plasma parameters through the analysis of individual line intensities is the aim of the investigator. A contrasting requirement is for oscillator strength data to aid the understanding of lower resolution spectra consisting of blended emission or absorption lines. For this a "blanket" approach is sometimes adopted in which *ab initio* or very approximate computations are applied to understand absorption or emission around cosmic sources. In such computations data for transition arrays are sometimes included to which various perturbations and configuration interactions contribute very large errors. With some care the blanket approach can, taking advantage of the sum rules (see Ref 1 pp 424) provide a meaningful analysis but its limitations are severe. Not only is the wavelength distribution of the oscillator strength uncertain but also the effects of specific resonances obscure. The data which either has or can in principle be computed with higher accuracy become overlaid with a mass of inferior data. It is therefore worthwhile excluding as far as possible the data with larger error bars and focussing on a selection of the most reliable calculations. A matter of grave concern is the facile misuse of large computers to archive large quantities of low grade material into data banks which the uninformed may use.

With oscillator strengths it is not possible to define error—bars for large numbers of values as each individual value is affected in many ways: by different contributions from the various types of dependancy detailed above and especially by any level mixing effects. With these limitations in mind the user can critically select the particular values required for a problem and try as far as possible to avoid values (such as small oscillator strengths associated with poorly defined mixed energy levels, or those with small cancellation factors) before resorting to detailed line intensity analysis. Now that atomic structure codes and fast computers are generally available, important cases can be rerun to generate output revealing accuracy information (eg: eigenvectors, mixing and cancellation). This can allow inspection in greater detail before application to expensive projects. For those posessing the necessary atomic—structure codes input of the reported optimised Slater parameters can facilitate duplication of printed data which could also have been archived in a data bank. It should also be realised that the accuracy of published oscillator strengths can be improved whenever new observational energy level data becomes available to be included in the least—squares fitting procedure. Frequently the data available at any one time is lacking in important multiplets, or even all terms of a particular multiplicity, and quite generally important interacting configurations may be unclassified. Furthermore the availability of increased computing power permits improved models with more configuration interaction. The reduction of error—bars is thus an evolving process.

The crucial test for calculated oscillator strengths must come from comparison with laboratory measurements. Oscillator strengths, lifetimes and transition probabilities are now being measured by a number of new, more reliable, techniques. Some, such as laser—fluorescence, have excellent precision. Beam—foil and beam—gas measurements are constantly improving and supplying data for multiply—ionised atoms which are lacking for many ions. Historically experimental oscillator strength measurements have been contradictory but as techniques improve data for some ions converge to agreement. Fortunately for users, the National Bureau of Standards, Washington has maintained critical data compilations(27—29) and bibliographies for both measured and calculated oscillator strengths which are an invaluable aid in an area of science so full of uncertainties and yet so vital as a key to our understanding of astrophysical observations.

REFERENCES

(1) Cowan, R. D. "The Theory of Atomic Structure and Spectra" (Author's description of HFR) (Univ. California Press, Berkeley and Los Angeles, 1981).
(2) Cowan R. D, J. Opt. Soc. Am. 58, 808, 924, (1968)
(3) Fawcett, B.C., Rutherford Appleton Laboratory Report RL—83—030 (Revised Edition). (1984).
(4) Cowan, R.D., Fawcett, B.C., Grant, I,P., Rose, S.J., Rutherford Appleton Laboratory Report RAL—85—098. (1985)
(5) Fawcett, B.C., Rutherford Appleton Laboratory Report RAL—87—051.(1987)
(6) Froese Fischer C, (1972) Comput Phys Commun. 4, 107 and *ibid* (1974) 7, 236 (Author's description of MCHF).
(7) Froese Fischer C, (1977) "The Hartree—Fock method for Atoms", Publ. J. Wiley, NY. Author's description of MCHF).
(8) Grant, I. P. Mckenzie B. J. Norrington P. H. Computer Physics Comm. 21, 207 (1980) *ibid* 21, 233. (Author's description of MCDF)
(9) Dyall, K.G. Grant, I.P. Johnson, C.T. Parpia F.A. and Plummer, E.P., Comput. Phys. Commun. 55, (1989) 425 (Author's description of GRASP)
(10) Parpia, F.A. Grant, I.P. Dyall, K.G. and Fischer, C.F. (In preparation) (Author's description of GRASP2).
(11) Desclaux, J. P., Computer Phys. Commun. 9, 31 (1975)

(12) Klapisch, M, C. R. Acad. Sci. B **265**, 914 (1967)
(13) Hibbert, A., Computer Phys. Commun. **9**, 141–172 (1975)
(14) Eissner, W. Jones M. and Nussbaumer H., Comput. Phys. Commun. **8**, 230 (1974).
(15) Fawcett, B.C., Physica Scripta Vol. **34**, 331–336 (1986).
(16) Fawcett, B.C. Rutherford Appleton Laboratory Report RAL–87–114 (1987).
(17) Fawcett, B.C. Cowan, R. D. (1988) J. Quant. Spectrosc. Radiat. Transfer **40**, 15–20 (1988).
(18) Blume M. and Watson R. E., Proc. R. Soc. A 270, 127 (1962), **271**, 565 (1963).
(19) Fawcett, B.C. Atomic Data and Nuclear Data Tables **41**,182–255 (1989).
(20) Bromage, G.E., Cowan, R.D., and Fawcett, B.C., Physica Scripta **15**, 177–182 (1977).
(21) Bromage, G.E., Cowan, R.D., Fawcett, B.C., Mon. Not. R. astr. Soc. **183**, 19–28 (1978).
(22) Fawcett, B.C., Atomic Data and Nuclear Data Tables **35**, 185–202 (1986).
(23) Fawcett, B.C., Atomic Data and Nuclear Data Tables **35**, 203–222 (1986).
(24) Fawcett, B.C., Atomic Data and Nuclear Data Tables **36**, 129–149 (1987).
(25) Fawcett, B.C., Atomic Data and Nuclear Data Tables **36**,151–166 (1987).
(26) Cowan, R.D., Bromage, G.E., Fawcett, B.C.,. Mon. Not. R. astr. Soc. **210**, 439–442 (1984).
(27) Wiese W. L. and Fuhr J. R., J. Phys. Chem. Ref. Data **4**, 263 (1973)
(28) Younger S. M., Fuhr J. R., Martin G. A., and Wiese W. L., J. Phys. Chem. Ref. Data **7**, 495 (1978)
(29) Fuhr J. R., Martin G. A., Wiese W. L. and Younger S. M., J. Phys. Chem. Ref. Data **10**, 305 (1981)
(30) Fawcett, B.C., Atomic Data and Nuclear Data Tables , "Computed Oscillator Strengths for Ce II and Landé–g values" **46** (1990).
(31) Fawcett, B.C. and Wilson, M. (Accepted 1990) Atomic Data and Nuclear Data Tables "Computed Oscillator Strengths and Lifetimes of Low–lying Levels of Yb+".
(32) Fawcett, B.C., (Accepted 1990).Physica Scripta ." Landé–g values for Fe II applied to error determination for oscillator strengths and data depending on eigenvectors"
(33) Fawcett, B.C., Atomic Data and Nuclear Data Tables **37**, 333–364 (1987).
(34) Fawcett, B.C., Atomic Data and Nuclear Data Tables **40**,1–8 (1988).
(35) Fawcett, B.C., (1984). Atomic Data and Nuclear Data Tables **30**, 1–26 (1977).
(36) Fawcett, B.C., Atomic Data and Nuclear Data Tables **33**, 479– 482 (1985).
(37) Fawcett, B.C., Atomic Data and Nuclear Data Tables **43**, (1989).
(38) Fawcett, B.C. and Mason, H.E. Laboratory Report RAL–89–010 (1989).
(39) Fawcett, B.C. and Mason, H.E. Atomic Data and Nuclear Data Tables **43**, 245–258 (1989), *ibid.* **47** (1990).
(40) Eissner W. and Seaton, M. J., J. Phys. B. Atom. Molec Phys. **5**, 2187 (1972).
(41) Saraph, H. E., Computer Phys. Commun. **3**, 256 (1972)
(42) Berrington, K. A., Burke P. G., Dourneuf, M. L., Robb, W. D., Taylor, K. T. and Lan, Vo Ky. Computer Phys. Commun. **14**, 367–412 (1978)
(43) Fawcett, B.C. and Mason, H.E. Atomic Data and Nuclear Data Tables (Accepted 1990).
(44) Fawcett, B.C., Atomic and Nuclear Data Tables **22**, 473–489 (1978). "Oscillator strengths for $2s^22p^n-2s2p^{n+1}$ and $2s^2p^{n+1}-2p^{n+2}$ transitions and for $2s^22p^n$. "Forbidden" transitions".
(45) Bromage, G.E., and Fawcett, B.C., Mon. Not. R. astr. Soc. **178** pp. 605–610 (1977). "The $2p^2-2p3d$ transition array of Fe XXI and isoelectronic ions".
(46) Bromage, G.E., and Fawcett, B.C., Mon. Not. R. astr. Soc. **178** pp. 591–598 (1977). "The $2p^4-2p^33d$ transition array of Fe XIX and isoelectronic ions".
(47) Bromage, G.E., Fawcett, B.C., Mon. Not. R. astr. Soc. Vol.**179** pp. 638–689. "The $2p^3-2p^23d$ transition array in Fe XX and isoelectronic ions".
(48) Fawcett, B.C., Atomic Data and Nuclear Data Tables **30**, 423–455 (1984). "Calculated oscillator strengths for n = 2–2 and n = 2–3 transitions in Be I, B II, C III and N IV".

(49) Fawcett, B.C., Atomic Data and Nuclear Data Tables **31**, 495–549 (1984). "Calculated wavelengths, oscillator strengths and energy levels for n = 2–2 and 2–3 transitions in F–like ions Mg IV to Ni XX and also 3–3 transitions for Mg IV, Al V and Si VI".

(50) Fawcett, B.C., Atomic Data and Nuclear Data Tables **34**, 215–260 (1986). "Calculated wavelengths and oscillator strengths of n = 2–2 and 2–3 transitions for ions in the O I–like isoelectronic sequence between Mg V and Ni XXI".

(51) Fawcett, B.C., Atomic Data and Nuclear Data Tables **37**,367–410 (1987). "Calculated wavelengths oscillator strengths and energy levels for allowed 2–2 and 2–3 transitions for ions in the C–like isoelectronic sequence between F IV and Ni XXIII".

(52) Fawcett, B.C., Atomic Data and Nuclear Data Tables **37**,411–418 (1987). "Oscillator strengths of allowed transitions for C I, N II and O III".

(53) Fawcett, B.C., Physica Scripta. **30**, 326–334 (1984). "Oscillator strengths for Al IV to Ar IX and revised energy levels of Ar IX".

(54) Fawcett, B.C., Jordan, C., Lemen, J.R., and Phillips, K.J.H., Rutherford Appleton Laboratory Report RAL–86–094. (1986). "New spectral line identifications in high temperature flares". with Appendix: Fawcett, B. C. " "Tables of Wavelengths and Oscillator Strengths for n=2–4 Transitions in Fe XXI, Fe XXII and Fe XXIII and Energy Levels"

(55) Bhatia, A. K., Fawcett, B.C., Lemen,J.R, Mason, H.E. and Phillips, K.J.H., Mon. Not. R. astr. Soc. (1989). "A comparison of theoretical and solar–flare intensity ratios for the Fe XIX X–ray lines".

Atomic Structure Calculations in Breit-Pauli Approximation

W. Eissner

Dept of Applied Mathematics and Theoretical Physics

The Queen's University

Belfast BT7 1NN, UK

1 Introduction

Results are presented of relativistic effects in one- and two-electron atomic systems using the low-Z Breit-Pauli (BP) approximation. In the structure calculations we include all one-body Breit-Pauli terms, i.e. ordinary spin-orbit coupling and the two non-finestructure contributions. Non-finestructure two-body terms are consistently ignored, whereas mutual spin-orbit effects — mutual spin-orbit coupling and spin-other-orbit coupling — are included in the atomic structure type calculations presented here but ignored in the collision-type approach. For the latter we use the BP variant of the R-matrix suite of programs (RM-BP, Scott and Taylor 1982), for the former the SUPERSTRUCTURE code (SSTRUCT, Eissner et al. 1974). Apart from level energies results to various degree of approximation are presented for radiative transitions.

2 Formulation

In their classic *Quantum Mechanics of One- and Two-Electron Atoms* Bethe and Salpeter give a full account of the Breit equation and their low-Z Pauli form as approximations of a fully relativistic theory. In this paper only a few points of interest in deriving the BP terms are sketched. In another chapter of this volume Ian Grant deals with the state-of-the art for relativistic treatment of many-electron atomic structure and excitation.

2.1 Schrödinger and Dirac equation

An approach to atomic structure in Breit-Pauli approximation for N electrons in the electric field of a nucleus with Z protons starts with suitably bounded solutions

$$\Psi = \Psi(\gamma S L M_S M_L | \vec{r}_1, \ldots \vec{r}_N) \tag{1}$$

of the time-independent Schrödinger equation

$$\left[\sum_{i=1}^{N} \left\{ -\frac{\hbar^2}{2m_0} \Delta_i - \frac{Ze}{r_i} + \sum_{j>i} \frac{e}{r_{ij}} \right\} \right] \Psi = E\Psi, \tag{2}$$

i.e. of a quantized non-relativistic energy expression H_{nr}, where the symbols have the ususal meaning of rest mass, electric charge, angular momentum quantum numbers etc. Antisymmetrization of Ψ is readily achieved after expanding as a product of single-particle wave functions, and it turns the coupled second-order differential into integro-differential equations.

Including effects of the Bohr magneton $\mu_B = \frac{e}{m}\frac{\hbar}{2c}$ along with all the other terms of relative order α^2 in the finestructure constant

$$\alpha = \frac{e^2}{\hbar c} \qquad (\approx 1/137.036\,0) \tag{3}$$

as a perturbation leads to wave functions in intermediate coupling (IC),

$$\Psi = \Psi(\Gamma SLJM_J|x_1,\ldots x_N), \tag{4}$$

which are eigenvectors of a Breit-Pauli hamiltonian H_{BP} (x stands for position $\vec{r}$ and spin $\vec{\sigma}$ of an electron). Such a hamiltonian can be derived in Dirac theory by retaining terms of absolute order α^2 and α^4, having exploited the low-velocity matrix representation

$$D(\gamma^0) = \frac{1}{i}\begin{pmatrix} I & 0 \\ 0 & -I \end{pmatrix}, \qquad D(\gamma^\mu) = \begin{pmatrix} 0 & -i\sigma^\mu \\ i\sigma^\mu & 0 \end{pmatrix} \tag{5}$$

of the underlying Clifford algebra. These 4×4 matrices are composed of the Pauli matrices σ^μ and the associated identity I. We note that

$$\mathrm{Ry} = \frac{\alpha^2}{2} m_0 c^2 \qquad (= 13.59\,\mathrm{eV}) \tag{6}$$

is the binding energy of the non-relativistic hydrogenic electron in the ground state; equally useful in subsequent conversions proves

$$a_0 = \frac{\hbar}{m_0 c}/\alpha \qquad (= 0.529 \cdot 10^{-8}\,\mathrm{cm}), \tag{7}$$

which relates the Bohr radius a_0 to the Compton wave length of the electron.

The one pervading one-body contribution to BP order, though, follows simply — as the third term after those of order $m_0 c^2$ and Ry — on expanding $m = \frac{m_0}{\sqrt{1-\beta^2}}$ for $\beta = \frac{v}{c} \ll 1$: this 'mass-velocity' term gives rise to one-body integrals ϵ_{nl}^m, which are the expectation values of an operator $f(m)$ with radial component

$$\frac{\alpha^2 \mathrm{Ry}}{4}\left(\frac{p a_0}{\hbar}\right)^4 = \frac{\alpha^2 \mathrm{Ry}}{4}\frac{\partial^4}{\partial(r/a_0)^4}. \tag{8}$$

Such a readily available quantity is of help when verifying reduced units used in some publication — the topic of the subsequent Section 2.2. The quantity $f(m)$ will fall into place in Section 2.4.

Other one-body terms follow straight from the Dirac hamiltonian $i\gamma^\mu \partial_\mu + \frac{\alpha}{e}A_\mu - \frac{m_0 c}{\hbar}$, and the two-body terms after suitably simplifying the $1/r_{ij}$ interaction. But as this paper is concerned with computational results we cast physics aside, summarising such preliminaries only in the form of number equations in Section 2.3 — which still bring out all the kinematic and scaling properties.

2.2 Observables as reduced variables

For computational work one extracts pure number equations, choosing the most natural or convenient unit for one quantity of each canonical pair — and phase invariance settles the case for the canonical conjugate! Using ALGOL notation for brevity we define the following dimensionless quantities instead of length, energy and angular momentum, parenthetically adding consequential redefinitions:

$$r := \frac{r}{a_0} \qquad \text{(hence wave numbers } \tfrac{p}{\hbar} \text{ transform as } k := k \cdot a_0)$$

$$E := \frac{E}{\mathrm{Ry}} \qquad \text{(hence } t := \frac{t}{\tau_0} \quad \text{where } \tau_0 = \frac{2 a_0}{\alpha c} = 4.839 \cdot 10^{-17} \cdot \mathrm{sec}) \tag{9}$$

$$d := \frac{d}{\hbar}$$

In tables though physical rather than reduced quantities will still be displayed for clarity.

Except perhaps for the last redefinition in (9) this is only superficially similar to the procedure in special relativity: there the speed of light c may be absorbed in quantities like mc^2 or $\beta = \frac{v}{c}$. But in such a case the new quanties loose nothing of their physical nature, nor do constraints such as phase space invariance apply. Although relevant in a Dirac approach we need not be concerned with such niceties, as Breit-Pauli treatment is distinctly not Lorentz-invariant.

2.3 Methods and codes

Use of spherical spatial coordinates yields Hartree-Fock type radial equations for the non-relativistic problem, the size being determined by the specified configurations $C = \prod (n_k l_k)^{q_k}$. Then wave functions and operators factorize, and the angular portions can be worked out by angular momentum algebra and Racah techniques. All that remains to be solved are radial equations. Two approximations are adopted in the following:

- in SSTRUCT, an atomic structure code in intermediate coupling described by Eissner et al., one starts off with independent particle radial functions satisfying

$$\left[\frac{d^2}{dr^2} - \frac{l(l+1)}{r^2} + \frac{2\mathcal{Z}_{\text{eff}}(\lambda_l \cdot r; Z, N)}{r} + \epsilon_{nl}\right] P_{nl}(r) = 0 \qquad (10)$$

in a statistical model (SM) potential $V^{\text{SM}}(r) = \frac{\mathcal{Z}_{\text{eff}}(\lambda_l \cdot r)}{r}$ with scaling parameters λ_l or perhaps even λ_{nl} (usually varied according to some functional $\mathcal{F}$ but in this paper kept at their default value 1.0 — which is tolerable for high residual charge $z = Z - (N-1)$).

- RM-BP is an electron-ion collision code, modelled on the ordinary non-relativistic R-matrix approach (Burke and Robb 1978, Berrington et al. 1978), where the coupled radial integro-differential equations (ID) for the 'colliding' electron are expanded as NRANG2 suitable basis functions within the range $r=$RA to which the target charge distribution is essentially confined. The equations that couple channels of a particular target state and collisional partial wave for some selected angular symmetry have often been outlined; Burke and Eissner (1983) write

$$\left[\frac{d^2}{dr^2} - \frac{l(l+1)}{r^2} + \frac{2Z}{r} + k_i^2\right] F_i(r) = 2 \sum_j^{\text{NCHF}} \left[V_{ij}(r) F_j(r) + \int_0^\infty dr' W_{ij}(r, r') F_j(r')\right]$$
$$+ \sum_{nl} \lambda_{i,nl} P_{nl}(r) \delta_{l, l_i}; \qquad (11)$$

the direct potential V is expanded in the usual way as spherical multipole potentials $y_\lambda(nl, n'l'; r)$, in the exchange term $\int WF$ orbitals P_{nl} and F_j swap places, and the Lagrange sum with parameters $\lambda_{j,nl}$ arises from orthogonality conditions imposed upon F_i for algebraic convenience. For a given angular symmetry the number NCHF increases with that of included target states. Formal extension to a BP hamiltonian is straightforward, Clebsch-Gordan coefficients providing for IC, but the number NCHF of ID's more than doubles over the case of LS coupling: e.g. inclusion of hydrogenic 1s–3d, i.e. of 9 target levels, leads to more than twenty channels for many sets of total angular momentum J and parity π. Methods and scope have been described by Scott and Taylor (1982) and general features and constraints discussed by Jones (1971). In the R-matrix approach one first provides a complete fundamental basis within the range RA. In the outer region beyond RA techniqes employed in codes for the OPACITY PROJECT and summarized by Berrington et al. (1987) have been implemented in the new version of the code. In this paper bound state solutions $F_i^{J^\pi}(r) = P_{nlj\,J^\pi}(r)$ are sought, which means solving an eigenvalue problem for channel energies k_i^2 associated with channels $i = (k_i l_i)$.

2.4 The Breit-Pauli hamiltonian and scaling properties

To varying degree both approaches treat Breit-Pauli terms as perturbations. We start with the full BP hamiltonian, by adding a relativistic correction operator H_{rc} — with the usual abbreviations for c[ontact], d[arwin], m[ass], o[rbit] and s[pin], and a prime indicating 'other' — to the non-relativistic hamiltonian:

$$H_{BP} = H_{nr} + \sum_{i=1}^{N}\{f_i(m) + f_i(d) + f_i(so)\} + \frac{1}{2}\sum_{i \neq j}^{N}\{g_{ij}(so+so') + g_{ij}(ss') + g_{ij}(css') + g_{ij}(d) + g_{ij}(oo')\}. \tag{12}$$

While accounting for all the one-body terms,

$$\begin{aligned} f_i(m) &= -\frac{\alpha^2}{4}\nabla_i^4 \\ f_i(d) &= -\frac{Z\alpha^2}{4}\nabla_i^2(1/r_i) \\ f_i(so) &= \frac{Z\alpha^2}{r_i^3}\vec{l}_i \cdot \vec{s}_i, \end{aligned} \tag{13}$$

among the BP two-body operators only the finestructure (FS) terms are retained here:

$$\begin{aligned} g_{ij}(so+so') &= -\alpha^2\left\{\left(\frac{\vec{r}_{ij}}{r_{ij}^3} \times \vec{p}_i\right)\cdot(\vec{s}_i + 2\vec{s}_j) + \left(\frac{\vec{r}_{ij}}{r_{ij}^3} \times \vec{p}_j\right)\cdot(\vec{s}_j + 2\vec{s}_i)\right\} \\ g_{ij}(ss') &= 2\alpha^2\left\{\frac{\vec{s}_i \cdot \vec{s}_j}{r_{ij}^3} - 3\frac{(\vec{s}_i \cdot \vec{r}_{ij})(\vec{s}_j \cdot \vec{r}_{ij})}{r_{ij}^5}\right\} \end{aligned} \tag{14}$$

where $\vec{r}_{ij} = \vec{r}_i - \vec{r}_j$. There is little point worrying about the omitted 2-body terms as long as one has not pushed ordinary configuration expansion (CI) — since interelectron correlation scales as $1/Z$ — sufficiently far for BP terms to matter.

A further approximation has been made in the current collision type RM-BP calculations, where the two-body finestructure terms $g(so+so')$ and $g(ss')$ are also omitted, which clearly is not a serious omission for sufficiently high ionisation stages — apply scaling laws!

Having written $\frac{1}{2}\sum_{j \neq i}$ rather than $\sum_{j>i}$ in the two-body sum — a difference of no great account as long as one deals with indistinguishable particles — smoothes the way towards the following verifying discussion. If one takes particle j for the 'fixed' nucleus then the mutual *spin–orbit* term $g_{ij}(so)$ equals the one-body $f_i(so)$ on changing electric charge sign and magnitude (while center-of-mass considerations take care of factors two). Since we assume the 'nucleus' j be at rest i experiences no *other* orbit and $g_{ij}(so')$ vanishes. So does the companion term $g_{ji}(so) = 0$, whereas spin–other-orbit term $g_{ji}(so')$ — on changing μ_B to protonic μ_p or in general to the magnetic moment of the nucleus — becomes the magnetic contribution to the hyperfine structure (HFS). However HFS is ignored in this paper.

The all-important scaling laws follow at a glance, on observing that the length r scales as the inverse of Z: consider the hydrogenic radial equation in $\rho = Zr$, i.e. after absorbing Z. Hence all the one-body BP terms scale as $\alpha^2 Z^4$ and — a feature familiar from H_{nr} — the two-body contributions one power less in Z. This has obvious repercussions on the scaling of the BP components in IC wavefunctions Ψ: expand perturbatively, observing in particular that the expectation value of $f(so)$ vanishes for half-filled subshells nl, and equally take note that the energy denominator will scale as Z rather that Z^2 if there is no change in the principal quantum number for the dominant configurations.

The hydrogenic solution suggests and quantum defect theory establishes the relation

$$E_{nl} = -\frac{z^2}{\nu_l^2} \tag{15}$$

as a powerful tool for expressing the energy of an electron bound by the potential of a 'core' or 'target' with residual charge number z Of course quantum defects $\mu = n - \nu$ of a series will be modified at the low-n side by BP terms, and for H-like ions they will no longer by zero.

Table 1. Lowest levels in H-like ions computed with the SSTRUCT code; also shown are the non-finestructure contributions that have been included: the m[ass-velocity] operator and — contributing to s-orbitals only and very much reducing the mass correction — the Darwin term, as they are indicative of the magnitude of the BP contribution in E_{nlj}.

Bound states of H-like ions in BP approximation

H-like nlj	Z = 13		Z = 26		Z = 52	
	$\dfrac{E_{nlj} - E_{1s}}{\text{Ry}}$	$\dfrac{\epsilon_{nl}^{m+d}}{\text{Ry}}$	$\dfrac{E_{nlj} - E_{1s}}{\text{Ry}}$	$\dfrac{\epsilon_{nl}^{m+d}}{\text{Ry}}$	$\dfrac{E_{nlj} - E_{1s}}{\text{Ry}}$	$\dfrac{\epsilon_{nl}^{m+d}}{\text{Ry}}$
1s	0	−.380	0	−6.083	0	−97.335
$2s_{\frac{1}{2}}$	127.0121	−.119	511.2254	−1.901	2097.635	−30.417
$2p_{\frac{1}{2}}$	127.0118	−.055	511.2020	−0.887	2096.118	−14.195
$2p_{\frac{3}{2}}$	127.1070		512.7341		2121.144	
$3s_{\frac{1}{2}}$	150.5611	−.042	606.3481	−0.675	2493.210	−10.815
$3p_{\frac{1}{2}}$	150.5610	−.023	606.3405	−0.375	2492.783	−6.008
$3p_{\frac{3}{2}}$	150.5889		606.7798		2499.300	
$3d_{\frac{3}{2}}$	150.5889	−.008	606.7788	−0.135	2499.239	−2.163
$3d_{\frac{5}{2}}$	150.5983		606.9290		2501.643	
E_{1s}	−169.3808 Ry		−682.1152 Ry		−2803.2886 Ry	
ν_{1s}	.99888		.99551		.98213	

3 The low-lying levels in the H and He isoelectronic sequence

Favourite among ions with a value Z amenable to BP treatment are those of iron, which is of great astrophysical interest because of of its high stellar abundance. As we wish to explore trends along a sequence, in particular the scaling behaviour, aluminium with half the charge number Z of iron becomes a convenient next choice, also because of astropysical interest in it. The value $Z = 13$ makes it likely that two-body FS terms — ignored in the RM calculations — are still small compared to one-body spin-orbit coupling (as borne out in Table 3).

Table 1 summarizes the salient features of H-sequence atomic structure. Doubling Z to 52 helps to highlight trends even when somewhat exceeding

$$\alpha^2 Z_{\text{eff}}^3 \ll 1, \tag{16}$$

the condition for the BP approximation to be valid. Clearly hydrogenic ϵ^{m+d} scales rigorously as Z^4. The bottom line of the table conveys a message that will be useful when assessing structure of He-like ions: BP terms affect quantum defects by values in the order of 0.005 at $z \sim 25$ and even less for smaller z, and this deviation also drops fast towards the series limit.

Selective results from extensive RM-BP bound state calculations for He-like Fe and Al are compiled in tables 2; for comparison effective quantum numbers derived from observation and (up to obitals 3d) from SSTRUCT calculations have been included. In IC calculations the dominant SL readily serves as a label on levels with the same symmetry J^{π}. Thus the 2-electron symmetries $^1P_1°$, $^3P_1°$ and $^3D_1°$ can combine linearly to 3 distinct levels 1^- in an RM-BP calculation based on the 9 target states in table 1. Likewise the 4 symmetries $^3P°$, $^1D°$, $^3D°$ and $^3F°$ make up the IC complex 2^-. No attempt was made to fill gaps. In particular in the 9-state case those bound levels not associated with transitions presented in Section 4.1 have been omitted.

Effective quantum numbers or quantum defects converge remarkably well with increasing correlation. Most noteworthy is the quality of the one-channel result for the ^1S ground state, but then $\{A^{+X} 1s\}1's \; ^1S$ — close enough to unrestricted Hartree-Fock — with a H-like positive ion A^{+X} is a big improvement over $A^{+(X-1)}1s^2 \; ^1S$. Higher levels $\{A^{+X}1s\}nl \; ^{2S+1}L_J$ in simple 1-state calculations behave increasingly poorly, and the finestructure splittings are way off.

Table 2a. Effective quantum numbers $\nu(1snl)$ of Fe XXV: derived from S[STRUCT] $\lambda_l = 1$. calculations (as well as making use of $E_{1s\,2s}$ from Table 1 and R_∞), computed by RM-BP including 1, 4, and 9 levels, and derived from o[bserved] wavelengths (again with $R_\infty = 109\,737.3$ cm^{-1}) as presented by Fuhr et al. (1981). RM-parameters in the 1, 4 and 9 state calculations were as follows: NRANG2 $= 10, 20, 20$ and RA $= 0.5$, 1.0, 4.0.

Bound states of He-like Fe XXV

$J^\pi =$		0+	1+		0−	1−		2−	
n		$^1S_0$	$^3S_1$	$^3D_1$	$^3P_0^\circ$	$^3P_1^\circ$	$^1P_1^\circ$	$^3P_2^\circ$	$^3F_2^\circ$
1	S	0.9812							
	1	0.9790							
	4	0.9806	—	—	—	—	—	—	—
	9	0.9808							
	o	0.9814							
2	S	1.9868	1.9727		1.9855	1.9865	2.0025	1.9870	
	1	1.9962				1.9653	1.9827		
	4	1.9805	1.9665			1.9804	1.9967		—
	9	1.9820				1.9820	1.9981	1.9896	
	o	1.9842	1.9691		1.9831	1.9840	1.9992	1.9907	
3	S	2.9915		3.0062	2.9915	2.9924	3.0072	2.9991	
	4	2.9803	2.9677	2.9954		2.9811	2.9965		—
	9	2.9766				2.9767	2.9933	2.9849	
	o	2.9816	2.9688			2.9821	2.9966		
4	4	3.9792	3.9671	3.9942		3.9802	3.9954		
	9	3.9793				3.9788	3.9934	3.9878	3.9975
	o	3.9803	3.9679			3.9811	3.9954		
5	4	4.9780	4.9660	4.9929		4.9790	4.9940		
	9	4.9787				4.9781	4.9927	4.9872	4.9969
	o	4.9796	4.9674			4.9805	4.9948		

Al XII bound states

$J^\pi =$		0+	1−	
n		$^1S_0$	$^3P_1^\circ$	$^1P_1^\circ$
1	S	0.96930		
	1	0.96917		
	4	0.96915	—	—
	9	0.96914		
	o	0.96913		
2	1	1.9782	1.9837	2.0049
	4	1.9842	1.9835	2.0046
	9	1.9854	1.9836	2.0047
3	1	2.9126	2.9843	3.0062
	4	2.9841	2.9836	3.0032
	9	2.9842	2.9834	3.0032

Table 2b. $\nu(1snl)$ of He-like Al XII, calculated as in table 2a. RM-parameters for 1, 4 and 9 levels were as follows: NRANG2 $= 20$ all and RA $= 0.9, 2.2, 4.0$ (as even the diffuse orbitals have dropped at this last radial value a smaller RA might have been more efficient for iso-electronic Fe in table 2a — as follows from $\rho = Zr$-invariance). Fernley et al. (1987) discuss the trend of the quantum defect $\mu = n - \nu$ along the sequence and why its value for $^1P^\circ$ is unusually small and can even be slightly negative — see their Fig. 2 in particular.

4 Radiative transitions

Electric quadrupole (E2) and magnetic dipole (M1) transitions come into their own only in intermediate coupling, even though electric multipole transitions formally exist already in LS coupling. Moreover it turns out that variants of the familiar electric dipole (E1) operator are also affected. As IC wavefunctions contain admixtures of order α^2 radiative operators *must* also be expanded up to BP order. We add them in M1 transitions, but in the velocity formulation they are still left out.

Table 3. He-like transition data computed with SSTRUCT (also for KUTSS=0 — no 2-body FS terms) and with RM-BP (n.b: no 2-body FS terms included!); NBS data for Al are by Wiese et al. (1969), Fe data by Fuhr et al. (1981).

E1 decay of the second and third excited terms

of He-like			Al ($Z=13$)	Fe ($Z=26$)
$1\,^1S_0 - 2\,^3P_1$			7.80 Å	1.8593 Å
SSTRUCT $\lambda_l = 1.$	gf	L	0.00185	0.613
		V	0.00189	0.636
	S	L	0.000 048	0.000 375
KUTSS=0	S	L	.000 041	.000 350
		V	.000 042	.000 364
RM-BP 1-level	S	L	4.445(−5)	0.000 378
		V	5.204(−5)	0.000 405
4-level	S	L	1.018(−4)	.000 424
		V	8.105(−5)	.000 426
NBS	S		—	0.000 421
$1\,^1S_0 - 2\,^1P_1$			7.75 Å	1.8502 Å
SSTRUCT	gf	L	0.759	0.711
		V	0.759	0.729
	S	L	.019 37	.004 33
KUTSS=0	S	L	.019 38	.004 35
		V	.019 39	.004 46
RM-BP 1-level	S	L	.016 86	.004 40
		V	.018 57	.004 57
4-level	S	L	.019 45	.004 31
		V	.018 80	.004 43
NBS	S		.0.019 2	0.004 48

4.1 Electric dipole transitions

In the long wavelength low intensity approximation the probability for spontaneous emisssion by E1 radiation,

$$A^{E1}_{i' \leftarrow i} = 2.6774 \cdot 10^9 \text{sec}^{-1} \cdot \frac{(E_i - E_{i'})^3}{g_i} \cdot S^{E1}(i, i'), \tag{17}$$

may be expressed in terms of the line strength

$$S^{Ek}_L(i', i) = S(i, i') = \left| \langle i' \| R^{[k]} \| i \rangle \right|^2, \tag{18}$$

where the linestrength amplitude $\langle i' \| R^{[1]} \| i \rangle$ is Wigner-Eckart reduced from matrix element $\langle i'|R_\kappa|i \rangle$, R_κ being any spherical tensor component $\vec{e}_\kappa \cdot \vec{R}$ (with polarization 'direction' $\vec{e}_\kappa$) of

$$\vec{R} = \sum_{j=1}^{N} \vec{r}_j . \tag{19}$$

Subscript L indicates 'length'. On replacing

$$r \longrightarrow \frac{2}{E_i - E_{i'}} \left\{ \frac{\partial}{\partial r} + \frac{(l(l+1) - l'(l'+1))}{2r} \right\} \tag{20}$$

in the radial length integrals $\langle n'l'|r_j|nl \rangle$ that contribute to $\langle i' \| R \| i \rangle$ one obtains the velocity form $S_V(i, i')$. Applying hypervirial relations once more leads to the acceleration form, and so forth.

We also consider oscillator strengths f^{abs} or f^{emi}; so as to avoid the need to distinguish between the absorption and emission oscillator strength one often prefers the symmetric (gf) value, by multiplying with the statistical weight g of the initial state:

Table 4. Hydrogenic gf in Length and Velocity formulation and their difference.

gf values of H-like ions

hydrogenic		$Z=8$	13	26	39	52
$1s_{\frac{1}{2}} - 2p_{\frac{1}{2}}$	L	.2766	.2752	.2682	.2565	.2400
	V	.278	.278	.278	.278	.278
	Δgf	.001	.003	.010	.021	.038
$-2p_{\frac{3}{2}}$	L	.5530	.5498	.5346	.5097	.4757
	V	.554	.554	.550	.544	.535
	Δgf	.001	.004	.015	.034	.059
$-3p_{\frac{1}{2}}$	L	.0521	.0510	.0460	.0384	.0292
	V	.0525	.0522	.0504	.0477	.0441
	Δgf	.0004	.0012	.0044	.0093	.0148
$2s_{\frac{1}{2}} - 3p_{\frac{1}{2}}$	L	.5758	.5694	.5391	.4923	.4337
	V	.578	.575	.561	.539	.510
	Δgf	.002	.006	.022	.047	.076

$$(gf)_{i',i} = (gf)_{i,i'} = \frac{|E_i - E_{i'}|}{3} \cdot S(i,i'). \tag{21}$$

Before turning to the main problem of BP corrections in the velocity form of electric multipole operators we discuss general features displayed in table 3: (i) as predicted it makes sense to omit $g(so+so')$ in highly ionized species; (ii) S for spin-allowed transitions scales as Z^{-2} in accordance with (18); (iii) spin-forbidden transitions obey no simply scaling pattern.

Dwelling briefly on singlet-triplet mixing: it suffices to look at the state vector

$$|2\,^3P_1\rangle \simeq |1s2p\,^3P_1\rangle + \frac{\langle 1s2p\,^1P_1|f(so) + g(so+so')|1s2p\,^3P_1\rangle}{\langle 1s2ps\,^1P_1|H_{BP}|1s2ps\,^1P_1\rangle - \langle 1s2p_T\,^3P_1|H_{BP}|1s2p_T\,^3P_1\rangle}|1s2p\,^1P_1\rangle, \tag{22}$$

whose second component alone causes E1 transitions $\langle 1\,^1S_0\|R\|2\,^3P_1\rangle$. Propertionality with $\alpha^2 Z^8$ for S at the low-Z end of the sequence changes quite drastically to Z^2 with increasing charge, when numerator and denominator in (22) are both controlled by BP terms. Only for the denominator does it look very different near the neutral end, which usually scales as Z or Z^2. In this case however terms of order α^0 come about only if triplet and singlet orbital functions $\langle r|2p_T\rangle$ and $\langle r|2p_S\rangle$ are allowed to be different, an effect scaling as $1/Z$ because described by correlation configurations associated with some other complex.

Turning to the operator problems we note that (20) is based on the commutator for a *non*relativistic hamiltonian (2) with spherical components r^κ: thus such an operator can give the same radiative results as the length operator only in LS coupling, and this only for *exact* wave functions (1); approximate wave functions yield differences depending upon the choice of gauge for the electromagnetic four-potential **A**.

Let us start with transitions for hydrogen-like ions, i.e. with wave functions that are eo ipso exact. Table 4 displays the differences Δ in gf, which obviously scale as $\alpha^2 Z^2$. No simple manipulation of the transition energies in (17) or (20) can remedy this situation.

Table 5 contains radiative data for our two He-like ions from recent test runs of the RM-BP code. Corresponding one-channel results are not good enough higher up the series to be worth showing. It can be seen that 4 'frozen cores' give answers of comparable quality to those obtained with 9 levels, i.e. including $n=3$ orbitals among the hydrogen-like target functions. The trend from length to velocity line strengths is qualitatively and it seems quantitatively reminiscent of the lesson drawn from table 4 and echoed in the SSTRUCT results.

For the spin-forbidden E1 transition $n\,^1S_0 \leftarrow n\,^3P_1$ an analytic answer has been given by Drake (1976), who supports it by numerical results for the decay to $1s^2\,^1S$ as well as to $1s2s\,^1S$ in neutral helium. Drake shows that the transition amplitude in the length form describes both the 'indirect' part arising from spin-orbit mixing (22) and the 'direct' part, which contains spin-dependent relativistic corrections to the radiative operator.

Table 5. S for He-like Al and Fe; — stands for not computed, - for consistency too poor (and not printed by RM-BP code).

Line Strength S in length (first entries) and velocity formulation

He-like method	Al XII			Fe XXV			
	SSTRUCT $\lambda_l = 1.0$	RM-BP NRANG2=20 RA=4.0		SSTRUCT $\lambda_l = 1.0$	RM-BP NRANG2=20 RA=1.0	NRANG2=20 RA=1.7	
transition $1s\,n's - 1s\,np$	$1s^2 - 1s3d$	9 levels		$1s^2 - 1s3d$	4 levels	9 levels	NBS
$1\,^1S_0 - 2\,^3P_1$	4.8(−5)	8.024(−5)		3.75(−4)	4.235(−4)	4.078(−4)	4.21(−4)
	4.9(−5)	6.563(−5)		3.89(−4)	4.259(−4)	4.108(−4)	B(=±10%)
$^1P_1$	.019 372	.019 21		.004 33	.004 308	.004 328	.004 28
	.019 36	.018 83		.004 43	.004 429	.004 431	B
$- 3\,^3P_1$	9.(−6)	4.093(−5)		6.2(−5)	6.779(−5)	9.011(−5)	8.8(−5)
	1.(−5)	2.723(−5)		6.8(−5)	7.389(−5)	9.097(−5)	E(>50%)
$^1P_1$	.003 427	3.652(−3)		6.82(−4)	7.075(−4)	6.990(−4)	7.15(−4)
	.003 471	3.245(−3)		7.40(−4)	7.417(−4)	7.270(−4)	B
$- 4\,^3P_1$	—	1.716(−6)		—	2.449(−5)	2.394(−5)	3.0(−5)
	—	2.850(−6)		—	2.668(−5)	2.613(−5)	E
$^1P_1$	—	8.178(−4)		—	2.492(−4)	2.476(−4)	2.49(−4)
	—	1.150(−3)		—	2.609(−4)	2.607(−4)	B
$- 5\,^3P_1$	—	1.008(−6)		—	1.178(−5)	1.147(−5)	1.4(−5)
	—	1.510(−6)		—	1.268(−5)	1.254(−5)	E
$^1P_1$	—	4.109(−4)		—	1.192(−4)	1.165(−4)	1.17(−4)
	—	5.461(−4)		—	1.232(−4)	1.229(−4)	B
$2\,^1S_0 - 2\,^3P_1$	4.42(−4)	5.743(−4)		.003 342	.003 883	.018 42	9.5(−5)
	2.01(−2)	—		—	—	—	D(=±50%)
$^1P_1$	.187 079	.185 7		.039 69	.038 28	.038 91	.0417
	.298 2	—		.066 44	—	—	B
$- 3\,^3P_1$	1.82(−4)	7.499(−4)		.001 242	.001 071	.001 820	.0121
	1.69(−4)	4.670(−4)		.001 212	.001 269	.001 653	B (sic)
$^1P_1$	.059 3	.063 75		.012 12	.010 35	.012 16	.0122
	.057 5	.054 70		.012 22	.012 17	.012 06	B

The answer in the velocity form equals that in the length form to full BP order if the length 'vector' (19) is replaced as follows:

$$\vec{r}_1 + \vec{r}_2 \longrightarrow \frac{2}{E(2\,^3P) - E(n\,^1S)}\left\{\vec{\nabla}_1 + \vec{\nabla}_2 + \alpha^2\frac{\sqrt{2}}{4}\left[Z\left(\frac{\vec{r}_1}{r_1^3} - \frac{\vec{r}_2}{r_2^3}\right) + 2\frac{\vec{r}_{12}}{r_{12}^3}\right]\right\}, \quad (23)$$

i. e. by adding BP terms of relative order $\alpha^2 Z^2$ and $\alpha^2 Z$ to the non-relativistic velocity operator (20). Naturally enough IC wavefunctions (4) — as approximated in (22) — have to be used to evaluate the matrix elements of order α^0, and LS wavefunctions (2) elsewhere.

4.2 Magnetic dipole transitions

Magnetic dipole transitions have become the most fascinating topic in Breit-Pauli work ever since Gabriel and Jordan (1969) correlated observed sharp lines in the solar soft X-ray spectrum with M1 decay of the first excited level $1s2s\,^3S_1$ of abundant He-like ions, challenging the previous view that this level can only decay by two-photon transitions — an assumption incompatible with sharp lines! Embedding the structural properties in a far broader context — alongside J/ψ, the particle that lead the way to the fourth flavour of quarks — Sucher (1978) tells the story of Drake's pivotal role in unravelling conflicting concepts.

Table 6. M1 decay of He-like 1s2s $^3S_1$ — the first excited level!
ex: extrapolated to the non-relativistic hydrogenic limit $A(2\,^3S_1) \to Z^{10} \times 1.7346 \cdot 10^6$/sec; Marrus and Mohr (1978) expand further, including the non-relativistic $\mathcal{O}(1/Z)$ due to configuration mixing as well as terms $\mathcal{O}(\alpha^2 Z^2)$ of Breit-Pauli order.

1s2s $^3S_1$ decay Z		observed τ/sec	Hata and Grant 1981 τ/sec	SSTRUCT τ/sec	A^{M1}*sec	Drake 1971 A^{M1}	ID
2	He	9.1±3(+3)	8.88(+3)	8.53(+3) 1.61(4)	1.172(−4) 6.208(−5)	1.272(−4)	ID SM
4	Be			2.47	4.141(−1)	5.618(−1)	SM
8	O			1.10(−3)	9.104(2)	1.044(3)	SM
14	Si			2.98(−6)	3.358(5)	3.563(5)	SM
16	S	7.06±.86(−7)	7.00(−7)	7.41(−7)	1.349(6)	1.408(6)	SM
18	Ar	2.02±.20(−7)	2.08(−7)	2.18(−7)	4.581(6)	4.709(6)	SM
22	Cr	2.58±.13(−8)	2.66(−8)	2.73(−8)	3.661(7)	3.656(7)	SM
23	V	1.69±.07(−8)	1.69(−8)	1.72(−8)	5.799(7)	5.751(7)	SM
26	Fe	4.8±.6(−9)	4.81(−9)	4.85(−8)	2.062(8)	2.002(8)	SM
36	Kr	1.70(−10)	1.71(−10)	1.65(−10)	6.049(9) 6.342(9)		SM ex

In order to compute M1 transition probabilities

$$A^{M1}_{fi} = 3.5644 \cdot 10^4 \cdot \sec^{-1} \frac{(E_i - E_f)^3}{g_i} \cdot S^{M1}(i,f) \tag{24}$$

one constructs the reduced matrix element that makes up the line strength

$$S_{fi} = \left| \langle f \| Q/\mu_B \| i \rangle \right|^2 \tag{25}$$

in the usual manner from spherical tensor operator components $\vec{e}_\kappa \cdot \vec{Q}$ of

$$\vec{Q} = \sum_{j=1}^{N} \left\{ \vec{Q}^I_j + \sum_{j<i} \vec{Q}^{II}_{ij} \right\}. \tag{26}$$

We write Drake's (1971) one-body parts like Eissner and Zeippen as

$$\vec{Q}^I_j = \mu_B \Bigg((\vec{l}_j + 2\vec{s}_j)\left\{1 + \frac{\alpha^2}{2}\left[\left(\frac{\partial^2}{\partial r_j^2} - \frac{l(l+1)}{r_j^2}\right) - \frac{\mathcal{E}^2}{20}r_j^2\right]\right\} \tag{27}$$
$$+ \frac{\alpha^2}{2}\left[\vec{p}_j \times (\vec{p}_j \times \vec{s}_j) + 2\vec{r}_j \times (\vec{r}_j \times \vec{s}_j)\left(\frac{\mathcal{E}^2}{20} - \frac{Z}{r_j^3}\right) + \frac{\mathcal{E}}{2}\vec{s}_j\right]\Bigg).$$

where $\mathcal{E}$ is essentially the photon energy. It so happens that the two-body component Q^{II} does not contribute to our particular transition 1s2s $^3S_1 \to$ 1s^2 ^1S.

The matrix element of the 'ordinary' M1 operator of order α^0 vanishes, because this term in (27) is purely angular and the state $|1s^2\,^1S\rangle$ — while orthogonal to $|1s2s\,^3S_1\rangle$ — is entirely isotropic. Therefore the terms of BP order in (27) remain the sole cause for the 1s2s $^3S_1$ decay in He-like ions; in fact they give rise to 14 distinct radial integrals. The transition therefore scales as Z^{10}: 6 powers coming from the energy factor and 4 from $\sim r^{-2}$ in (26) when squaring.

Table 6 summarizes results in a number of approaches. While showing good results further down the table when only modestly complex configuration expansions involving pure SM orbitals are employed, it comes not unexpectedly that neutral helium requires a much greater effort. The label 'ID' in the trailing comment column marks a special approach, in which collision type bound orbitals 1s, 2s and 2p have been included in a 28-configution expansion; these orbitals have been obtained in runs of the electron collision code IMPACT (Crees et al. 1978) and are of the kind dealt with in tables 2 obeying (11) — this way one builds up the He atom from He$^+$ orbitals, which provides 'frozen cores', and the IMPACT orbitals. Eissner and Zeippen obtained this result when testing part of the SSTRUCT extension in 1981. With similar output from RM one would now get the same good answer. For comparison the table contains a second entry for neutral He: 28 configurations composed entirely of ordinary SM orbitals still give a poor result. For all ions correlation configurations had been included.

In half-filled shells nl the one- and two-body BP terms in (26) compete in (25) to the same orders in α and Z with components due to the ordinary $(\vec{l}+2\vec{s})$ (Eissner and Zeippen 1981), which can lead to destructive interference in the transition amplitudes. This affects the calibration when using line intensity ratios in such ions for electron density diagnostics.

5 Conclusion

The paper grew out of a particular task, namely to provide bench marks before extending the new version of Scott and Taylor's RM-BP code so as to include E2 and M1 radiative processes, all consistent up to Breit-Pauli order as discussed during a CECAM Workshop in 1989 (Berrington et al.). The results shown in this paper give confidence that the code will be performing reliably in its main task: to provide intermediate coupling data for electron-ion collsion processes and for photoionization on a large scale — analytically the extension of the bound state regime into the continuum!

In the context of the Conference the results give insight to separate physical processes contributing in relativistic structure calculations.

Acknowledgements

The computations for this paper were done on the IBM 3090 180 S supplied by IBM under a Joint Study Contract with The Queen's University of Belfast; the R-matrix Breit-Pauli code has been set up within the Alpha Project exploiting the hardware eXtended Architecture. Throughout this work, especially when the machine was becoming operational, the staff at the Queen's Belfast Computer Centre has been most helpful. It is a great pleasure to thank Keith Berrington, who provided a prototype version of the RM-BP code that for the first time interfaces with the OPACITY PROJECT asymptotic package. Last but not least I would like to thank Philip Stafford for stimulating discussions.

References

Berrington K A, Burke P G, Butler K, Seaton M J, Storey P J, Taylor K T, and Yu Yan, 1987, J. Phys. B **20**:6379–97.

Berrington K A, Burke P G, Le Dourneuf M, Robb W D, Taylor K T, and Vo Ky Lan, 1978, Computer Phys. Commun. **14**:367–412.

Berrington K A, Eissner W, and Zeippen C J *in* "The R-matrix Method Applied to Atomic and Molecular Processes," P G Burke and M Le Dourneuf M eds: to appear *in CECAM 1990 Annual Report.*

Bethe H A and Salpeter E E, Handb. Phys. **35** (1957).

Burke P G and Eissner W, *in Atoms in Astrophysics*, P G Burke, W B Eissner, D G Hummer and I C Percival eds, PLENUM New York 1983.

Burke P G and Robb W D, 1975, Adv. At. Mol. Phys. **11**:143–214.

Crees M A, Seaton M J, and Wilson P M H, 1978, Computer Phys. Commun. **15**:23–83.

Drake G W F, 1971, Phys. Rev. A **3**:908–15.

Drake G W F, 1976, J. Phys. B **9**:L169–71.

Eissner W, Jones M, and Nussbaumer H, 1974, Computer Phys. Commun. **8**:270–306.

Eissner W and Zeippen C J, 1981, J. Phys. B **14**:2125–37.

Fernley J A, Taylor K T, and Seaton M J, 1987, J. Phys. B **20**:6457–76.

Fuhr J R, Martin G A, Wiese W L, and Younger S M, 1981, J. Phys. Chem. Ref. Data **10**:305–565.

Gabriel A H and Jordan C, 1969, Mon. Not. R. astr. Soc. **145**:241–8.

Hata J and Grant I P, 1981, J. Phys. B **14**:2111–24.

Jones M, 1971, Phil. Trans. Roy. Soc. (Lond.) **4**:1422–39.

Marrus R and Mohr P J, 1978, Adv. At. Mol. Phys. **14**:181–224.

Scott N S and Taylor K T, 1982, Computer Phys. Commun. **25**:347–87.

Sucher J, 1978, Rep. Prog. Phys. **41**:1781–838.

Wiese W L, Smith M W, and Miles B M, *Atomic Transition Probabilities* Vol. II: Sodium through Calcium, NSRDS-NBS 22 (1969).

Discussion

Professor Patrick Sandars' comment: Your hydrogenic result $E_{2s_{1/2}} \neq E_{2p_{1/2}}$ is at variance with Dirac theory: when expanding in powers of α^2, the two level positions must be the same to all orders.

Answer: Yes, they must — and they would if the SSTRUCT code retained terms up to order α^2 in the *eigenvalues* of the H matrix. However expectation values of order α^2 are retained in the *matrix elements*, and the situation is made worse in the hydrogenic case because off-diagonal matrix elements contain nothing of order α^0. Look at the submatrix for $J = \frac{1}{2}$ (which of course reduces further according to parity), arranged in the same order as in table 1:

$$\langle H_{J=\frac{1}{2}}\rangle = \begin{pmatrix} -\frac{Z^2}{1^2} - \alpha^2 Z^4 \cdot 0.25 & -\alpha^2 Z^4 \cdot 0.1440 & 0 & \cdots \\ -\alpha^2 Z^4 \cdot 0.1440 & -\frac{Z^2}{2^2} - \alpha^2 Z^4 \cdot 0.0781 & 0 & \cdots \\ 0 & 0 & -\frac{Z^2}{2^2} - \alpha^2 Z^4 (.0364 + .0417) & \cdots \\ -\alpha^2 Z^4 \cdot 0.0842 & -\alpha^2 Z^4 \cdot 0.0461 & 0 & \cdots \\ \vdots & \vdots & \vdots & \ddots \end{pmatrix}$$

Indeed it displays the same diagonal values for $2s_{\frac{1}{2}}$ and $2p_{\frac{1}{2}}$ (for mass plus Darwin term cf table 1, which does not give the spin-orbit component; therefore the expectation value $\langle n'l|f(\text{so})|nl\rangle$ shows up as a separate second entry within the parentheses of our H matrix). However the off-diagonal elements $\langle n'l|f(K)|nl\rangle$ spoil the tidyness in the eigenvalues E_{nlj} on diagonalizing H. Perturbation theory yields eigenvalues

$$E_{2s_{\frac{1}{2}}} = -\frac{Z^2}{4} + \epsilon_{2s}^{m+d} + \frac{\alpha^2 Z^4}{\frac{3}{4}Z^2}\langle 1s|f(m) + f(d)|2s\rangle + \text{similar terms from other } ns, \text{ where } n > 2,$$

$$E_{2p_{\frac{1}{2}}} = -\frac{Z^2}{4} + \epsilon_{2p_{\frac{1}{2}}}^{m+so} + 0 + \text{terms of relative order } \alpha^2 Z^2 \text{ from other } np, \text{ where } n > 2.$$

Thus the off-diagonal terms give rise to admixtures two orders lower in Z than the BP terms. Numerical cumulative integration errors, which can be estimated from deviations in the non-relativistic answers $-Z^2/n^2$ and the likes, are smaller than the BP effects; in our IBM double precision calculations relative errors remain below 10^{-7} (which coincides with the accuracy parameter specified as ITOL = 7 that controls the eigenvalue problem (10)), at least for all the ground states. Only at $Z = 52$ may the uncertainties grow by nearly 2 orders for the more oscillating and diffuse orbitals; this is insignificant compared to the rise of the BP terms at such a big value of Z. Another source of error is neatly avoided: apart from the obvious measure of distributing the fourth derivative in $f(m)$ evenly over the two orbital functions the resulting second derivatives of $\langle r|nl\rangle = P_{nl}(r)$ are available from (10) as a simple expression. A final remark: for s-orbitals the integrand at $r = 0$ in $\langle n's|f(m)|ns\rangle$ is finite, and the integral can become fairly sensitive — to say nothing of $\langle f(d)\rangle$, which apart from a factor Z is entirely given by the slopes of the orbitals at $r = 0$: see (13). One checks the code for stable answers on varying the step length of the integration grid.

RELATIVISTIC CALCULATIONS of PARITY NON-CONSERVING EFFECTS in ATOMS

A.C. Hartley, P.G.H. Sandars

Clarendon Laboratory
Parks Rd.
Oxford
OX1 3PU

Abstract

Atomic parity non-conserving (PNC) effects are produced by electronic weak interactions and lead to macroscopic mirror asymmetries. These PNC effects can be measured experimentally but in order to extract fundamental information about the weak force it is necessary to have accurate atomic calculations of the appropriate PNC E1 transition matrix elements. In this paper we give an over-view of relativistic calculations of PNC effects in atoms according to successive approximation schemes. From the combination of atomic PNC experiment and theory we derive an accurate value for $\sin^2\Theta_w$, the fundamental electro-weak mixing angle, which is in excellent agreement with the current world average.

1. Introduction

The Parity of a system describes its behaviour under reflection through the origin. This transformation is formally equivalent to a mirror reflection in a plane through the origin plus a 180° rotation about an axis perpendicular to that plane. Rotational invariance of physical systems means that a parity non-conserving (PNC) system is one that exhibts a mirror asymmetry. Of the four fundamental forces only the weak force is known to be parity non-conserving.

The effect of weak interactions in the atomic environment leads to small mirror asymmetries such as the rotation of plane polarised light in an atomic vapour. These effects, though extremely small are measurable because of their PNC signature. Recent advances in experimental technique have enabled high precision measurements of atomic PNC phenomena. Atomic PNC effects have a Z^3 enhancement so that experiments are performed on heavy atoms: currently, atomic PNC experiments have been perfomed on Cs, Tl, Pb and Bi (for a recent review of atomic PNC experiments see Hinds 1989). The fact that these are all heavy atoms means that a theoretical treatment has to be fully relativistic.

The most important atomic weak interaction is described in the 'standard model' (Glashow 1970, Weinberg 1967, Salam 1968, see e.g. Mandl and Shaw for an expostion) by the exchange of an intermediate Z^0 vector boson between the nucleus and an electron. The atomic Hamiltonian for this is given by

$$h^{pnc} = \frac{G_F Q_w}{2\sqrt{2}} \rho(r) \gamma_5 \qquad (1.1)$$

where $G_F = 2.22 \times 10^{-14}$ au gives the small size of the interaction, $\rho(r)$ is a nuclear density, normalised to unity, Q_W is the weak charge (Bouchiat and Bouchiat 1974) given (without radiative corrections) by

$$Q_w = -N + Z(1 - 4\sin^2\theta_w) \qquad (1.2)$$

where N and Z are the number of neutrons and electrons respectively, $\sin^2\theta_w$ is the fundamental parameter for the standard electro-weak model, and γ_5 is the matrix

$$\gamma_5 = \begin{pmatrix} 0 & 1 \\ 1 & 0 \end{pmatrix}. \qquad (1.3)$$

When this matrix acts on a relativistic Dirac two component spinor it exchanges the top and bottom components so that the resulting orbital has the opposite nominal parity to the unperturbed orbital. This gives the PNC nature of the atomic weak interaction Hamiltonian.

The presence of atomic PNC weak interactions means that quantities which are normally parity forbidden can now occur. For example the 6s -> 7s Cs E1 transition is parity forbidden. In the presence of a PNC interaction the orbitals become parity mixed so that there is then a non-zero amplitude for this transition due to the opposite parity of the admixture:

$$<\tilde{7s}_{1/2}| \text{-r} |6\tilde{s}_{1/2}> = <7s_{1/2}^{pnc}| \text{-r} |6s_{1/2}> + <7s_{1/2}| \text{-r} |6s_{1/2}^{pnc}> \qquad (1.4)$$

where the ~ denotes a parity mixed orbital. The 6s -> 7s transition amplitude, which is proportional to Q_w, can be measured experimentally. If the value of Q_w can be extracted then a value for $\sin^2\theta_w$ can be obtained. However, in order to extract Q_w it is necessary to have a value for the PNC E1 transition matrix element. These have to be calculated separately and this is the task of atomic PNC theory calculations. The most accurate PNC measurments have been performed on Cs (Noecker et al. 1988) and as this is a single valence system the theory is relatively easy. We shall accordingly concentrate on Cs and also Tl which can be treated as a single valence system as well.

2. The Hierarchy of Approximations in MBPT

In this section we shall discuss the hierarchy of successive approximations that are normally made in an *ab initio* many-body perturbation theory (MBPT) calculation. We shall first discuss normal, parity conserving calculations before including PNC effects as well. As discussed by e.g. Bouchiat and Piketty (1986) the parity conserving physical quantities which are most relevant to a PNC calculation are E1 transition matrix elements and the hyperfine interaction constants. We shall accordingly concentrate on these.

The atomic Hamiltonian that we shall use is given by

$$H = \sum_i h_D(i) + \sum_{i>j} \frac{1}{r_{ij}} \qquad (2.1)$$

where

$$h_D = c\underline{\alpha}\cdot\underline{p} + mc^2(\beta - 1) + V_{nuc}(r) \qquad (2.2)$$

is the standard Dirac single particle Hamiltonian and $V_{nuc}(r)$ is the nuclear potential energy. As yet there has been little work with the Breit interaction included. In practice, instead of the inter-electronic Coulomb interaction an approximate electronic potential is used as a starting point and MBPT is then applied. The Dirac Hartree-Fock potential is often used as an initial approximation.

2.1 The Dirac Hartree-Fock Equations

The standard starting point for a MBPT calculation is the Dirac Hartree-Fock (DHF) equations. For a single valence system it is most convenient to use a V_{n-1} DHF potential where only the N-1 core electrons contribute to the electronic potential. The DF equations are given by

$$(h_D + V_{HF}) | a > = \epsilon_a | a > \qquad (2.3)$$

where

$$V_{HF} | a > = \sum_{b.core} < b | g_{12} | b > | a > \qquad (2.4)$$

and

$$g_{12} = \frac{1}{r_{12}} (1 - P_{12}) \qquad (2.5)$$

where P_{12} is the usual exchange operator. These differential equations can be solved by standard finite difference techniques.

2.2 The Random Phase Approximation

The next level of approximation in our hierarchy is to include certain single particle excitations to all orders. The resulting equations are often referred to as the Random Phase Approximation (RPA). The RPA equations are closely related to the Time Dependent HF equations (TDHF) for a time dependent perturbation.

There are two ways of viewing the RPA equations. One of the simplest is to consider a perturbation δV which we can include directly in the DHF Hamiltonian. If we expand the equation up to first order in δV for the wave functions and the eigenvalues we obtain:

$$(h_D + \delta V + V_{HF} + \delta V_{HF}) | a + \delta a > = (\epsilon_a + \delta \epsilon_a) | a + \delta a > \qquad (2.6)$$

Taking terms linear in the perturbation and rearranging we obtain, for the hyperfine interaction for example

$$(\epsilon_a - h_D - V_{HF}) | \delta a > = h^{hyp} | a > + V^{hyp}_{RPA} | a > - | a >< a | h^{hyp} + V^{hyp}_{RPA} | a > \qquad (2.7)$$

where $V^{hyp}_{RPA} = \delta V_{HF}$ is the change in the self-consistent potential due to the perturbation h^{hyp} acting on the other core orbitals.

For a time dependent perturbation, such as a radiation field, we have to include an 'ω term' on the left-hand side of the equation. There are now two possible excitations to be solved for : $| \delta a^+ >$ and $| \delta a^- >$. For an electric field we have:

$$(\epsilon_a \pm \omega - h_D - V_{HF}) | \delta a^\pm > = -r | a > + V^\pm_{RPA} | a > \quad - \text{orthog.} \qquad (2.8)$$

where 'orthog.' refers to the orthogonality terms that ensure that the excitations are orthogonal to the unperturbed orbital. $V^\pm_{RPA}$ is given by

$$V^\pm_{RPA} | a > = \sum_{b.core} \left(< \delta b^\pm | g_{12} | b > + < b | g_{12} | \delta b^\mp > \right) | a >. \qquad (2.9)$$

In all these RPA type of equations the excitations have to be split up into their different allowed angular momentum channels and each solved for separately.

To obtain the matrix element including the RPA correction, the operator simply has the appropriate RPA potential added to it. For example, for the hyperfine interaction one requires the expectation value over the valence orbital:

$$< v | h^{hyp} + V_{RPA}^{pnc} | v > . \qquad (2.10)$$

For an E1 transition matrix element between valence states i and f the matrix element

$$< f | -r + V_{RPA}^{+} | i > . \qquad (2.11)$$

is required.

Physically the V^{+}_{RPA} potential gives the shielding correction to the E1 transition matrix element. In Figure 1 we give some of the MBPT diagrams that the RPA equations represent. It is straightforward to show that equations (2.7) to (2.11) will generate all the MBPT RPA E1 diagrams to all orders when the equations are solved to self-consistency. The inclusion of the ω term on the LHS of equation (2.8) gives the correct energy denominator for the diagrams. It is interesting to note that although equations (2.6) and (2.8) allow admixtures of the other core orbitals b into the solution for core correction $| \delta a >$ and $| a^{\pm} >$, which is not allowed in the MBPT diagrams, the admixture of b into the excitation of a is exactly cancelled by the admixture of a into the excitation of b as long as the perturbation is Hermitian.

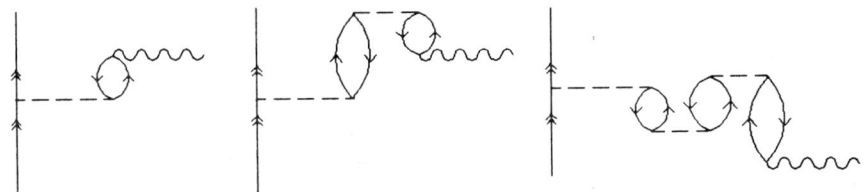

Figure 1. Some MBPT RPA diagrams

These RPA equations have been solved by several groups (e.g. Johnson *et al.* 1986, Dzuba *et al.* 1984,1985, Hartley and Sandars 1990, Hartley and Martensson-Pendrill 1990) and the results are generally consistent. We shall present the answers later.

2.3 Lowest Order Correlation

All the equations used so far have been coupled differential equations that have been solved using standard finite difference techniqes. However, for the calculation of correlation it is necessary to include two particle effects and evaluating these using differential equations leads to severe difficulties (see e.g. Lindroth 1987,1988, Lindroth *et al.* 1987). One alternative solution is to use finite basis set techniques. This has been used by several groups in the past (Johnson *et al.* 1987, 1988$_a$, Dzuba *et al.* 1984,1985,1987$_a$). The method described here was used by Hartley and Martensson-Pendrill (1990). It involves deriving a basis set by diagonalising a matrix representation of the DHF Hamiltonian (eq. 2.3) as used by Salomonson and Oster (1989). This involves taking an atom in a finite space and representing the space on a discrete logarithmic radial grid. The radial part of the DHF Hamiltonian (2.3) can then be represented as a matrix in this space.

$$\begin{pmatrix} V & D^T \\ D & V - 2mc^2 \end{pmatrix} \begin{pmatrix} F \\ G \end{pmatrix} = \epsilon \begin{pmatrix} F \\ G \end{pmatrix} \qquad (2.12)$$

where

$$\begin{pmatrix} F & G \end{pmatrix} = \begin{pmatrix} f_1 \ f_2 \ f_3 \ldots f_N & g_1 \ g_2 \ g_3 \ldots g_N \end{pmatrix}.$$

The differential operator submatrix, D, is obtained by using finite difference numerical formulae and thus has contributions along a band a few points wide along the diagonal. The direct part of the DHF potential is obtained using DHF core orbitals obtained by solving equation (2.3) using standard differential equation techniques. This contributes along the diagonal of the potential submatrix V. The exchange part of the potential introduces off diagonal elements in V. Diagonalising this matrix to obtain the eigenfunctions and eigenvalues results in an ortho-normal basis set that is complete on the matrix space. For a given size of 'box' the grid size can be reduced and the results extrapolated to derive a zero grid size answer. This can then be repeated for larger and larger box sizes and the results extrapolated to the physical result. In practice for a grid of 120 points and a box from r=exp(-12) to r=40 a.u. we found little appreciable change after extrapolation.

With this basis set it is possible to construct any MBPT diagram by explicit summation. The four lowest order correlation corrections are given in figure 2. Their effect can be expressed in terms of a non-local correlation potential

$$\sum (v) = \sum_{s \geq r}^{exc} \sum_{b}^{core} \frac{<b|g_{12}|sr><sr|g_{12}|b>}{\epsilon_v + \epsilon_b - \epsilon_s - \epsilon_r} + \sum_{s}^{exc} \sum_{a \geq b}^{core} \frac{<ba|g_{12}|s><s|g_{12}|ab>}{\epsilon_b + \epsilon_a - \epsilon_s - \epsilon_v}$$

(2.13)

which depends, through ϵ_v, on the orbital which it acts upon.

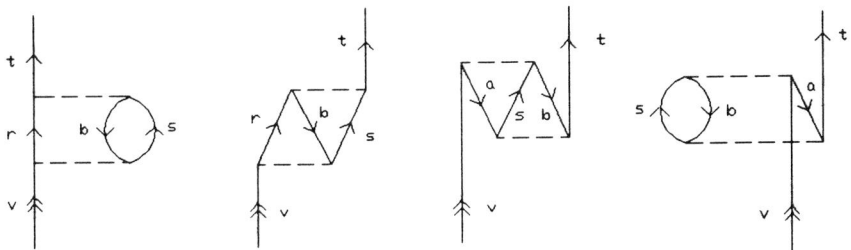

Figure 2. Lowest order correlation corrections

These four diagrams give the lowest order corrections of a DHF orbital towards a Brueckner orbital (BO) which has no single particle corrections to it to any order. The DHF orbital plus these lowest order corrections are often referred to as approximate BO's.

The lowest order BO energy corrections are given by

$$\delta\epsilon_v = <v|\sum(v)|v>$$

(2.14)

and the lowest order BO wave function corrections given by

$$|\delta v> = \sum_{i \neq v} |i><i|\frac{\sum(v)|v>}{\epsilon_v - \epsilon_i}$$

(2.15)

so that the BO corrections to the E1 transition matrix elements and the hyperfine interaction constants are given by

$$< v | h^{hyp} | \delta v > + < \delta v | h^{hyp} | v > \qquad (2.16)$$

and

$$< f | -r | \delta i > + < \delta f | -r | i > \qquad (2.17)$$

respectively. It is straighforward to include the RPA corrections to the BO contributions by including the RPA correction potentials in the above expressions.

The results for all the contributions up to this level of approximation for Cs and Tl are given in Tables 1 to 3. From this it can be seen that the lowest accuracy, which is for the hyperfine constants, is at 5% for Cs but only around 10% for Tl. The lowest order BO calculations by W. Johnson et al. (1987) did not include the shielding corrections; their calculation also only summed explicitly up to $l = 6$ in their partial wave summation and extrapolated to obtain the higher l contributions. In the results presented here (Hartley and Martensson-Pendrill 1990) explicit summation was performed up to $l = 10$ before extrapolation. It was found that the higher $l > 6$ values contributed more than is estimated from extrapolation. In their calculations for Tl matrix elements, Dzuba et al. (1987) had large basis set errors due to incompleteness which accounts for the large discrepancies with the numbers presented here. This problem has been corrected in their latest calculations.

2.4 Beyond the Lowest Order

To calculate even the lowest order BO corrections take a large amount of computing resources. To go beyond lowest order is no easy task. Blundell et al. (1988) have performed an explicit calculation of all third order MBPT corrections for Cs valence energies using a finite basis set constructed out of B splines. The accuracy obtained was 0.4% or better for the energies. However, this took an enormous amount of computing time and in general order-by-order calculations quickly become prohibitively large.

2.4.1. The Coupled Cluster Equations

One of the main alternatives to order-by-order methods is to use all-order equations. As discussed earlier, the RPA equations give the contributions of certain single particle diagrams to all orders. It is also possible to derive all order equations for two or more particle effects such as the Coupled Cluster (CC) equations (see e.g. Lindgren and Morrison 1986). These equations generate diagrams such as in Figure $3_{a,b}$. If a Hermitian formulation for deriving the CC equations is used (Lindgren 1990) then certain additional diagrams are included as shown in Fig $3_{c,d}$. These CC equations have been solved non-relativistically for the lighter alkali metals for single and two particle excitations (e.g. Martensson-Pendrill et al. $1990_{a,b}$, Salomonson and Ynnerman 1990) but the accuracy for some quantites is still at the few percent level. It is only when certain three particle diagrams such as in Fig. $3_{d,e}$ are included that, for Na, the agreement goes to 0.01% for the eigenvalues and <0.2% for the hyperfine constants (Salomonson and Ynnerman 1990). Work is currently in progress on complete CC calculations for Cs and Tl.

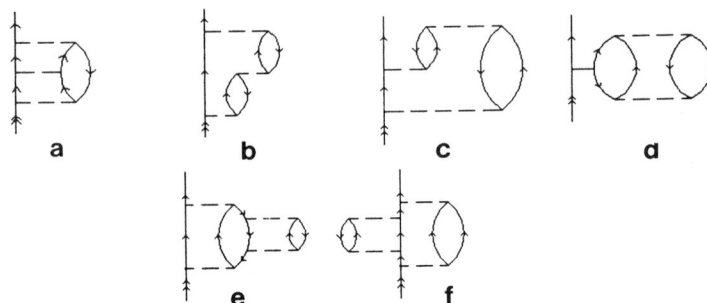

Figure 3. Coupled Cluster diagrams: a and b are included in the standard CC equations, c and d are only included in the Hermitian formulation.

Table 1

Cs energies

		$6s_{1/2}$	$6p_{1/2}$	$7s_{1/2}$	$7p_{1/2}$
DHF		-0.12737	-0.08562	-0.05519	-0.04202
δBO	a	-0.01782	-0.00694	-0.00422	-0.00213
	b	-0.01774	-0.00691	-0.00420	
	c	-0.01601	-0.00652	-0.00374	-0.00197
	e	-0.01771	-0.00698	-0.00429	-0.00217
DHF+δBO	a	-0.14519	-0.09256	-0.05941	-0.04415
+3rd order MBPT	d	-0.14318	-0.09238	-0.05844	
'Russian Bubbles'	e	-0.14295	-0.09232	-0.05865	-0.04392
Experiment	f	-0.14310	-0.09217	-0.05865	-0.04393

a: Hartley and Martensson-Pendrill (1990)
b: Johnson et al. (1987)
c: Dzuba et al. (1985)
d: Blundell et al. (1988)
e: Dzuba et al. (1989$_b$)
f: Moore

Tl energies

		$6p_{1/2}$	$6p_{3/2}$	$7s_{1/2}$	$7p_{1/2}$
DHF		-0.19968	-0.16693	-0.09618	-0.06505
δBO	a	-0.03536	-0.03077	-0.00923	-0.00465
	b	-0.03302	-0.02904	-0.00896	-0.00453
DHF+δBO	a	-0.23504	-0.19770	-0.10541	-0.06970
'Russian Bubbles'	c	-0.22786	-0.19229		
Experiment	d	-0.22446	-0.18896	-0.10382	-0.06882

a: Hartley and Martensson-Pendrill (1990)
b: Dzuba et al. (1987$_a$)
c: Dzuba et al. (1988)
d: Moore

Table 2

Cs hyperfine interaction constants (MHz) for ^{133}Cs with $g_I = 0.7377208$ n.m.

		$6s_{1/2}$	$6p_{1/2}$	$7s_{1/2}$	$7p_{1/2}$
DHF		1427	161.0	392.0	57.63
δRPA		+293	+40.6	+79.7	14.01
δBO		+645	+73.2	101.6	20.18
δBO-RPA		+40	+5.9	+3.4	+1.29
Total	a	2405	280.7	576.7	93.11
	b	2366		572	
	c	2296	292.3	541.1	94.31
	d	2302	290.1	542.6	93.69
Experiment	e	2298	291.9(1)	546(3)	94.34(3)

a: Hartley,Martensson-Pendrill (1990)
b: Johnson et al. (1987)   : no BO-RPA corrections included
c: Dzuba et al. (1989$_b$)   : 'Russian bubble method'
d: Dzuba et al. (1989$_b$)   : 'Russian bubbles' with semi-emprical fit to energy levels
e: Arimondo (1977)

Tl hyperfine interaction constants (GHz) for ^{205}Tl with $g_I = 3.2764268$ n.m.

		$6p_{1/2}$	$6p_{3/2}$	$7s_{1/2}$	$7p_{1/2}$
DHF		17.57	+1.302	+7.63	+1.958
δRPA		+4.31	- 3.381	+3.15	+0.061
δBO		+2.33	+0.226	+2.37	+0.155
δBO-RPA		-0.15	-0.032	-0.09	-0.054
Total	a	24.06	-1.885	+13.06	+2.120
	b	21.77	-1.919	+12.47	+2.069
	c	21.32	+0.600	+12.56	+2.039
Experiment	d	21.31	+0.265	+12.5	+2.13(6)

a: Hartley and Martensson-Pendrill (1990)
b: Dzuba et al. (1987$_a$)   : approximate BO potential
c: Dzuba. et al. (1987$_a$)   : includes κ diagonal internal BO substitutions
d: Lurio and Prodell (1956); Gould (1956); Pollack and Wong (1971); Flusberg et al. (1976)

Table 3

Cs Reduced E1 transition matrix elements

		$6s_{1/2} \to 6p_{1/2}$	$6s_{1/2} \to 6p_{1/2}$	$6s_{1/2} \to 6p_{1/2}$	$6s_{1/2} \to 6p_{1/2}$
DHF		-5.278	+4.414	-0.370	-11.012
δRPA		+0.303	+0.037	+0.133	+0.088
δBO		+0.584	-0.106	-0.094	+0.857
δBO-RPA		+0.036	-0.010	+0.012	+0.003
Total	a	-4.355	+4.335	-0.319	-10.064
	b	-4.393			
	c	-4.45	+4.21	-0.261	-10.31
	d	-4.50	+4.22	-0.260	-10.32
	e	-4.49	+4.25	-0.275	-10.29
Experiment	f	-4.52(3)	+4.24(1)	-0.285(2)	-10.21(6)

a: Hartley, Martensson-Pendrill (1990)
b: Johnson et al. (1987) : no δBO-RPA corrections included
c: Dzuba et al. (1985)
d: Dzuba et al. (1989$_b$) : screened correlation 'Russian Bubbles'
e: Dzuba et al. (1989$_b$) : 'Russian Bubbles + energy fit
f: Bouchiat and Pikety (1986), J. Guena, private communication quoted therein

Tl Reduced E1 transition matrix elements

		$7s_{1/2} \to 6p_{1/2}$	$7s_{1/2} \to 7p_{1/2}$	$7s_{1/2} \to 6p_{3/2}$
DHF		+2.049	-6.618	+3.965
δRPA		-0.144	+0.218	-0.355
δBO		-0.243	+0.493	-0.544
δBO-RPA		+0.009	+0.044	+0.012
Total	a	+1.671	-5.863	+3.078
	b	+1.75	-5.88	+3.24
	c	+1.72		+3.18
Experiment	d	+1.82(3)	-5.94(6)	+3.27(7)

a: Hartley, Martensson-Pendrill (1990)
b: Dzuba et al. (1987$_a$) : approximate BO potential
c: Dzuba et al. (1987$_b$) : includes structural radiation
d: Gallagher and Lurio (1964) ; James et al. (1985)

2.4.2. The Russian Bubble Method

The solution of the CC equations requires large computer resources of both time and memory. Rather than evaluating all these diagrams, many of which are negligibly small or cancelling, the Russian group at Novosibirsk identified the most important terms beyond lowest order correlation. These propagator diagrams (Fig. 4_a) represent the shielding of the Coulomb interaction by the core in the lowest order correlation correction. These shielded correlation terms were then chained together as in Fig 4_b.

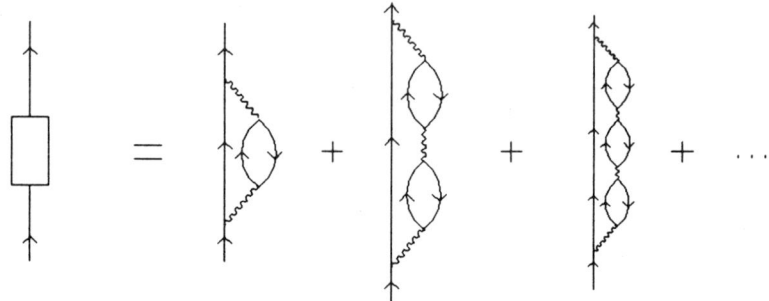

Figure 4a. The chaining of the bubble diagrams to give the screening of the correlation

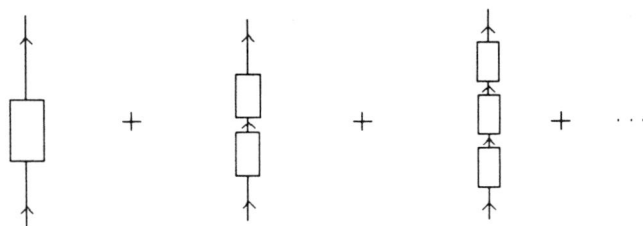

Figure 4b. The chaining of the screened correlation diagram

It is not easy to calculate the chaining of the core 'bubbles' of Fig 4_a by conventional MBPT as the different time-orderings of the Coulomb interaction lead to non-trivial energy denominator terms. The alternative approach adopted by the Russians was to use the Feynmann propagator method that is normally used for QED calculations. This involves using fermion propagators; its different treatment of the energy denominator terms by using contour integrals enables the calculation of the bubble diagrams to all orders.

This method has been applied to the calculation of energy levels in Cs and Tl (Dzuba et al. 1989_a, 1988 respectively) giving accuracies of <0.2% (see Table 1). It has also been used to calculate E1 matrix elements and hyperfine values for Cs to give an accuracy of better than 1% (Dzuba et al. 1989_b) (see Tables 2 and 3). This represent the most accurate *ab initio* MBPT calculation performed on Cs to date.

3. The Inclusion of PNC Effects in MBPT Calculations

As first pointed out by Sandars (1977) it is possible to include the PNC interaction directly in the DHF Hamiltonian. By separating out the terms of one power of G_F one obtains the PNC DHF equations

$$(\epsilon_a - h_D - V_{HF}) | a^{pnc} > = (h^{pnc} + V^{pnc}_{HF}) | a > \qquad (3.1)$$

The PNC DHF orbital $| a^{pnc} >$ has the opposite parity but the same j value as the orbital $| a >$ (i.e. it has the opposite sign κ value). If one now considers a parity mixed basis set consisting of each orbital plus its PNC correction then it is straighforward to show that this set is still ortho-normal to first order in G_F:

$$< \tilde{a} | \tilde{b} > = < a + a^{pnc} | b + b^{pnc} > = < a | b > + < a^{pnc} | b > + < a | b^{pnc} > = \delta_{ab} \qquad (3.2)$$

where the two PNC overlaps cancel due to the Hermiticity of the PNC DHF Hamiltonian. Thus it is possible to perform MBPT with this parity-mixed basis in the standard way. To pick out the required PNC corrections to an E1 transition one takes the standard MBPT E1 diagrams constructed with a parity-mixed basis set and expands them in powers of G_F. The PNC corrections are all first order in G_F. This amounts to taking a parity conserving diagram and replacing each DHF orbital one by one by its PNC DHF counter-part. This is equivalent to performing MBPT with two perturbations: the PNC interaction and the E1 interaction.

3.1 PNC Corrections

PNC HF

The lowest order PNC E1 transition matrix element is obtained by replacing each valence orbital in the transition by its PNC HF counter-part. Thus we have for the Cs 6s -> 7s PNC E1 transition

$$< 7s_{1/2}^{pnc} | -r | 6s_{1/2} > + < 7s_{1/2} | -r | 6s_{1/2}^{pnc} > \qquad (3.3)$$

The values for this are given at the top of Table 4 for Cs and Tl PNC E1 transitions.

PNC RPA

The next PNC term is found from the PNC RPA equations. There are two parts to this. The first one is the standard electric field shielding which can be included by replacing the dipole operator $-r$ in the above by $-r + V^+_{RPA}$. The second type is where the PNC orbital occurs in the shielding diagram so that we have a correction of the type

$$< 7s_{1/2} | V^{+pnc} | 6s_{1/2} > \qquad (3.4)$$

The results are also given in Table 4. It is interesting to note that this correction is very small for Cs but larger than the standard shielding correction for Tl.

PNC+BO

When considering PNC corrections with BO corrections there are several types of correction:
i). PNC DHF valence orbital plus BO valence orbital. Thus we get the terms :
$$< 7s_{1/2}^{pnc} | -r | \delta 6s_{1/2} > + < \delta 7s_{1/2} | -r | 6s_{1/2}^{pnc} >$$
ii) DHF valence orbital plus 'external' PNC BO. This is where the PNC interaction occurs on an external orbital of the valence BO correction (Fig $5_{a,b}$).
$$< \delta 7s_{1/2}^{pnc} | -r | 6s_{1/2} > + < 7s_{1/2} | -r | \delta 6s_{1/2}^{pnc} >$$
iii) DHF valence orbital plus 'internal' PNC BO. This is as above except that the PNC interaction occurs on one of the three internal orbitals in the BO correction (Fig 5_c).
iv) Structural Radiation: this is where the photon occurs on an internal BO line. There can be both external and internal PNC corrections for this (Fig 5_d).

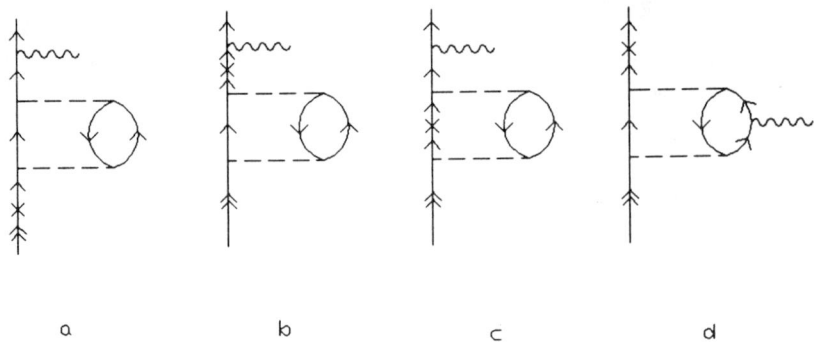

Figure 5. PNC BO corrections: a) BO corrections, b) external PNC-BO corrections, c) internal PNC-BO corrections, d) structural radiation.

The values for the PNC transitions are given in Table 4. The shielding potential terms can easily be included with these BO corrections as well as the V^{+pnc} shielding term. These terms are also given in the table.

The large discrepancy for Cs between the results of Table 4 (Hartley et al. 1990) and Dzuba et al. 1987$_b$ is due to basis set errors in the latter's calculation. This has been corrected for in later calculations (Dzuba et al. 1989$_c$).

3.2 Beyond Lowest Order PNC BO Corrections

As yet a full PNC Coupled Cluster calculation has not been performed though work is in progress in Indiana (W. Johnson, private communication). The most comprehensive PNC calculation performed to date is by Dzuba et al . (1989$_c$) using the 'Russian Bubble Method'. In this calculation they evaluate the 'Russian Bubble' diagrams for the BO and external PNC-BO corrections. The internal PNC-BO corrections and the structural radiation corrections are evaluated to lowest order only.

4. Results and Discussions

The list of all the PNC contributions so far discussed are given in Table 4 for the PNC transitions considered. If these values are to be used for an atomic determination of $\sin^2\Theta_w$ then it is necessary to estimate the accuracy of the PNC values. The usual approach it to examine the accuracy of experimentally known values calculated to the same approximation as the PNC terms. As first discussed by Bouchiat and Piketty (1986) the geometric mean of the appropriate hyperfine constants is a better yardstick than the individual hyperfine constants for assessing the accuracy of PNC calculations as it has the same dependence on the wavefunction density at the nucleus as the PNC Hamiltonian. The calculated values of experimentally known E1 transition matrix elements and hyperfine interaction constants in the approximation of lowest order BO + shielding corrections have been given in Tables 1 to 3. However internal BO and structural radiation corrections were not included in these whereas they are included for the PNC calculations. For some quantities, such as the Tl p3/2 hyperfine constant these contriubutions make a large difference, but for most of them they are very small. We can thus estimate an accuracy of 4% for the Cs 6s -> 7s PNC E1 transition and about 10% for the Tl 6p1/2-7p1/2 and 6p1/2 -> 6p3/2 transitions to this level of approximation.

As mentioned above, the most complete PNC calculation on Cs is that of Dzuba et al. 1989$_c$. Since their calculations of the experimentally known E1 transition matrix elements and hyperfine constants using their approach are accurate to 1% or less (Dzuba et al. 1989$_b$) then this accuracy is assumed for the PNC calculation as well.

An alternative approach to PNC calculations has recently been used by Hartley and Sandars (1990).

Table 4. PNC Results (Taken from Hartley, Lindroth and Martensson-Pendrill 1990)

	Cs 6s -> 7s	Tl $6p_{1/2}$ -> $7p_{1/2}$	Tl $6p_{1/2}$ -> $6p_{3/2}$
PNC HF	+0.927	-9.99	-39.74
+V$^+$	-0.035	+7.97	-1.75
+V^{+pnc}	-0.002	-7.63	+11.13
	———	———	———
Total	+0.890	-9.65	-30.36
BO +PNC-BO	+0.056	+2.25	+6.92
V$^+$ + V^{+pnc} on BO + PNC-BO	-0.013	-0.39	+0.49
Internal PNC BO	+0.003	-0.04	-2.94
Structrual Radiation	+0.003	-0.2	-0.9
Normalisation	-0.006	+0.13	+0.6
	———————	———————	———————
	+0.933	-7.90	-26.2
Dzuba et al. 1987$_b$ (basis set errors)	+0.90	-7.9	-27.0
Johnson et al. 1988$_c$ (no V$^+$, etc. terms on BO,BO-PNC)	+0.951		
Russian Bubbles Dzuba et al. 1989$_c$	+0.91(1±0.01)		

Table 5. Atomic PNC Results

Theory Results on Cesium	E1PNC ($\times 10^{-11}$ iea$_0$Q$_W$/(-N))
Dzuba et al. (1989)	0.91(1±0.01)
Hartley and Sandars (1990)	0.904(1±0.02)
Hartley et al. (1990)	0.933(1±0.04)
Johnson et al. 1986	0.951(1±0.05)
Bouchiat and Piketty (1986)	0.935±0.02±0.03

They used a semi-empirical correlation potential in the relevant RPA type of equations which when combined with a semi-empirical correction factor leads to an accuracy of 2% for the PNC E1 transition matrix element.

4.1 An Atomic Determination of $\sin^2\theta_w$

In Table 5 we give the latest theory calculations in units of Q_w. Combining these with experimental measurements of $E1^{pnc}$ allows us to extract a value for Q_w. Using the expression for Q_w from Marciano and Sirlin 1983 which includes radiative corrections:

$$Q_w = -0.9793 N + Z(0.9793 - 3.8968 \sin^2\theta_w) \quad (4.1)$$

and taking the best value theory from Dzuba et al. (1989$_c$) together with the most accurate experimental measurement on Cs (Noecker et al. 1988) we obtain an atomic physics determination of $\sin^2\theta_w$ given by

atomic physics: $\sin^2\theta_w = 0.226 \pm 0.007$ (exp) ± 0.004 (theory) \quad (4.2)

where the first error is due to experiment and the second to theory. If we compare this with the current world average (Langacker 1989) of

world average: $\sin^2\theta_w = 0.2264 \pm 0.0054$ (exp)

then we see that there is excellent agreement.

Conclusions

An atomic physics determination of $\sin^2\theta_w$ has a great deal more importance then merely confirming the world average value ! It is dependent on interactions at a much lower momentum transfer, Q, than high energy particle physics measurements and represents an important test of electro-weak theory in this energy range. Perhaps of even greater interest is the fact that an atomic physics value could be sensitive to different parameters from particle physics, such as the mass of the top quark, for example, so that more accurate atomic experiment and theory could allow determinations of fundamental physical significance. Also, with out doubt, the need for accurate atomic calculations of PNC quantities has been a great spur to atomic structure calculations which have improved tremendously over the last decade. It is now possible to calculate energies for single valence systems to fractions of a percent and it is hoped to achieve this level of accuracy for other quantities in the near future as well. There is still plenty of work to be done though: we have to master heavy atoms with more than one valence electron yet !

References

Arimondo E, Inguscio M, Violino P, 1977, Rev. Mod. Phys., 49:31
Bouchiat MA, Bouchiat C, 1974, J. de Physique, 35:899
Bouchiat C, Piketty CA, 1986, Europhys. Lett., 2:511
Blundell SA, Johnson WR, Sapirstein J, 1988, Phys. Rev. A, 38:4961
Dzuba VA, Flambaum VV, Shushkov OP, 1984, J. Phys. B, 17:1953
Dzuba VA, Flambaum VV, Silvestrov PG, Shushkov OP, 1985, J. Phys. B, 18:597
Dzuba VA, Flambaum VV, Silvestrov PG, Shushkov OP, 1987, J. Phys. B, 20:1399
Dzuba VA, Flambaum VV, Silvestrov PG, Shushkov OP, 1987$_b$, J. Phys. B, 20:3297
Dzuba VA, Flambaum VV, Silvestrov PG, Shushkov OP, 1988, Phys. Lett. A, 131:461
Dzuba VA, Flambaum VV, Shushkov OP, 1989$_a$, Phys. Lett. A, 140:493
Dzuba VA, Flambaum VV, Kraftmakher Ya, Shushkov OP, 1989$_b$, Phys. Lett. A, 142:373

Dzuba VA, Flambaum VV, Shushkov OP, 1989$_c$, Novosibirsk pre-print 89-111
Flusberg A, Mossberg T, Hartman SR, 1976, Phys. Lett., 55A:403
Glashow SL, et al., 1970, Phys Rev. D, 2:1285
Gallagher A, Lurio A, 1964, Phys. Rev., 136:A87
Gould G, 1965, Phys. Rev., 101:1828
Hartley AC, Martensson-Pendrill A-M, 1990, to appear in Z. Phys. D
Hartley AC, Lindroth E, Martensson-Pendrill A-M, 1990, submitted to J. Phys. B
Hartley AC, Sandars PGH, 1990, to appear in J. Phys. B
Hinds EA, 1989, Atomic Phys. 11, Proceedings of 11th Internatioanl Conference on Atomic Physics, World Scientific Pub.
James JV, Wang CC, Chuan Guo, 1985, Phys. Rev., A32:643
Johnson WR, Guo DS, Idrees M, Sapirstein J, 1986, Phys. Rev A, 34:1043
Johnson WR, Idrees M, Sapirstein J, 1987, Phys. Rev A, 35:3220
Johnson WR, Blundell S, Sapirstein J, 1988$_a$, Phys. Rev A, 37:307
Johnson WR, Blundell S, Liu ZW, Sapirstein J, 1988$_c$, Phys. Rev A, 37:1395
Langacker P, 1989, Phys. Rev. Letts., 18:1920
Lindgren I, 1990, submitted to J. Phys B
Lindgren I, Morrison J, 1986, Atomic Many-Body Theory (Springer series on Atoms and Plasmas)
Lindroth E, 1987, Physica Scripta, 36:485
Lindroth E, 1988, Phys. Rev. A, 37:316
Lindroth E, Heully J-L, Lindgren I, Martensson-Pendrill A-M, 1987, J. Phys. B, 20:1679
Lurio A, Prodell AG, 1956, Phys. Rev., 101:79
Mandl F, Shaw G, Quantum Field Theory, Wiley-Interscience Publication
Marciano WJ, Sirlin A, 1983, Phys. Rev. D, 27:552
Martensson-Pendrill A-M, Pendrill L, Salomonson S, Ynnerman A, Warston H, 1990$_a$, to be submitted
Martensson-Pendrill A-M, Ynnerman A, 1990$_b$, to appear in Physica Scripta
Moore CE, Atomic Energy Levels
Noecker MC, Masterson BP, Wieman CE, 1988, Phys. Rev. Lett., 61:310
Pollack SP, Wong EY, 1971, Am. J. Phys., 39:1388
Salam A, 1968, Elementary Particle Phys., ed. Svartholm, Stockholm
Salomonson S, Ynnerman A, 1990, to be submitted
Salomonson S, Oster P, 1989, Phys. Rev. A, 40:5559, 40:5548
Sandars PGH, 1977, J. Phys. B, 10:2983
Weinberg S, 1967, Phys Rev. Lett., 19:1264

HIGH PRECISION RELATIVISTIC ATOMIC STRUCTURE CALCULATIONS USING THE FINITE BASIS SET APPROXIMATION

H.M. Quiney

Department of Theoretical Chemistry
University of Oxford
5, South Parks Road
Oxford,   OX1 3UB

1. Relativistic Basis Set Methods

Most relativistic atomic structure calculations use some sort of finite dimensional single–particle basis. The well–known computer program of Desclaux and co–workers (Desclaux, 1975) and the GRASP^2 package produced by the Oxford Atomic Theory Group (Dyall et al, 1990) are the direct decendants of earlier non–relativistic programs which employ finite difference methods and whose origins may be found in work pioneered by Hartree and collaborators. The current status of the published finite difference programs and their application to bound–state problems and to electron–ion collision models is reviewed in the lecture by Ian Grant in this volume.

Relativistic many–body problems have been the subject of active theoretical interest for the last decade. The Oxford Atomic Theory Group became involved in the development of practical methods for relativistic many–body perturbation theory in 1984, when we considered the relativistic extension of a simple model problem using finite basis set methods to represent the complete Dirac spectrum (Quiney et al, 1985). It was clear from this model calculation that a properly constructed square–integrable basis could be used to evaluate relativistic perturbation theory diagrams to high accuracy order–by–order, including all contributions from the positive– and negative–energy continua, without the quadrature problems which are encountered in brute–force finite difference calculations. Perturbation sums are evaluated using the basis set method simply by adding a finite number of discrete contributions from square–integrable wavepackets, and features of the relativistic perturbation theory which arise because of the filled negative–energy sea are readily incorporated simply by counting states. We have continued to pursue an algebraic approach, in which basis sets are constructed from analytic functions. This has allowed us to control the accuracy of our numerical procedures and to demonstrate that a sequence of calculations converges to a well–defined limit as the dimension of the basis is enlarged towards completeness. This approach does not, of course, represent a unique viewpoint on relativistic many–body perturbation theory using basis sets. Analytic methods have been developed by Drake and Goldman (1981, 1988) using different functional forms to those implemented in our codes. Piecewise polynomial basis functions have been employed by Johnson and his collaborators at Notre Dame in highly successful calculations of many–body and parity–non–conserving effects in the caesium atom (Johnson et al, 1988) and a finite difference discretization procedure has been developed at Göteborg which has been applied successfully to relativistic many–body problems (Salomonson and Öster, 1989). We have persevered with our analytic approach because of its success in treating the atomic many–body problem, the direct applicability of the method to relativistic molecular models and the computational efficiency of the scheme when implemented on vector–processing machines (see the articles by Ishikawa, Wilson and Dyall et al in this volume for molecular applications). The method has also proved useful in the mathematical analysis of the relativistic finite basis set approximation (Grant, 1989).

This talk is about the design of practical finite basis set methods for atoms. We have named our prototype computer program SWIRLES, to honour one of the pioneers of relativistic atomic structure theory (Swirles, 1935). The general strategy is familiar to quantum chemists who perform molecular calculations using finite basis sets of Slater– or Gaussian–type functions. In relativistic atomic structure calculations, however, we emphasize the importance of a complete representation of the spectrum and of the systematic approach towards completeness of the basis. This should be compared with the widespread quantum chemical practice of using energy optimized basis sets which may not be enlarged in any systematic fashion and which often suffer from computational linear dependence.

The first version of SWIRLES is able to calculate bound–state energy levels and expectation values for closed–shell, average–of–configuration and simple open–shell problems. These calculations are comparable in both speed and accuracy with those produced by the finite difference program GRASP for any element up to about $Z = 100$. Unlike GRASP, the low–frequency Breit interaction may be included in the construction of the self–consistent field potential as a matter of routine with negligible computational overhead. This is possible because of the extensive use of recurrence relations in the evaluation of intermediate quantities for two–electron basis set integrals. The matrix structure of the Dirac–Fock equations and the subsequent diagrammatic expansion of the correlation energy are translated into a sequence of matrix multiplications which are built around library subroutines written in CRAY Assembler Language (CAL). The most CPU–intensive part of the calculation which transforms basis set integrals into reduced matrix elements runs at a substantial fraction of the maximum rate per processor on the CRAY X–MP/416, so that the remaining bottlenecks in the calculation are connected with some essential scalar manipulations and tests in the integral generation which ensure the accuracy of the calculation. The tensor algebra machinery introduced and refined by Grant (1970, 1988) for the Dirac–Fock problem has been adopted throughout SWIRLES. The heart of the program is a module called XKABCD. The input to this module consists of four sets of variables which define the basis set parameters, quantum numbers and expansion coefficients for the orbitals belonging to four distinct atomic symmetry types. All basis set integrals are generated as they are required and partial transformations performed on the lists to reduce storage. The output of XKABCD is a complete list of reduced matrix elements of all allowed tensor orders for both the Coulomb and low–frequency Breit interactions. It is then a simple matter to combine the reduced matrix elements together following the index structure prescribed by perturbation theory, multiply these contributions by the appropriate angular momentum coupling coefficient and add the resulting contribution to the diagrammatic expansion at a given order.

The program is structured for an order–by–order approach to the many–body problem. Starting with a many–body diagram at a given order, the angular and radial parts of the diagram are separated and the angular parts decomposed and evaluated according to the standard Yutsis–Vanagas–Levinson rules (Lindgren and Morrison, 1986). SWIRLES evaluates the radial parts and the two contributions are recombined to form the many–body diagrammatic components. Fourth–order in many–body perturbation theory represents a practical upper limit to this approach because of the complexity of the higher-order diagrams and the rapid growth in the number of terms. Another approach, which is popular in non-relativistic atomic and molecular theory, has been adapted to relativistic MBPT by Johnson et al (1988) and by the group at Chalmers (Salomonson and Öster, 1989), who sum some classes of diagram through all-orders in perturbation theory using the coupled-cluster ansatz (Lindgren and Morrison, 1986). Within an approximation which includes all single- and double-excitations in the correlation of electron pairs, the coupled-cluster approximation is a powerful and relatively straightforward technique which is able to recover the bulk of electron correlation corrections. A relativistic coupled-cluster extension of SWIRLES is currently under development which exploits the special features of our basis functions and much that is already known from non-relativistic quantum chemical studies. Order-by-order methods are attractive from a conceptual point of view because many-body perturbation theory diagrams represents a direct transcription of the standard rules of QED, but the practical problems associated with the complexity and poor convergence of the diagrammatic expansion may make coupled-cluster approximations the method of choice in many applications.

It is planned to interface the next version of SWIRLES with the angular modules of GRASP^2, to produce a general atomic structure program for multiconfigurational, coupled-cluster and many-body perturbation theory calculations of complex atoms using analytic basis functions. We are currently investigating methods for the *ab initio* evaluation of quantum electrodynamic (QED) corrections in atoms. In view of the current precision of atomic spectra and the interest in QED effects in many-electron systems, more sophisticated computational techniques are necessary and crude approximations to the Lamb shift in many-electron systems are no longer adequate. Our preliminary work towards the calculation of radiative corrections in atoms is the subject of the second part of this talk.

2. Spinor Basis Functions

The first calculation of relativistic atomic structures using the analytic basis set approximation was made by Kim (1967) while he was a graduate student with Roothaan in Chicago. He had to contend with two main problems which appeared to undermine the matrix approach to the Dirac-Fock problem. Firstly, he commented that the computers of the day were too slow and too short on memory to make relativistic calculations of meaningful accuracy. Computer resources are now more powerful and more readily available than was imaginable in 1967, so this first problem is of no concern, at least for atoms. For molecules, however, computing power is still a problem and we have yet to see relativistic *ab initio* molecular orbital programs applied to chemically interesting systems containing heavy atoms. Clearly this situation is changing, and we have heard encouraging reports at this meeting from a number of groups working towards *ab initio* quantum chemistry. Kim noticed another more enduring problem which became the subject of active research throughout the 1980's. Variational calculations based on the Dirac one-electron operator have a tendency to "collapse". The distinction between positive- and negative-energy solutions is lost, the spectrum exhibits wild fluctuations as the basis set parameters are adjusted and the correct non-relativistic limit of the spectrum is not recovered if the numerical value of the speed of light is allowed to tend to infinity. In Kim's calculations, this problem manifested itself in the instability of relativistic basis set optimization procedures.

The situation was reviewed by Kutzelnigg (1984) and a number of methods have since been developed which reduce or eliminate the problem. I will consider only those methods which avoid variational collapse through a special choice of basis set; the alternative techniques, most of which are based on corrections to the Pauli limit of the Dirac equation, are now so numerous that they require careful consideration in their own right.

Kim chose as his radial basis functions modified exponential functions of the form

$$f^m_{\kappa i}(r) = r^{\gamma+m} \exp(-\zeta_i r) \qquad [2.1]$$

where κ is the usual relativistic angular quantum number, $\gamma = [\kappa^2 - (Z\alpha)^2]^{\frac{1}{2}}$, m is a positive integer, Z is the nuclear charge, α is the fine–structure constant and $\{\zeta_i\}$ contains a chosen set of exponential parameters. Throughout the following discussion, we will ignore the spin–angular dependencies of the spinors, which are treated algebraically. There was no explicit connection between the functions belonging to the large– and small–component basis sets and it was found that satisfactory answers could only be obtained by imposing the additional constraint that solutions should satisfy the relativistic virial theorem. Later, it was demonstrated by Lee and MacLean (1982) that quantum chemical calculations may be performed using standard non–relativistic methods if the large– and small–component basis sets are related by the Pauli condition

$$\{f^S(r)\} \simeq \{\sigma.p\, f^L(r)\} \qquad [2.2]$$

where the superscripts L and S refer to matched pairs of large– and small–component basis functions, respectively, and the operator $\sigma.p$ provides the off–diagonal coupling of the spinor components corresponding to the action of the relativistic kinetic energy operator $c\alpha.p$ in the absence of an external field and in the Pauli limit of the Dirac equation. This basis set prescription is now widely known as kinetic balance and avoids the worst aspects of variational collapse. It is not a sufficient condition to avoid all symptoms of variational collapse. Finite matrix approximations of Dirac single–particle states do not provide upper bounds to the exact eigenvalues; small "variational failures" of the order of an atomic unit are observed using kinetically balanced functions in heavy point–nuclear ions. For example, the solutions of the point–nuclear Dirac equation are singular at $r = 0$ and have non–integer power law behaviour. Such analytic behaviour may not be reproduced by the direct application of the kinetic balance approximation because it leads to singular matrix elements, particularly for the symmetry–type $\kappa = +1$ and to small variational failures. The use of a uniform proton distribution reduces the size of the bounds failure but does not eliminate it completely for orbitals of $\kappa = +1$ symmetry type (Ishikawa et al, 1986). It has been suggested that the use of a Gaussian distribution for the nuclear potential eliminates the variational problem entirely (Aerts, 1986), but I know of no published numerical evidence which establishes this when the kinetic balance prescription is observed rigorously. This approximation has the very important practical feature that no new integral types are required if Gaussian–type primitive functions are chosen as a basis for kinetically balanced molecular calculations (Aerts, 1986; Dyall et al, this volume). The disadvantage is that there is no physical basis in the choice of a single Gaussian function to model the nuclear potential, and the possibility remains that a more realistic empirical model of the nucleus such as the Fermi distribution would reintroduce small variational collapse problems. This area requires further study, but these comparatively minor details need not invalidate the kinetic balance approach.

Our experience of relativistic calculations using the finite difference approximation indicates that we must look carefully at the boundary conditions on orbital models, particularly at the nucleus. In finite difference calculations, the boundary conditions are an integral part of the solution algorithm, but in non–relativistic basis set calculations they seem to appear only incidentally, when we examine how well our calculation satisfies the "cusp condition" at the nucleus after the wavefunctions have been determined. We have found that the boundary conditions form an integral part of any valid relativistic finite basis set procedure. The analysis of kinetic balance provided by Stanton and Havriliak (1984) indicates that these small bounds failures introduce spurious errors $O[(Z\alpha)^4]$. It is likely be some time before quantum chemical calculations are able to overcome the formidable technical difficulties which arise if we wish to evaluate all important corrections $O[(Z\alpha)^2]$, which represents the level of approximation we require in order to achieve a chemical accuracy of 1 milli-hartree. The attractive feature of the kinetic balance approach is that it allows the recycling of most of the technology of quantum chemistry without substantial modification. This is particularly true of the algorithms for the generation and manipulation of two-electron matrix elements and methods to deal with the many-body problem and the construction of basis sets, which have been major technical preoccupations of computational quantum chemists for almost thirty years. It is likely that the only real problem in performing *ab initio* relativistic calculations of molecular properties to chemical accuracy will be found in the numerical complexity of these calculations, rather than in any deep problem of principle.

We may place the relativistic basis set approximation in a well-defined mathematical context if we combine the idea that the large- and small-component basis sets for one-body model equations must be related through the action of a operator coupled with the requirement that physical boundary conditions determined by the algebraic form of the external field must be be imposed on the space of trial functions. This is equivalent to the familiar result that a complete basis set may be constructed from the discrete solutions of a differential operator, subject to appropriate boundary conditions. This is the essence of the Rayleigh-Ritz method which exploits the fact that the complete set of solutions obtained through the above prescription forms a basis in which to expand any arbitrary function which satisfies the same boundary conditions. This leads us to a discussion of S- and L-spinor basis sets for relativistic atomic calculations.

It is not widely appreciated that the completeness of Slater- or Gaussian-type basis sets in non-relativistic studies arises because these functions satisfy self-adjoint differential equations of Sturm-Liouville type. Another example is the less widely used non-relativistic Sturmian basis pioneered by Hylleraas (1930) and popularized by Rotenberg (1970) in high-precision atomic calculations. The functions which we term S-spinors and L-spinors have as their non-relativistic (Pauli) limits kinetically balanced sets of Slater functions and Sturmian functions, respectively, and have the same range of applicability in atomic calculations. The S- and L-spinors are also solutions of a self-adjoint differential equation: a generalized Sturm-Liouville form of the Dirac equation. The first demonstration that S-spinors, which are based on the discrete solutions of the point-nuclear Dirac equation, exhibit no trace of variational collapse was provided by Quiney (1987) and Grant and Quiney (1988). The validity of S-spinor and L-spinors expansions in the context of Sturm-Liouville theory has been discussed by Grant (1989) and by Drake and Goldman (1988). The latter authors call L-spinor functions "relativistic Sturmian functions" but there is no difference up to a multiplicative constant. The L-spinor functions form a complete basis for an exact expansion of the Dirac-Coulomb Green's function, a fact which is implicit in the work of Wichmann and Kroll (1956) and explicit in the work of Hostler (1982, 1987) and Manakov *et al* (1984). The radial L-spinor functions have the general form

$$f^T_{\kappa n}(r) = C_{n\kappa}\, r^\gamma \exp(-\lambda r)\left\{ -(1-\delta_{n0})L^{2\gamma}_{n-1}(2\lambda r) \pm \frac{(N_n - \kappa)}{(n+2\gamma)} L^{2\gamma}_n(2\lambda r)\right\}$$
[2.3]

where the upper sign implies T = L, and the lower, T = S and the labels L and S designate the large and small components, respectively. The parameter λ

is the same for all n and fixed κ but is an otherwise arbitrary real, positive constant. For practical purposes, λ is usually taken to be of the same order of magnitude as the nuclear charge, Z. $C_{n\kappa}$ is a normalization constant, N_n is the relativistic "apparent" principal quantum number, and $L_n^{2\gamma}(x)$ is the generalized Laguerre polynomial. The S-spinor functions are formed by choosing a set of exponential parameters, $\{\lambda_i\}$, and the functional form of the L-spinor function with the smallest value of n permissible for symmetry-type κ. For $\kappa < 0$ the value of n = 0 is chosen; for $\kappa > 0$ we choose n = 1. This forms a set of nearly-nodeless radial functions which are sufficiently flexible to represent the spectrum for realistic mean-field potentials but which uses relatively simple functional types. The general form of the unnormalized S-spinor radial basis is

$$f_{\kappa i}^T(r) = \begin{cases} r^\gamma \exp(-\lambda_i r) & \text{for } \kappa < 0 \\ (A_\kappa^T + \lambda_i r) r^\gamma \exp(-\lambda_i r) & \text{for } \kappa > 0 \end{cases} \quad i = 1, 2, \ldots N(\kappa) \quad [2.4]$$

The superscript T takes the values L (large) and S (small) for a given angular symmetry-type, κ, and radial dimension, $N(\kappa)$. The contraction coefficients A_κ^T are readily obtained from the explicit functional forms of the generalized Laguerre polynomials which appear in the L-spinor prescription. This reasonably simple basis possesses all the properties we require to avoid variational collapse problems, and provides a set which may, in principle, be extended towards completness in the limit $N(\kappa) \to \infty$. For $\kappa = +1$, for example, the required contraction coefficients are

$$A_\kappa^L = (2 - N)(2\gamma + 1)/(2N - 2)$$

$$A_\kappa^S = -N(2\gamma + 1)/(2N - 2) \quad [2.5]$$

where

$$\gamma = (1 - Z^2/c^2)^{\frac{1}{2}} \text{ and } N^2 = 2 + 2\gamma \quad [2.6]$$

Each basis set has particular advantages depending on the situation. The S-spinor is analogous to the non-relativistic Slater spin-orbital, and is a particularly good choice for calculations involving heavy, many-electron systems. It is this basis which has been adopted in the radial modules of the program SWIRLES. The L-spinor functions take advantage of the orthogonality and recursive properties of Laguerre polynomials, and is an ideal choice for high-precision calculations on few-electron systems. The orthogonality properties of L-spinors which result in sparse and banded matrices greatly reduce the effects of computational linear dependence in the basis, so that radial basis sets of several hundred functions per angular symmetry type may be used without problem.

The basis set parameters for S-spinor calculations must be chosen in a systematic fashion which ensures that the basis becomes complete in a well-defined limit and which controls the effects of computational linear dependence. The rather special choice of function ensures that much of the experience gained in the construction of systematic sequences of even-tempered basis functions may be transferred directly to relativistic electronic structure calculations. The even-tempered prescription

$$\lambda_i = \alpha_N \cdot \beta_N^{i-1} \quad i = 1, 2, \ldots, N \qquad [2.7]$$

generates a suitable basis of dimension N if the parameters α_N and β_N satisfy

$$\alpha_N \to 0, \; \beta_N \to 1 \text{ and } (\beta_N)^N \to \infty \qquad [2.8]$$

in the limit $N \to \infty$. In practice, if the value of any parameter β in a sequence of calculations is close to unity, the basis becomes computationally linearly dependent as is evidenced by a striking growth in the magnitude of the orbital expansion coefficients and one or more of the eigenvalues of the basis set Gram matrix approaching zero. As a rough guide, the value of β should not be less than about 1.30 for a set of ten or more functions per radial symmetry type. The value of α merely scales the exponent values. In principle, basis set optimization using this scheme is very simple for finite sets, because there are only two free parameters for a fixed value of N. In practice, however, the regular spacing of the exponents and the high quality of a moderately large even-tempered set introduces deep multiple minima in the energy surface, which may cause the optimization procedure to sample regions where linear dependence is a problem. For atoms, optimization is pointless, since it is always easier and more profitable to extend the basis by increasing its dimension. For molecules, we contend that a strong case can be made for the use of large, flexible even-tempered sets, rather than compact energy-optimized sets, because of the need to use a richly detailed virtual spectrum in many-body calculations.

3. Many-Body Calculations Using S-Spinors

Self-consistent field calculations of the mercury ground state obtained using SWIRLES are presented in this volume by Ian Grant. These calculations demonstrate that the finite basis set approximation is competitive in both speed and accuracy with the finite difference approach embodied in the GRASP² program and that new areas of research are suggested. The Breit interaction may be included in the self-consistent field procedure as a routine part of any atomic structure calculation. Interesting and significant modifications of the orbitals energies and shapes are found if the potential includes low-frequency Breit contributions, but the total energy is not significantly changed from the first-order result. It has been been demonstrated by Lindroth et al (1989) that the self-consistent treatment of the Breit interaction leads to important contributions in calculations of the enhancement factor between the electric dipole moment of the electron and the corresponding moment in an atom. The most important consequence of treating the Breit interaction self-consistently is the substantial simplification in the relativistic many-body theory which results from the cancellation of some classes of diagram through all orders of perturbation theory, rather than the magnitude of the effects which are due to the modification of the orbitals. It is likely that calculations of transition probabilities will be sensitive to self-consistent Breit modifications because of the significant changes which are observed in the orbital mean radii, particularly for core electrons. We hope to examine this question in the near future.

The potential of this method is nowhere more apparent than in the calculation of many-body effects. Such calculations have been reported previously within the Dirac-Coulomb approximation (Quiney, 1987, 1988; Quiney et al, 1987, 1989, 1990), but here we construct a no-virtual-pair relativistic many-body perturbation theory based on a Dirac-Fock-Breit reference function, and incorporating the Coulomb-gauge low-frequency form of the Breit interaction in the evaluation of second-order pair correlation corrections (see Quiney et al, 1990).

Table 1.

Relativistic corrections to the electronic energy of Ne^{8+}

(1) Hartree–Fock energy = −93.861 094 117 a.u.

(2) Dirac–Fock–Breit energy = −93.970 669 651 a.u.

(3) Relativistic shift [(2) − (1)] = −0.109 575 534 a.u.

(4) Total relativistic shift from all sources (Drake, 1989)
= −0.110 829 364 a.u.

(5) Relativistic correlation energy [(4) − (3)]
= −1253.8 × 10^{-6} a.u.

Second–order MBPT contributions with sign reversed

$\kappa_r \kappa_s$	Nonrelativistic	Relativistic	Difference
$s\ s$	14104.3	14472.1	367.8
$\bar{p}\ \bar{p}$	8274.9	8593.3	320.4
$p\ p$	16549.7	16908.9	359.2
$\bar{d}\ \bar{d}$	1499.0	1575.7	76.7
$d\ d$	2248.6	2313.4	64.8
$\bar{f}\ \bar{f}$	442.7	475.2	32.5
$f\ f$	590.2	616.7	26.5
$\bar{g}\ \bar{g}$	170.1	187.2	17.1
$g\ g$	212.6	226.5	13.9
$\bar{h}\ \bar{h}$	76.9	86.9	10.0
$h\ h$	92.3	100.5	8.2
$\bar{i}\ \bar{i}$	38.8	45.0	6.1
$i\ i$	45.3	50.4	5.1

The relativistic many–body shift from all sources, including the low–frequency Breit interaction is −1277 μ–hartree using the S–spinor basis set approximation, which compares well with the value −1254 μ–hartree derived from Drake's calculation at this order of approximation. The agreement is poor, however, with the value of −1851 μ–hartree obtained by Lindroth (1988) using pair equation methods, and with similar estimates obtained using multiconfigurational expansions (Gorceix et al, 1987). Similar agreement with Drake, at the 2% level, has been found for He–like beryllium and zinc. The evidence suggests that relativistic many–body corrections are well–represented by low–order many–body perturbation theory if the Breit interaction is treated self–consistently throughout a calculation. The residual many–body corrections are dominated by higher–order non–relativistic contributions involving many interactions, in which relativistic effects contribute comparatively little. The most likely source of the differences between the current calculations and those of Lindroth (1988) is the self-consistent use of the low-frequency Breit interaction in the former and the perturbative use of the Gaunt interaction in the latter. The differences between the basis set calculation and multiconfigurational expansions may be numerical rather than physical in origin, since the set of numerical configuration state functions is comparatively small due to the computational cost involved in high-precision MCDF calculations. It is difficult to define "relativistic shifts" in MCDF calculations because the method is founded on the variational principle and different basis sets are used in the non-relativistic, Dirac-Coulomb and Dirac-Breit models.

Many-body corrections evaluated with the low-frequency forms of the Coulomb- and Feynman-gauge interactions in no-pair theory may be substantially different, and some comment seems necessary to explain why I have chosen to use the low-frequency Breit interaction rather than the simpler Gaunt interaction. The recent discussion of this problem appears to have been prompted by Sucher (1989), who indicated that the Feynman-gauge form of the interaction should not be used in no-virtual-pair calculations of many-electron atomic structures because the leading order relativistic effects are incompletely determined. Lindroth and Mårtensson-Pendrill (1989) reported the remarkable result that gauge independence between the two forms of the interaction is restored in the low-frequency limit only at the expense of including negative-energy intermediate states in the Gaunt result, which is a domain in which this approximate interaction is not valid. Further analysis of this problem by Lindgren (1990) indicates that when a single-photon potential of the type used in atomic structure calculations is employed within the no-pair approximation, the Coulomb-gauge Breit operator represents the entire interaction, correct to $O(\alpha^2)$, but when the Gaunt interaction is used we must include the lowest-order retardation correction, which contributes to irreducible crossed-photon Feynman diagrams, in order to restore gauge independence. This fundamental work is of practical importance in atomic and molecular calculations. It is very tempting to use the Gaunt interaction in self-consistent field and many-body calculations because no new integral types are encountered in basis set approximations, and because the retardation of the electromagnetic interaction is known to contribute only about 10% of the relativistic shift at the mean-field level of approximation. Such calculations would not, however, include all important relativistic many-body effects unless we use the interaction in negative-energy intermediate states where the Gaunt interaction is not valid, or we include the leading-order retardation correction. In either case, the apparent simplifications which we sought by using the Gaunt interaction are lost, and it seems clear that within a theory which neglects virtual-pair production, the use of the Coulomb-gauge Breit interaction is preferable in calculations of relativistic MBPT.

The presentation of relativistic many-body theory in this paper has had a essentially non-relativistic flavour. The general strategy of the calculations has followed closely the practices of quantum chemistry because so much is known about computational algorithms in this field. There seems to be little point in calculating relativistic or quantum electrodynamical corrections unless the non-relativistic parts of the calculation are satisfactory and the numerical procedures are carefully defined. The use of an analytic basis ensures that we may make a direct connection with very high precision calculations in the non-relativistic limit of the theory, so that the major features of the problem are accurately reproduced. Accurate relativistic corrections to total electronic energies are readily obtained if we adopt a theory based on Dirac spinors instead of non-relativistic spin-orbitals. The use of a two-body interaction operator which includes both the instantaneous Coulomb and frequency-independent Breit terms in both the specification of the self-consistent field atomic potential and in the evaluation of no-virtual-pair many-body perturbation theory diagrams leads to a theory in which, in principle, all $O[(Z\alpha)^2]$ corrections to the non-relativistic limit are recovered. Residual many–body corrections are due mainly to non–relativistic contributions which arise because of the truncation of the perturbation expansion or from uncalculated terms in the multipole expansion of the potential.

The clear challenge to atomic and molecular theory is to develop methods which enable the reproduction and interpretation of spectra to an accuracy comparable to that of current experiments. Continual improvements in experimental technique have been the motivation for the development of theoretical methods to model spectroscopic transitions in many–electron systems which are of interest to astrophysics and plasma physics and for the establishment of relativistic quantum chemistry. Advances in the fundamental models of physical interactions since the 1950's have led to experiments which are designed to measure parity non–conservation in atomic states caused by the weak electron–nucleon force and the possible existence of atomic electric dipole moments which would arise though the violation of symmetry under both parity and time reversal. An overview of parity violation in atoms is presented in the lecture by Hartley and Sandars in this volume. The other area in which relativistic many–body is essential is in the interpretation of data which aims to test the theory of quantum electrodynamics in strong external fields and in many–electron systems. Contributions which are attributable to QED corrections, particularly those known collectively as the Lamb shift, are now the largest source of uncertainty in high–precision calculations of few–electron systems. In the second part of this talk we examine the extension of the theory to the calculation of terms which are characteristic of relativistic electron theory and the theory of quantum electrodynamics, including retarded interelectron potentials, the polarization of the vacuum and electron self–interactions and describe our progress in developing numerical methods to deal with these problems.

4. Quantum Electrodynamics

When we consider effects which are essentially relativistic or quantum electrodynamical in origin, the "follow–your–nose" approach (Sucher, 1989) which has guided us until now lands us in trouble: the equations which govern the propagation of Dirac particles and the effective interactions between these particles may not be deduced by naive generalization of non–relativistic quantum mechanics. The underlying problem is that the notion of a "single–particle system" cannot be given physical meaning in relativistic atomic structure theory because of the presence of the filled negative–energy sea: valid relativistic models of atoms, molecules and solids are always infinitely–many–body problems which are most conveniently discussed within quantum field theory. So far, we have considered a simplified physical model in which the negative–energy many–body contributions have been neglected. A classical condition has been imposed which conserves the number of particles throughout an arbitrary sequence of unretarded interactions.

Quantum electrodynamics, which is the theory of fundamental processes involving electrons (all leptons) and photons, must be brought to bear in order to resolve the true role of the negative–energy states in the Dirac theory and the many divergences which are encountered in calculations of electrodynamic interactions (Schweber, 1961; Mandl and Shaw, 1984; Itzykson and Zuber, 1980). The development of methods which generate a discrete representation of the Dirac spectrum suggests the exciting possibility that the rules of QED may be transcribed into practical computational algorithms which are suitable for modelling complex systems. The extent to which we may append the residual effects of QED for many–particle systems to our simplified models of atomic and molecular physics raises difficult questions of both theory and practice. In this section, we discuss some relevent features of QED which arise in atomic calculations when we venture "beyond the no–virtual–pair approximation".

4.1. Electrons

The Green's function for the non–relativistic Kepler problem in quantum mechanics is known in closed form (Schwinger, 1964). The relativistic Green's function for the Dirac–Coulomb equation known only as a partial wave expansion, both for the linear (spinor) (Wichmann and Kroll, 1956) and second–order (scalar) forms (Hostler, 1982; 1987). Apart from many subtle problems of theory, the difficulties which arise in the accurate construction and manipulation of numerical representations of electron Green's functions (or propagator functions) present the main obstacle in the development of relativistic QED calculations (Lieber, 1989).

In non–relativistic quantum mechanics, where the scalar wavefunction, ψ, satisfies the Schrödinger equation

$$H\psi = i\hbar \frac{\partial \psi}{\partial t} \qquad [4.1]$$

the evolution of the wavefunction over a small time interval Δt is determined by the action of the operator $\exp(-iH\Delta t/\hbar)$. If the particle has the initial coordinates $(\mathbf{r}_1, t_1)$ then at some other time:

$$\psi(\mathbf{r}_2, t_2) = \int K(\mathbf{r}_2, t_2; \mathbf{r}_1, t_1) \psi(\mathbf{r}_1, t_1) d^3 \mathbf{r}_1 \qquad (t_2 > t_1)$$

$$\psi(\mathbf{r}_2, t_2) = 0 \qquad (t_2 < t_1). \qquad [4.2]$$

The kernel $K(\mathbf{r}_2, t_2; \mathbf{r}_1, t_1)$ is the propagator function which carries the wavefunction to later times. For earlier times the non–relativistic propagator function vanishes; the future evolves from initial conditions specified at previous times. If we can find a complete, orthonormal set of solutions of the time–independent Schrodinger equation, $H\varphi_n = \epsilon_n \varphi_n$, the superposition principle allows us to express the propagation kernel as a spectral expansion

$$K(\mathbf{r}_2, t_2; \mathbf{r}_1, t_1) = \sum_n \varphi_n(\mathbf{r}_2) \varphi_n^*(\mathbf{r}_1) \exp[-i\epsilon_n(t_2 - t_1)] \qquad (t_2 > t_1)$$

$$= 0 \qquad (t_2 < t_1)$$

$$[4.3]$$

This propagator forms the theoretical basis of quantum chemical calculations based on many–body perturbation theory. It is implicit in most non–relativistic calculations that we introduce a finite–dimensional representation of the complete spectrum in the expansion of the above propagator. Many–body perturbation theory may be developed from this starting point by combining propagators and states with instantaneous interactions and performing integrations over the time–interval $(t_2 - t_1)$. The resulting energy denominators which appear in the perturbation theory expressions are characteristic of the finite dimensional representation rather than the original physical model. This procedure is equivalent to a slight modification of the physical model of a free atom which is removed in a well–defined limit. If we use a finite set of analytic basis functions which decay exponentially, we may imagine that we have introduced a weak confining potential which discretizes the spectrum at all energies. In the case of finite–difference or spline calculations the atom is placed in a box, whose radius is systematically increased until its presence is undetectable in the results of the calculation. In either case, these are realistic and controllable modifications of the model, since the idea of a free atom is itself an approximation to the experimental conditions. Both the finite basis and finite difference approximations are familiar techniques in the solution of integral equations (Hochstadt, 1973).

Construction of an approximate relativistic electron propagator using finite spectral representations requires special care. The Schrödinger equation is replaced by the Dirac equation which is, for an electron moving in a time-independent external four-potential with components A_μ,

$$(i\gamma_\mu \partial^\mu - A_\mu \gamma^\mu - mc)\psi(x) = 0 \qquad [4.4]$$

where $\psi(x)$ is the solution four-vector and the four-dimensional matrices γ_μ satisfy

$$\gamma_\mu \gamma_\nu + \gamma_\nu \gamma_\mu = 2g_{\mu\nu}. \qquad [4.5]$$

The metric tensor, $g_{\mu\nu}$, is defined by $g_{00} = 1$; $g_{11} = g_{22} = g_{33} = -1$. If we omit the external field, the electron propagator is known exactly for free-particles. If we absorb the time-independent external field into the specification of the zero-order problem, as we wish to do in bound-state problems in order to reduce the strength of the perturbation, the model which describes the time-evolution of the system is known in QED as the Furry bound-state interaction picture (Furry, 1951; Schweber, 1964). There are two fields; one describes the localized creation and annihilation of electrons in a time-independent external field and the other defines the photon field. The interaction Hamiltonian, which specifies the interaction between the matter and radiation fields, constitutes the entire perturbation.

When we attempt to extend the idea of an electron propagator to relativistic particles, our instincts tempt us simply to replace the eigenvalues of the finite-dimensional representation of the Schrödinger equation with those of the Dirac equation, the scalar wavefunctions φ_n with the spinors ψ_n and the complex conjugates of the wavefunctions with the adjoint vectors, defined by

$$\overline{\psi}_n = \psi_n^* \gamma_0 \qquad [4.6]$$

This naive prescription violates the physical requirements of the Dirac hole theory of the positron, in which of the negative-energy states are filled with electrons according to the Pauli Exclusion Principle and may not be used as a basis in which to expand the propagator for $t_2 > t_1$. We could just forget about the negative-energy states entirely, but then we have an incomplete representation of the propagator. This is the origin of the no-virtual-pair approximation which was been widely adopted in atomic calculations. It is stressed that such a theory leads to qualitative agreement with experiment for processes in atoms and molecules which arise through low-energy interactions and the deficiencies of the theory may be recovered, if required. With care (Mårtensson-Pendrill and Lindroth, 1989; Lindgren, 1990), the no-virtual-pair approximation yields all effects correct to $O(Z\alpha)^2$. In high-energy interactions, where pair-creation is possible, the negative-energy states fulfil a vital role in ensuring the convergence of the theory, because many of the linear divergences of the unquantized theory are softened to logarithmic divergences in hole theory. Both the no-virtual-pair propagator and the propagator which ignores the Pauli Exclusion Principle in negative-energy intermediate states are solutions of the homogeneous Dirac equation and their difference is the propagator for times $t_1 > t_2$. The hole-theory of the positron appears in propagator theory as a boundary condition on allowable solutions of propagation equation. The relativistic propagator for electrons, valid for all time intervals, is (Feynman, 1949):

$$K(\mathbf{r}_2,t_2;\mathbf{r}_1,t_1) =$$

$$+ \sum_n \psi_n(\mathbf{r}_2)\bar{\psi}_n(\mathbf{r}_1)\exp[-i\epsilon_n(t_2-t_1)]\theta(t_2-t_1) \qquad (t_2 > t_1)$$

$$- \sum_n \psi_n(\mathbf{r}_2)\bar{\psi}_n(\mathbf{r}_1)\exp[-i\epsilon_n(t_2-t_1)]\theta(t_1-t_2) \qquad (t_1 > t_2)$$

[4.7]

The function $\theta(\tau)$ takes the value $\theta(\tau) = 1$ for $\tau > 0$ and $\theta(\tau) = 0$ for $\tau < 0$. In this theory, positrons appear not as holes in an unobservable infinite negative–energy sea but as electrons propagating backwards in time. Notice that in extending the definition to $(t_1 > t_2)$ a minus sign has been introduced, which is consistent with results from the original formulation of hole theory (Weisskopf, 1939). In a finite basis set calculation, we replace the sums over a complete basis by finite–dimensional sums over a numerical basis according to the above rule, where the energy character of each of the square–integrable approximations is determined by its eigenvalue. From a computational viewpoint, this allows us to make a connection with the theory of quantum electrodynamics. In principle, we may construct approximations to the propagator and employ the rules of QED to translate Feynman diagrams into perturbation sums. The correct application of boundary conditions on the allowable forms for basis functions now becomes crucial, because the separation of positive– and negative–energy states is part of the physical model. Different formulae arise depending on whether the intermediate states are of positive– or negative–energy and some combinations of states are excluded when subsequent time–like integrations are performed.

The electron propagator function is conveniently represented as a contour integral over the Dirac Green's function, $G(x_2,x_1;z)$, in the complex energy plane and the boundary conditions of hole theory are embodied in the direction of the contour and in the poles and branch cuts in the complex energy plane. The conventional notation is:

$$-\frac{1}{2} S_F^e(x_2,x_1) = \frac{1}{2\pi i}\int_{C_F} dz\, G(x_2,x_1;z)\exp[-iz(t_2-t_1)]\gamma^0$$

$$= -K(x_2,x_1) \qquad [4.8]$$

The contour of integration passes from $z = (-\infty - i\delta)$ through the origin of the complex plane to $z = (+\infty + i\delta)$, where δ is a small positive quantity. The integrand is analytic and holomorphic throughout the complex plane, except at isolated simple poles and branch cuts on the real line, and the contour may be deformed to any convenient path. The hole theory propagator is obtained from the contour integral representation by the application of Cauchy's Theorem and the evaluation of the residues at the simple poles of the Dirac Green's function.

Calculations of QED effects are a particularly severe test of the basis function method because any slight numerical pollution of either branch, which would go unnoticed in no–virtual–pair theory, is amplified when we calculate electromagnetic matrix elements involving negative–energy states. A part of the self–energy calculation, which we discuss later, involves sums which resemble well–known sum rules. The contribution from negative–energy intermediate states dominates the calculation.

4.2. Photons

In non-relativistic quantum mechanics, static electrons interact through an instantaneous Coulomb mechanism depending only on the inverse of the spatial separation between the particles. In quantum electrodynamics, we require a function which acts only on a relativistically invariant space-time interval $\Delta s_{12} = 0$ (Feynman; 1949, 1962), defined by:

$$(\Delta s_{12})^2 = c^2(t_1 - t_2)^2 - (\mathbf{r}_1 - \mathbf{r}_2)^2 \qquad [4.9]$$

It seems reasonable to replace the instantaneous Coulomb interaction

$$\frac{1}{r_{12}} \delta(t_1 - t_2) \qquad [4.10]$$

by a similar function which acts only on $\Delta s_{12} = 0$, so that the interaction is retarded by the finite speed of propagation

$$\frac{1}{r_{12}} \delta(\Delta s_{12}) \qquad [4.11]$$

Again, the "follow your nose approach" is wrong, because the Fourier transform of this interaction contains both positive and negative frequencies: we observe only photons with positive frequency. The negative-frequency components are removed by employing the result (Feynman, 1949), for times $t_2 > t_1$:

$$\delta_+(x) = \frac{1}{\pi} \int_0^\infty \exp(-i\omega x)\,d\omega = \lim_{\epsilon \to 0} \frac{1}{i\pi}\left(\frac{1}{x - i\epsilon}\right) = \delta(x) + \frac{1}{i\pi x} \qquad [4.12]$$

This is averaged with the corresponding result for times $t_1 > t_2$ to obtain the static interaction valid for all time intervals t_{12}

$$\delta_+(\Delta s_{12}^2) = \frac{1}{2r_{12}}[\delta_+(ct_{12} - r_{12}) + \delta_+(-ct_{12} - r_{12})] \qquad [4.13]$$

where

$$\Delta s_{12}^2 = c^2 t_{12}^2 - r_{12}^2 \qquad [4.14]$$

This relativistically invariant operator, which contains only positive frequency Fourier components, depends on the square of the space-time interval, Δs_{12}. So far, the interaction has been between static charges. To obtain the Lorentz gauge interaction between moving charges, we replace the classical velocity, v_i with the matrix operator, α_i. The complete relativistic electron-electron interaction is

$$(1 - \alpha_1 \cdot \alpha_2)\delta_+(\Delta s_{12}^2). \qquad [4.15]$$

This is the Feynman gauge operator; the Coulomb-gauge form is discussed by Lindgren (1990). When the Fourier transform of the Coulomb-gauge interaction operator is taken, a potential between electrons is obtained in which contributions from scalar and longitudinal modes are eliminated and the instantaneous Coulomb interaction is recovered with the frequency-dependent Breit interaction as a correction to the potential. We shall develop the theory in Feynman gauge, because this leads to a more convenient potential for self-energy calculations of hydrogenic ions and the results are, of course, gauge invariant.

The interaction operator may be expressed as the four-dimensional Fourier transform

$$\delta_+(\Delta s_{12}^2) = \int \frac{d^4k}{(2\pi)^4} \exp(-ik.x_{21}) \frac{1}{k^2 + i\epsilon} \qquad [4.16]$$

where the limit $\epsilon \to 0$ is understood. The problem with using this interaction is that it leads to divergent matrix elements for some processes. It is usual to introduce a relativistally invariant cutoff function into the integrand which eliminates the contributions from very large values of $|k|$. This parametric modification of the theory must be removed at the end of a calculation; if the result does not depend on the form of the cutoff then the result is considered to be valid. The one-parameter cutoff function which is commonly used in QED calculations leads to the modified interaction for photons

$$\delta_+(\Delta s_{12}^2) = \int \frac{d^4k}{(2\pi)^4} \exp(-ik.x_{21}) \left\{ \frac{1}{k^2 + i\epsilon} - \frac{1}{k^2 - \Lambda^2 + i\epsilon} \right\}$$

$$= -\frac{i}{2} D_F^\Lambda(x_2 - x_1) \qquad [4.17]$$

The second term in the braces cuts off the interaction for $k^2 \gg \Lambda^2$ and the modification may be removed in the limit $\Lambda^2 \to \infty$. The second equality defines the regulated Feynman photon propagator. The use of a parametrized cutoff functions provides an elementary example of the Pauli–Villars regularization prescription (Pauli and Villars, 1949) and is necessary because the intermediate quantities which are calculated in QED are frequently not those of physical relevence. The parameters which are attributed to the mass and charge of a free electron already contain radiative corrections summed through all orders in perturbation theory and no shift in the values of these parameters is allowed through QED interactions. In atomic systems, the QED effect which is largest in magnitude is the Lamb shift (Johnson and Soff, 1982), which is mainly due to the emission and reabsorption by a bound electron of a virtual photon; the self-energy. In an atomic calculation, we must subtract from the self-energy for the bound electron the corresponding self-energy for a free-electron wavepacket with the same probability distribution as the bound state, because this second counterterm has already been absorbed in the definition of the free electron mass. The unrenormalized self-energy of a free electron in state $|a>$, which is required later as a renormalization operator, is:

$$\Delta E_a(\Lambda) = \frac{\alpha}{\pi} \left[\frac{3}{4} \ln(\Lambda^2) + \frac{3}{8} \right] <\psi_a | \beta | \psi_a>$$

$$= \delta m(\Lambda) <\psi_a | \beta | \psi_a> \qquad [4.18]$$

which is clearly divergent in the limit $\Lambda \to \infty$. This is chosen as the renormalization counterterm so that the electromagnetic shift in the energy of a free electron vanishes identically. Since $\Delta E_a(\Lambda)$ is proportional to the expectation value of the relativistic matrix β, the operator corresponds to a renormalization of mass. For a bound electron there is a residual shift due to the external field. The process of renormalization obviously raises deep philosophical and mathematical difficulties. The difference between the unrenormalized self-energies of the bound and free-electron is a finite quantity which represents a change in the electron mass-energy due to binding and is called the electron self-energy. The unrenormalized quantities are logarithmically divergent with respect to the contributions from the radiation field and the regularization parameters serve only to manage the divergences on a temporary basis: they are unphysical features of the theory which are eliminated at the end of any valid calculation.

5. Self–Energy Calculations

The changing nature of the self–energy operator as different energy regions are sampled is the main origin of the numerical difficulties which are encountered in calculations of radiative corrections in atoms. A few sample calculations will now be examined to indicate the qualitative nature of these numerical problems.

At photon energies substantially less than the electron rest mass–energy, the intermediate states which contribute to the self–energy shift are predominantly non–relativistic in character. By making non–relativistic approximations for the self–energy of both a free electron and an electron in an external Coulomb field, Bethe (1947) derived the remarkable formula:

$$\Delta E_n = \frac{2\alpha}{3\pi c^3} \int_0^{K_{max}} \sum_m (E_m - E_n) \frac{<n|\mathbf{v}|m><m|\mathbf{v}|n>}{E_m - E_n + k} \qquad [5.1]$$

for the radiative shift in the energy of the non–relativistic state $|n>$. The sum over $\{|m>\}$ includes all intermediate states excluding the reference state, and the integration over k is interpreted as a principal value integral. The use of the velocity operator, $\mathbf{v}$, indicates that a low–frequency dipole approximation has been made in the evaluation of radiation matrix elements. Since the integral has an upper energy limit of $K_{max} \simeq mc^2$ a.u., the low–frequency approximation does not seem justified, but Bethe later showed (Bethe and Salpeter, 1957) that the frequency–dependent parts of the interaction for the bound– and free–electron self–energy problems cancel and the correct shift is obtained in the dipole approximation. Bethe obtained a value of 1040 MHz in this approximation, which compares favorably with the modern experimental value of 1058 MHz, considering the approximations involved. An additional correction of −16 MHz to the theoretical value is also required at this order of approximation for vacuum polarization.

The non–covariant sharp photon cutoff is clearly an unsatisfactory device which may not form part of a consistent theory. In Bethe's non–relativistic theory, however, it was justified on physical grounds by the argument that a fully relativistic calculation should be convergent (a finite shift is observed) which can only be true if the bound– and free–electron parts of the calculations are asymptotically equal in the high–energy region. This is a feature of the relativistic operator which we wish to exploit in calculations; in the very high–energy region where the photon energy is greater than the total electron energy, the effect of binding must become less important. As a second requirement it would be a great advance if a relativistic calculation could incorporate the qualitative features of Bethe's calculation; the observable self–energy shift arises mainly through non–relativistic intermediate states and the effect of retardation and of the negative energy states on the observed shift is small.

The relativistic self–energy shift formula is (Baranger, Bethe and Feynman, 1953):

$$\Delta E_n(\Lambda) = \mathrm{Re}\left\{ -i a \pi \int \bar{\psi}_n(x_2) \gamma_\mu S_F^c(x_2,x_1) \gamma^\mu \psi_n(x_1) D_F^\Lambda(x_2 - x_1) d^3x_2 d^3x_1 d(t_2 - t_1) \right. $$
$$\left. - \delta m(\Lambda) <\psi_n|\beta|\psi_n> \right\} \qquad [5.2]$$

If the spectral expansion of the electron propagator and the Fourier transform of the unregulated photon propagator are substituted into the above expression and the time–like integration performed, observing the distinction between earlier and later times in the propagators, then the second–order energy shift derived by Weisskopf (1939) using transverse radiation gauge within the Dirac hole–theory is obtained:

$$\Delta E_n = +\sum_m^+ \int \frac{d^3k}{4\pi k} \frac{<\psi_n| a_\mu \exp(-i\mathbf{k}\cdot\mathbf{r})| \psi_m><\psi_m| a_\mu \exp(+i\mathbf{k}\cdot\mathbf{r})|\psi_n>}{E_m - E_n + k}$$

$$-\sum_m^- \int \frac{d^3k}{4\pi k} \frac{<\psi_n| a_\mu \exp(-i\mathbf{k}\cdot\mathbf{r})| \psi_m><\psi_m| a_\mu \exp(+i\mathbf{k}\cdot\mathbf{r})|\psi_n>}{E_n - E_m + k}$$

[5.3]

This shift is ultimately renormalized by the subtraction of the divergent free–particle self–energy, $\delta m(\Lambda=\infty) < \psi_n |\beta| \psi_n >$. It is assumed that the sums above exclude the reference state but include both transverse photon polarizations, μ. It was within this hole–theory formulation that the first calculations of the self–energy shift were made by Kroll and Lamb (1949) and French and Weisskopf (1949), and its familiar structure suggests that it may be possible to use finite basis sets to approximate the above sums as we often do in electron correlation studies. No photon cutoff parameter appears in this formula. The very early calculations of the self–energy based on the above prescription avoided the use of integral cutoff functions either by performing the calculations for the bound and free electrons in parallel using perturbed plane wave approximations or by calculating the $2s$–$2p_{1/2}$ interval directly. These subtractions allow the elimination of divergent algebraic expressions which are common to both unrenormalized contributions.

A hint as to how we might proceed is provided by the paper of French and Weisskopf (1949), who define a "mass operator" which is used to renormalize the calculation of the Coulomb self–energy shift. Their operator is:

$$\Delta M = +\sum_m^+ \int \frac{d^3k}{4\pi k} \frac{a_\mu \exp(-i\mathbf{k}\cdot\mathbf{r})| \varphi_m><\varphi_m| a_\mu \exp(+i\mathbf{k}\cdot\mathbf{r})}{E_m^0 - H_0 + k}$$

$$-\sum_m^- \int \frac{d^3k}{4\pi k} \frac{a_\mu \exp(-i\mathbf{k}\cdot\mathbf{r})| \varphi_m><\varphi_m| a_\mu \exp(+i\mathbf{k}\cdot\mathbf{r})}{H_0 - E_m^0 + k}$$

[5.4]

where $\{\varphi_m\}$ are solutions of

$$H_0 \varphi_m = E_m^0 \varphi_m \qquad [5.5]$$

for the free–particle Dirac operator, H_0. The operator H_0 appears in the energy denominators to distinguish between positive– and negative–energy free–particle states. The expectation value of this operator is defined for any state which can be expressed as a linear combination of free–particle Dirac spinors.

This approach has been criticized because the relativistic invariance of the theory is not obvious at every stage of the calculation (Schwinger, 1958). The original algebraic method of solution involves expansions of the intermediate states and the evaluation of matrix elements over many effective operators; Kroll and Lamb

(1949), whose approach was similar to that of French and Weisskopf, list twenty-five operators, each of which is multiplied by fifty elementary integrals over the photon momentum parameter. The results of the two calculations named above are, however, identical to those obtained by the lowest-order corrections in modern QED, and it is possible that a similar strategy in a numerical treatment of the renormalization problem may be more fruitful than the straightforward implementation of the invariant cutoff methods described above. We shall see later that a brute force numerical approach is technically very demanding. The calculation of ΔE_n may be made to lowest order in the Coulomb interaction if a number of simplifying assumptions are made. I will summarize the features of the lowest-order calculation from Feynman (1962) because they provide a qualitative description of the more complete calculation which we ultimately require.

A feature of Bethe's non-relativistic calculation is that the mean excitation energy of intermediate states contributing to the shift is very much greater than the binding energy: for the non-relativistic s-states of hydrogen the mean excitation energy is about 17 Ryd, compared with 1 Ryd binding energy. For energies greater than this mean value, it is reasonable to treat the electron as free, and to include only the perturbation of intermediate states due to a single Coulomb scattering between the emission and reabsorption of the self-energy photon. This approximation becomes more accurate in the high energy region, where multiple Coulomb scatterings become less likely. The formula derived by Schwinger (1949) and Feynman (1949) for radiative corrections to scattering in an external potential, V, and in the limit of low momentum transfer

$$\Delta E_> = -\frac{\alpha^2}{3\pi}\left[\ln\left(\frac{mc}{2\mu}\right) + \frac{11}{24}\right]<\nabla^2 V> \qquad [5.6]$$

may be used for the energy shift due to photons with a lower cutoff energy μ. A second ultra-violet cutoff parameter Λ has been eliminated in the mass-renormalization process. An infra-red regulator is necessary in this approximation because there is a divergent probability that one or more real soft photons will be emitted in inelastic collisions. This apparent divergence is really only an artifact of making a free-particle approximation in the bound-state energy region, where it is not valid. For energies less than the cutoff μ, the relativistic scattering formula may be matched to the non-relativistic shift formula derived by Bethe. For an external Coulomb field,

$$<\nabla^2 V> = 4\pi Z |\psi(0)|^2 = 4Z(Z/n)^3. \qquad [5.7]$$

This is the origin of the well-known qualitative behaviour of the self-energy, which depends on the square of the amplitude at r = 0, inversely as the cube of the principal quantum number, n, and approximately as Z^4, neglecting the weak logarithmic dependence.

An average excitation energy is defined in such a way that the calculations for the low- and high-energy regions match smoothly. Finite basis sets have already been employed by Broad (1985) and by Drake and Goldman (1988) in the numerical evaluation of this excitation energy, which is characterized by the so-called Bethe logarithm. We have found that the relativistic scattering formula, [5.6], may readily be pasted onto the low-energy formula by evaluating sum-over-states in the latter using a non-relativistic finite basis set approximation, and integrating over dk using an infra-red cutoff parameter, λ. The two calculations match smoothly over a large energy range; a convenient place at which to join the results is at λ = mc a.u. The numerical value of the calculated radiative shift is not sensitive to the value of λ chosen, but λ should be large compared with the binding energy to justify the free-particle approximation and small compared with the rest mass energy so that the dipole approximation is valid. Within this basis set approach it is possible to reproduce the lowest-order contributions to the self-energy without explicit knowledge of the Bethe logarithm, so that the numerical values of 1051 Mhz obtained from earliest calculations of the 2s–2p interval are reproduced after correcting for vacuum polarization. A basis of 15 Slater functions is sufficient to achieve 5 figure accuracy in this calculation.

The contributions from negative-energy states and low photon energy may be estimated by replacing the energy denominators $E_m - E_n + k$ with the representative value $2mc^2$. Only matrix elements of the operator α contribute significantly to the unrenormalized sum:

$$\Delta E_{n,<}^{-} = \sum_m^{-} \int_0^\lambda k^2 \frac{dk}{k} \frac{<n|\boldsymbol{\alpha}|m><m|\boldsymbol{\alpha}|n>}{2mc^2}$$

$$\simeq \sum_m^{+/-} \int_0^\lambda k^2 \frac{dk}{k} \frac{<n|\boldsymbol{\alpha}|m><m|\boldsymbol{\alpha}|n>}{2mc^2}.$$

[5.8]

In the second expression, the sum over states includes both positive- and negative-energy intermediates, except the reference state, without restriction. This is a reasonable approximation because the sum over positive-energy states is smaller than the negative- energy sum by terms $O[(Z\alpha)^2]$. Closure is now invoked to evaluate the approximate shift

$$\Delta E_{n,<}^{-} \simeq \int_0^\lambda <n|\alpha^2|n> k^2 \frac{dk}{2mc^2 k} = \frac{3\lambda^2}{4mc^2}$$

This contribution is very much larger than the observable shift and is strongly divergent with increasing cutoff. Retardation attenuates this term at very high energies, but it is striking feature of the problem that the contribution to the unrenormalized self-energy is dominated by negative-energy intermediate states and is independent of the strength of the applied external field. The physical effect is masked by the approximations which are made in arriving at this crude order-of magnitude result. Since the physical effect is obtained by subtracting two unrenormalized terms $O(Z\alpha)$ to estimate an observable $O[(Z\alpha)^4\ln(Z\alpha)^{-1}]$ the numerical requirements of this problem are severe.

Although these results are interesting, they do not really contribute anything to the development of numerical methods for the relativistic calculation of QED effects in atoms, other than to provide qualitative information about the self-energy integrand. It is straightforward to improve the accuracy of the above non-relativistic calculation of the lowest-order shift using a large Sturmian basis, since the required dipole integrals may be calculated exactly, but this is really only an indirect way of calculating the Bethe logarithm, for which very accurate methods already exist. The fundamental problem is to develop fully relativistic methods, because it is relativistic atomic structure calculations to which radiative corrections are to be applied. A number of schemes exist which attempt to approximate Lamb-shifts using a scaled nuclear charge, but it is the arbitrariness inherent in these schemes which limits agreement between experiment and theory.

The Lamb–shift corrections for atoms are of greatest interest in strong fields so the incorporation of terms of higher order in the Coulomb interaction are required.

The practical aspects involved in the direct numerical evaluation of the relativistic self–energy shift in a bound electron state are now assessed. Configuration space representations of the electron and photon propagator functions are presented in forms which exploit known results from the theory of angular momentum coupling, recurrence relations for the generation of amplitudes and which are compatible with the basis set approach to relativistic atomic structure calculations. The standard against which new methods in this field are to be judged has certainly been set by Mohr (1974), who calculated self–energy shifts for hydrogenic ions using a faithful transcription of the rules of QED into numerical algorithms suitable for computation.

5.1. Expansion of the photon propagator

Three of the four integrations which define the photon propagator may be performed analytically, with the result

$$\frac{1}{2} D_F^{\Lambda=\infty}(x_2 - x_1) = -\theta(t_2 - t_1) \int_0^\infty \frac{dk}{(2\pi)^2} \exp(-ik(t_2 - t_1)) \frac{\sin(kR)}{R}$$

$$+ \theta(t_1 - t_2) \int_0^\infty \frac{dk}{(2\pi)^2} \exp(+ik(t_2 - t_1)) \frac{\sin(kR)}{R} \quad [5.9]$$

where $R = |r_2 - r_1|$. Combining this with a similar expression for the propagator for finite values of Λ and performing the time integrations with respect to a general harmonic factor $\exp(i\omega t)$ representing the time–dependence of the electron states, leads to the definition of an integrated interaction kernel

$$B^\Lambda(r_1, r_2; \omega) = \int_0^\infty dk \frac{\sin(kR)}{R} [\boldsymbol{a}_1 \cdot \boldsymbol{a}_2 - I]$$

$$\times \left\{ \frac{1}{\omega + k} - \frac{1}{\omega + k\sqrt{1 + \Lambda^2/k^2}} \frac{1}{\sqrt{1 + \Lambda^2/k^2}} \right\}. \quad [5.10]$$

No analytic expression for the integral over k could be found, either for the regulated or unregulated parts of the interaction. If a simple form were known, then an effective Breit–like potential would be recovered and the calculation of self–energy shifts greatly simplified. In subsequent calculations of the energy shift, the factor ω appears as an energy difference between the reference state and an intermediate state; if $\omega < 0$ the principal value of the integral is implied. In this notation, the energy shift is the limit $\Lambda \to \infty$ of:

$$E_n(\Lambda) = -\frac{a}{\pi} \sum_m{}^+ < \psi_n \mid (< \psi_m \mid B^\Lambda(r_1,r_2;E_m - E_n) \mid \psi_m >) \mid \psi_n >$$

$$+ \frac{a}{\pi} \sum_m{}^- < \psi_n \mid (< \psi_m \mid B^\Lambda(r_1,r_2;E_n - E_m) \mid \psi_m >) \mid \psi_n >$$

$$- \delta m(\Lambda) < \psi_n \mid \beta \mid \psi_n > \qquad [5.11]$$

The multipole expansion of the interaction is obtained by using the well-known results

$$\frac{\sin(kR)}{R} = k \sum_{\nu=0}^\infty (2\nu + 1) j_\nu(kr_1) j_\nu(kr_2) \; C^\nu(1) \cdot C^\nu(2) \qquad [5.12]$$

$$C^\nu(1) \cdot C^\nu(2) = \sum_q (-1)^q C_{\nu q}(1) C_{\nu-q}(2) \qquad [5.13]$$

$$a_1 \cdot a_2 \left[C^\nu(1) \cdot C^\nu(2) \right] = \sum_\lambda (-1)^{\nu+\lambda+1} \left[(X^{(1\nu)\lambda}(1) \cdot X^{(1\nu)\lambda}(2) \right] \qquad [5.14]$$

$$X_Q^{(1\nu)\lambda}(1) = [a \cdot C^\nu]_Q^\lambda \qquad [5.15]$$

A matrix element of the operator $g(k,R) = -(a_1 \cdot a_2 - I)\sin(kR)/R$ takes the explicit form

$$< an \mid g(k,R) \mid na > = -k \sum_{\nu=0}^\infty (2\nu+1) \sum_{q=0}^\infty (-1)^q$$

$$\times \left\{ < a \mid j_\nu(kr_1) C_q^\nu(1) \mid n >< n \mid j_\nu(kr_2) C_{-q}^\nu(2) \mid a > \right.$$

$$\left. + \sum_{\lambda=\nu-1}^{\nu+1} < a \mid X_q^{(1\nu)\lambda}(1) j_\nu(kr_1) \mid n > < n \mid X_q^{(1\nu)\lambda}(2) j_\nu(kr_2) \mid a > \right\}$$

$$[5.16]$$

The reduced matrix elements are reproduced here for completeness:

$$< a \mid j_\nu C_q^\nu \mid n > = (-1)^{j_a - m_a} \begin{pmatrix} j_a & \nu & j_n \\ -m_a & q & m_n \end{pmatrix} < a \parallel C^\nu j_\nu \parallel n > \qquad [5.17]$$

$$< a \mid X_q^{(1\nu)\lambda} j_\nu \mid n > = (-1)^{j_a - m_a} \begin{pmatrix} j_a & \nu & j_n \\ -m_a & q & m_n \end{pmatrix} < a \parallel j_\nu X^{(1\nu)\lambda} \parallel n > \qquad [5.18]$$

$$< a \parallel C^\nu j_\nu \parallel n > = \Pi^e(\kappa_a,\kappa_n;\nu) < j_a \parallel C^\nu \parallel j_n > J_{an}^\nu(k) \qquad [5.19]$$

$$< a \parallel j_\nu X^{(1\nu)\lambda} \parallel n > = -i\, \Pi^o(\kappa_a,\kappa_n;\nu) < j_a \parallel C^\lambda \parallel j_n > K_{an}^{\nu\lambda}(k) \qquad [5.20]$$

$$\langle j_a \| C^\nu \| j_n \rangle = (-1)^{j_a + \frac{1}{2}} [j_a, j_n]^{\frac{1}{2}} \begin{pmatrix} j_a & \nu & j_n \\ \frac{1}{2} & 0 & -\frac{1}{2} \end{pmatrix} \qquad [5.21]$$

$$\Pi^e(\kappa_a, \kappa_n; \nu) = \tfrac{1}{2}[1 - (-1)^{\ell_a + \ell_n + \nu}] \qquad [5.22]$$

$$\Pi^o(\kappa_a, \kappa_n; \nu) = 1 - \Pi^e(\kappa_a, \kappa_n; \nu) \qquad [5.23]$$

The notation of Grant (1988) has been adopted; $j_n(z)$ is a spherical Bessel function, as defined in Abramowitz and Stegun (1970). If we reserve the integration over k until later, as we did for the calculation in the style of Bethe (1947), the problem is reduced to the evaluation of the above angular factors and radial integrals of the form:

$$J^\nu_{an}(k) = \int_0^\infty [P_a(r)P_n(r) + Q_a(r)Q_n(r)] j_\nu(kr) dr \qquad [5.24]$$

$$K^{\nu\lambda}_{an}(k) = \frac{(\kappa_a - \kappa_n) I^+_\nu(k) + \nu I^-_\nu(k)}{(\nu(2\nu+1))^{1/2}} \quad \text{for } \lambda = \nu - 1$$

$$= - \frac{(\kappa_a + \kappa_n) I^+_\nu(k)}{[\nu(\nu+1)]^{1/2}} \quad \text{for } \lambda = \nu$$

$$= \frac{(\kappa_a - \kappa_n) I^+_\nu(k) + (\nu+1) I^-_\nu(k)}{[(\nu+1)(2\nu+1)]^{1/2}} \quad \text{for } \lambda = \nu + 1 \qquad [5.25]$$

$$I^\pm_\nu(k) = \int_0^\infty [P_a(r)Q_n(r) \pm Q_a(r)P_n(r)] j_\nu(kr) dr \qquad [5.26]$$

These expressions are finally combined to provide a convenient expression for an exchange matrix element of the interaction at fixed k

$$\langle an | g(k,R) | na \rangle =$$

$$- k(2j_n + 1) \sum_{\nu=0}^\infty (2\nu + 1) \begin{pmatrix} j_a & \nu & j_n \\ \frac{1}{2} & 0 & -\frac{1}{2} \end{pmatrix}^2 \Pi^e(\kappa_a, \kappa_n; \nu) |J^\nu_{an}(k)|^2$$

$$+ k(2j_n + 1) \sum_{\nu=0}^\infty (2\nu + 1) \Pi^o(\kappa_a, \kappa_n; \nu) \sum_{\lambda=\nu-1}^{\nu+1} \begin{pmatrix} j_a & \lambda & j_n \\ \frac{1}{2} & 0 & -\frac{1}{2} \end{pmatrix}^2 |K^{\nu\lambda}_{an}(k)|^2$$

$$[5.27]$$

The partial wave–expansion of the matrix elements of $-(1 - \alpha_1 \cdot \alpha_2) \sin(kR)/R$ may now be combined with the energy denominators which occur in the expression for the integrated interaction kernel.

The evaluation of these matrix elements requires the evaluation of spherical Bessel functions for a very large range of argument and order values. For values of k up to about mc a.u., we have found that the required argument values lie in the range $0 \leq kr < 500$ for Bessel functions in the order range $0 \leq \nu \leq 300$, in keeping with the conclusions of Mohr in the part of the calculation which he calls the "low–energy region". The convergence rate of the partial wave expansion deteriorates as the upper limit of k is increased, and the convergence is not smooth. The evaluation of a single point in the integrand over k closely resembles a calculation of the sum–rule for spherical Bessel functions

$$\sum_{n=0}^{\infty} (2n+1)|j_n(x)|^2 = 1 \qquad \text{for all } x \qquad [5.28]$$

A large number of terms contribute to the sum which are of similar magnitude and whose signs depend on a slowly oscillating phase factor. The magnitude of the terms which contribute significantly to the sum fall off rapidly once the order index is greater than a certain critical value which depends on the argument, $z = kr$.

For a fixed value of x, all spherical Bessel functions which contributes significantly to the energy shift formula may be evaluated using the recurrence relation

$$j_{n+1}(x) + j_{n-1}(x) = (2n+1)x^{-1}j_n(x) \qquad [5.29]$$

which is stable only in the direction of decreasing n. Using Miller's Algorithm, two arbitrary seed values for the Bessel functions are chosen and the recurrence applied for decreasing n until the value of $j_0(x)$ has been determined. The table of values is then rescaled to the value $j_0(x) = \sin(x)/x$. The starting point, n_{max}, for this procedure is determined using a number of empirical rules for different value ranges of x. For small values of x, the power series representation for the Bessel functions is useful up to the first stationary point for each value of n, which occurs, approximately, at $x = (2n+1)$. Since the values of the spherical Bessel functions are required at many values of the argument, x, and for many values of the index, n, an algorithm may be constructed which moves the recursive dependencies to outer loops, allowing the efficient generation of tables of function values. Subroutines have been constructed to evaluate the spherical Bessel and spherical Hankel functions which commonly occur in QED calculations which are stable, efficient and accurate over very large ranges of argument and parameter values.

5.2. Numerical representations of the electron propagator

We have chosen to make a spectral expansion of the Dirac Green's function using the L–spinor basis functions described earlier. A practical upper limit for the dimensions of the basis is of the order of 500 functions per angular symmetry–type. The dimension of the basis may not be extended indefinitely because the effects of computational linear dependence in the basis are manifested in very large expenaion coefficients and deteriorating orthogonality integrals between the approximate and exact solutions of the Dirac–Coulomb equation for bound–state functions. Compared with other basis set methods, however, this basis set is considered to be very large and is far more detailed than that required for the high–precision calculations of electron correlation discussed earlier.

Extensive use is made of stable recurrence relations in the construction of the Dirac matrix in a finite L–spinor representation. For a point–nuclear hydrogenic model, we require overlap, potential energy and kinetic energy matrix elements as described in the accompanying lecture by Grant for the S-spinor basis. In order to simplify the presentation, we define the auxiliary functions N_n, $H_{mn}(2\gamma)$ and $G_{mn}(2\gamma)$:

$$N_n = (n^2 + 2n\gamma + \kappa^2)^{\frac{1}{2}} \qquad [5.30]$$

$$H_{mn}(2\gamma) = \sum_{k=0}^{\min(m,n)} \frac{\Gamma(2\gamma + k)}{k!} \qquad [5.31]$$

$$G_{mn}(2\gamma) = \frac{\Gamma(2\gamma + m + 1)}{m!} \delta_{mn} \qquad [5.32]$$

and the related function $h_m(2\gamma)$:

$$h_m(2\gamma) = \frac{\Gamma(m+1)}{\Gamma(2\gamma+m)} \sum_{k=0}^{m} \frac{\Gamma(2\gamma+k)}{k!}. \qquad [5.33]$$

In practice, the L-spinor algorithms embodied in the computer programs that we have written are based on the function $h_m(2\gamma)$, but the presentation will be in terms of the functions $H_{mn}(2\gamma)$ and $G_{mn}(2\gamma)$ for clarity. The principal difference between the two formulations is the explicit use of the gamma function, $\Gamma(z)$, in the formulae for $H_{mn}(2\gamma)$ and $G_{mn}(2\gamma)$, which may be avoided by the use of the recurrence relation for $h_{mn}(2\gamma)$:

$$h_0(2\gamma) = 1$$

$$h_m(2\gamma) = 1 + \frac{m}{m + 2\gamma - 1} h_{m-1}(2\gamma), \quad m = 1,2,3,... \qquad [5.34]$$

which ensures that very large or very small intermediate quantities are never required in the construction of the Dirac matrix. The required matrix elements for symmetry-type κ are:

$$< f_m^T | f_n^T > = \begin{cases} 1 & n = m \\ \mp \left[\left(1 - \frac{\kappa}{N_n}\right)\left(1 + \frac{\kappa}{N_{n+1}}\right)\right]^{\frac{1}{2}} & |n - m| = 1 \\ 0 & |n - m| > 1 \end{cases}$$

$$< f_m^T | 1/r | f_n^T > = \frac{\lambda}{m} \left[1 + \frac{\kappa}{N_m}\right]^{\frac{1}{2}} \left[1 + \frac{\kappa}{N_n}\right]^{\frac{1}{2}} \left[\frac{(n-1)(n-2)\ldots m}{(n-1+2\gamma)(n-2+2\gamma)\ldots(m+2\gamma)}\right]^{\frac{1}{2}}$$

$$\times \left[1 \mp \frac{N_n - \kappa}{n + 2\gamma}\right] \left[h_m(2\gamma)\left\{1 \mp \frac{N_m - \kappa}{m + 2\gamma}\right\} - 1\right] \quad \text{for } m \leq n$$

$$< f_m^T | 1/r | f_n^T > = < f_n^T | 1/r | f_m^T > \quad \text{for } m > n$$

$$< f_m^S | \frac{d}{dr} + \frac{\kappa}{r} | f_n^L > = \frac{\lambda}{2} \left[\left(1 + \frac{\kappa}{N_m}\right)\left(1 + \frac{\kappa}{N_n}\right)\Gamma(m)\Gamma(n)\right]^{\frac{1}{2}} \times \sum_{k=1}^{10} T_k$$

$$< f_m^L | \frac{d}{dr} + \frac{\kappa}{r} | f_n^S > = < f_n^S | \frac{d}{dr} + \frac{\kappa}{r} | f_m^L > \qquad [5.35]$$

where

$$T_1 = 2(N_n + n + \gamma - 1) \; H_{m-1, n-1}(2\gamma)$$

$$T_2 = 2(N_n + n + \gamma - 1) \left[\frac{N_m - \kappa}{m + 2\gamma}\right] H_{m, n-1}(2\gamma)$$

$$T_3 = -2(n + \gamma - 1) \left[\frac{N_n - \kappa}{n + 2\gamma}\right] H_{m-1, n}(2\gamma)$$

$$T_4 = -2(n + \gamma - 1) \left[\frac{N_m - \kappa}{m + 2\gamma}\right] \left[\frac{N_n - \kappa}{n + 2\gamma}\right] H_{m, n}(2\gamma)$$

$$T_5 = -2(n + 2\gamma - 1) \; H_{m-1, n-2}(2\gamma)$$

$$T_6 = -2(n + 2\gamma - 1) \left[\frac{N_m - \kappa}{m + 2\gamma}\right] H_{m, n-2}(2\gamma)$$

$$T_7 = -G_{m-1, n-1}(2\gamma)$$

$$T_8 = -\left[\frac{N_m - \kappa}{m + 2\gamma}\right] G_{m, n-1}(2\gamma)$$

$$T_9 = -\left[\frac{N_n - \kappa}{n + 2\gamma}\right] G_{m-1, n}(2\gamma)$$

$$T_{10} = \left[\frac{N_m - \kappa}{m + 2\gamma}\right] \left[\frac{N_n - \kappa}{n + 2\gamma}\right] G_{m, n}(2\gamma). \qquad [5.36]$$

In the formulae above, the label T may take the values $T = L$ or $T = S$, for the large and small component basis functions, respectively. The upper sign of a pair $\mp$ is chosen for $T = L$ and the lower sign for $T = S$. The basis function labels m and n lie in the range $[0, N_b - 1]$, where N_b is the dimension of the spinor basis. If either subscript in the auxiliary functions $H_{mn}(2\gamma)$ and $G_{mn}(2\gamma)$ is less than zero, then the auxiliary function vanishes.

The above algorithm is sufficient to generate a point nuclear Dirac spectrum containing up to 500 positive– and 500 negative–energy discrete states per symmetry–type. The cost of constructing the matrix is negligible in this scheme, compared with the solution of the generalized matrix eigenvalue problem. Using the CRAY X–MP/416 at the Rutherford Appleton Laboratory, the spectrum for each value of κ costs less than two seconds of CPU time to generate.

The similarity between finite basis set and exact expansions of the Dirac Coulomb Green's function is now discussed. The use of an analytic basis allows us to carry over qualitative results from non–relativistic electron scattering models regarding the convergence of expectation values with increasing basis set dimension. Existing methods for modifying the Dirac Coulomb Green's function to that of the Hartree–Fock potential through iteration are also directly applicable.

The Dirac–Coulomb Green's function is known only as a partial wave expansion.

The radial Green's function, $G(r_2,r_1;z)$, for a potential $V(r_k) = -Z\alpha/r_k$ satisfies:

$$\begin{bmatrix} 1 + V(r_2) - z & -\frac{1}{r_2}\frac{d}{dr_2}r_2 + \frac{\kappa}{r_2} \\ \frac{1}{r_2}\frac{d}{dr_2}r_2 + \frac{\kappa}{r_2} & -1 + V(r_2) - z \end{bmatrix} \begin{bmatrix} G_\kappa^{11}(r_2,r_1;z) & G_\kappa^{12}(r_2,r_1;z) \\ G_\kappa^{21}(r_2,r_1;z) & G_\kappa^{22}(r_2,r_1;z) \end{bmatrix}$$

$$= I \frac{1}{r_2 r_1} \delta(r_2 - r_1). \quad [5.37]$$

The functions $G_\kappa^{ij}(r_2,r_1;z)$ are defined, for $r_1 > r_2$, by:

$$G_\kappa^{11}(r_2,r_1;z) = (1+z)\, \mathbb{Q}\,[(\lambda-\nu)M_{\nu-(1/2),\lambda}(2cr_2) - (\kappa-(\gamma/c))M_{\nu+(1/2),\lambda}(2cr_2)]$$
$$\times [(\kappa+(\gamma/c))W_{\nu-(1/2),\lambda}(2cr_1) + W_{\nu+(1/2),\lambda}(2cr_1)]$$

$$G_\kappa^{12}(r_2,r_1;z) = c\mathbb{Q}\,[(\lambda-\nu)M_{\nu-(1/2),\lambda}(2cr_2) - (\kappa-(\gamma/c))M_{\nu+(1/2),\lambda}(2cr_2)]$$
$$\times [(\kappa+(\gamma/c))W_{\nu-(1/2),\lambda}(2cr_1) - W_{\nu+(1/2),\lambda}(2cr_1)]$$

$$G_\kappa^{21}(r_2,r_1;z) = c\mathbb{Q}\,[(\lambda-\nu)M_{\nu-(1/2),\lambda}(2cr_2) + (\kappa-(\gamma/c))M_{\nu+(1/2),\lambda}(2cr_2)]$$
$$\times [(\kappa+(\gamma/c))W_{\nu-(1/2),\lambda}(2cr_1) + W_{\nu+(1/2),\lambda}(2cr_1)]$$

$$G_\kappa^{22}(r_2,r_1;z) = (1-z)\, \mathbb{Q}\,[(\lambda-\nu)M_{\nu-(1/2),\lambda}(2cr_2) + (\kappa-(\gamma/c))M_{\nu+(1/2),\lambda}(2cr_2)]$$
$$\times [(\kappa+(\gamma/c))W_{\nu-(1/2),\lambda}(2cr_1) - W_{\nu+(1/2),\lambda}(2cr_1)]$$

$$[5.38]$$

where

$$c = (1-z^2)^{\frac{1}{2}};\ \mathrm{Re}(c) > 0;\ \lambda = (\kappa^2 - \gamma^2)^{\frac{1}{2}};\ \nu = \gamma z/c;\ \gamma = Z\alpha \text{ and}$$

$$\mathbb{Q} = \frac{1}{(r_1 r_2)^{\frac{1}{2}}(4c^2 r_1 r_2)} \frac{\Gamma(\lambda-\nu)}{\Gamma(1+2\lambda)} \quad [5.39]$$

and $M_{ab}(x)$ and $W_{ab}(x)$ are the regular and irregular Whittaker functions, respectively (Abramowitz and Stegun, 1970). It is this form which was used by Mohr (1974) in his non-perturbative calculations of the electron self-energy in hydrogenic ions.

In order to make contact with the discrete expansion methods which have recently become popular in atomic physics calculations, it is convenient to consider an equivalent but less well-known representation of the Dirac-Coulomb radial Green's function, derived from the second-order Dirac equation. An integral transform is used to convert the product of regular and irregular Whittaker functions into a modified Bessel function:

$$\frac{\Gamma(\omega - \lambda)}{\Gamma(2\omega)} M_{\lambda,\omega-(1/2)}(z_1) W_{\lambda,\omega-1/2}(z_2) =$$

$$(z_1 z_2)^{\frac{1}{2}} \int_0^\infty \exp[-(z_1+z_2)\cosh(x)/2] [\coth(x/2)]^{2\eta} I_{2\omega-1}[(z_1 z_2)^{\frac{1}{2}}\sinh(x)] dx$$

[5.40]

An expansion for modified Bessel functions in terms of the orthogonal Laguerre polynomials is then adopted to replace the product of Whittaker functions by an expansion in a complete basis of exponentially weighted generalized Laguerre polynomials:

$$I_{2\omega-1}\left[\frac{2(uvz)^{\frac{1}{2}}}{1-z}\right] = (1-z)(uvz)^{(2\omega-1)/2}\exp\left[\frac{z(u+v)}{1-z}\right] \times \sum_{k=0}^{\infty} \frac{k! z^k}{\Gamma(2\omega+k)} L_k^{2\omega-1}(u) L_k^{2\omega-1}(v).$$

[5.41]

Four auxiliary functions $S_\kappa^0(r_2,r_1;z)$, $S_\kappa^1(r_2,r_1;z)$, $S_\kappa^+(r_2,r_1;z)$ and $S_\kappa^-(r_2,r_1;z)$ are defined:

$$S_\kappa^m(r_2,r_1;z) = (\rho\rho')^{\gamma-1}\exp[-(\rho+\rho')/2] \times \sum_{k=0}^{\infty} \frac{k! L_k^{2\gamma}(\rho) L_k^{2\gamma}(\rho')}{\Gamma(k+2\gamma+1)(k+\gamma-\eta+m)} \quad ; m=0,1$$

$$S_\kappa^\pm(r_2,r_1;z) = (\rho\rho')^{\gamma-1}\exp[-(\rho+\rho')/2] \times$$

$$\times \sum_{k=0}^{\infty} \frac{k! [L_k^{2\gamma-1}(\rho)L_k^{2\gamma}(\rho') \pm L_k^{2\gamma}(\rho)L_k^{2\gamma-1}(\rho')]}{\Gamma(k+2\gamma+1)(k+\gamma-\eta+m)}$$

[5.42]

where $\rho = 2\nu^{-1}r_1$, $\rho' = 2\nu^{-1}r_2$, $\epsilon = Z/m$, $\nu = (1-\epsilon^2)^{-1}$, $\eta = \alpha\nu Z\epsilon$.

These radial functions are then combined to form the discrete representation of the radial Dirac–Coulomb Green's function (Dimitriev et al, 1984):

$$G_\kappa^{11}(r_2,r_1;z) = g(1+\epsilon)[(\alpha\nu Z+\kappa)S_\kappa^1(r_2,r_1;z)+(\alpha\nu Z-\kappa-2\gamma-2\eta)S_\kappa^0(r_2,r_1;z)+S_\kappa^+(r_2,r_1;z)]$$

$$G_\kappa^{12}(r_2,r_1;z) = g(1-\epsilon^2)^{\frac{1}{2}}[(\alpha\nu Z+\kappa)S_\kappa^1(r_2,r_1;z)-(\alpha\nu Z-\kappa)S_\kappa^0(r_2,r_1;z)-S_\kappa^-(r_2,r_1;z)]$$

$$G_\kappa^{21}(r_2,r_1;z) = g(1-\epsilon^2)^{\frac{1}{2}}[(\alpha\nu Z+\kappa)S_\kappa^1(r_2,r_1;z)+(\alpha\nu Z-\kappa)S_\kappa^0(r_2,r_1;z)+S_\kappa^-(r_2,r_1;z)]$$

$$G_\kappa^{22}(r_2,r_1;z) = -g(1-\epsilon)\left[(a\nu Z+\kappa)S_\kappa^1(r_2,r_1;z) + (a\nu Z-\kappa+2\gamma+2\eta)S_\kappa^0(r_2,r_1;z)\right.$$
$$\left. - S_\kappa^+(r_2,r_1;z)\right]$$

[5.43]

where $g = 2m/\nu$.

From these parts, the complete Dirac–Coulomb Green's function,

$$G(r_2,r_1;z) =$$
$$\sum_{\kappa,m} \begin{bmatrix} G_\kappa^{11}(r_2,r_1;z)\chi_{+\kappa m}(\theta_2,\phi_2)\chi_{+\kappa m}^\dagger(\theta_1,\phi_1) & iG_\kappa^{12}(r_2,r_1;z)\chi_{-\kappa m}(\theta_2,\phi_2)\chi_{+\kappa m}^\dagger(\theta_1,\phi_1) \\ iG_\kappa^{21}(r_2,r_1;z)\chi_{-\kappa m}(\theta_2,\phi_2)\chi_{+\kappa m}^\dagger(\theta_1,\phi_1) & G_\kappa^{22}(r_2,r_1;z)\chi_{-\kappa m}(\theta_2,\phi_2)\chi_{-\kappa m}^\dagger(\theta_1,\phi_1) \end{bmatrix}$$

[5.44]

may be constructed by summing over all symmetry-types, $\{\kappa\}$, and all magnetic substates, $\{m\}$. The two-component spin-angular function, $\chi_{\kappa m}(\theta,\phi)$, is familiar from atomic structure theory:

$$\chi_{\kappa m}(\theta,\phi) = \begin{bmatrix} <\ell\ \tfrac{1}{2}\ m-\tfrac{1}{2}\ \tfrac{1}{2}\ |\ \ell\ \tfrac{1}{2}\ j\ m>\ Y_\ell^{m-\tfrac{1}{2}}(\theta,\phi) \\ <\ell\ \tfrac{1}{2}\ m+\tfrac{1}{2}\ -\tfrac{1}{2}\ |\ \ell\ \tfrac{1}{2}\ j\ m>\ Y_\ell^{m+\tfrac{1}{2}}(\theta,\phi) \end{bmatrix}$$

[5.45]

where $\ell = |\kappa|$ if $\kappa > 0$, $\ell = -\kappa-1$ if $\kappa < 0$ and $j = |\kappa|-\tfrac{1}{2}$.

Clearly, calculation of the Dirac–Coulomb Green's function and the subsequent determination of the self–energy shift depends primarily on the accurate evaluation of four radial auxiliary functions per angular symmetry–type, either in terms of the Whittaker functions, or as expansions in Laguerre polynomials. Mohr has already devised algorithms for the evaluation of Whittaker functions over the regions of the complex plane which are sampled by his contour integral technique. The second method based on Laguerre polynomial expansions is familiar from our work on L-spinor representations of the Dirac equation, which yields a finite-dimensional approximation to the above expansions of the radial auxiliary functions, and is formally equivalent to it in the limit of a complete set.

All complete orthogonal polynomial sets, such as the generalized Laguerre polynomials,

$$L_n^\alpha(z) = \sum_{k=0}^n (-1)^k \begin{bmatrix} n + \alpha \\ n - k \end{bmatrix} \frac{z^k}{k!},$$

[5.46]

satisfy an elementary three-term recurrence relation:

$$(n+1)L_{n+1}^\alpha(x) = [(2n+\alpha+1) - x]L_n^\alpha(x) - (n+\alpha)L_{n-1}^\alpha(x).$$

[5.47]

This recurrence is stable for increasing degree, n, and has been implemented in our computer codes by scaling the polynomial values by the weight function $\exp(-x/2)$; the starting values for polynomials of degree 0 and 1 are trivial. Using this simple scheme, which may readily be vectorized for efficient computation, it is possible to generate the radial L-spinor amplitudes or auxiliary radial Green's for any symmetry-type and for any value of the radial coordinate which is required in the subsequent numerical integrations. The accuracy and stability of the procedure was tested by evaluating numerically the orthogonality integrals over square–integrable pseudo–states; an accuracy of one part in 10^{12} is readily achieved using our numerical techniques.

The numerical integration scheme was based on the Romberg algorithm, extrapolating the Riemann sum to the limit of vanishing step–length. Numerical approximations to well–known relativistic sum–rules and perturbation sums were tested using this scheme, and found to be accurate to the required accuracy of 1 part in 10^7 with modest computational cost, using basis sets of up to 500 L–spinors per symmetry type. Obviously a more efficient scheme in most cases would be Gauss–Laguerre quadrature, but for testing we did not wish to be restricted to the matrix elements of very smooth, polynomial–like operators.

If we look more closely at this procedure, a feature is revealed which is not found in calculations of many–body corrections in atoms and molecules, which determine matrix elements by integrating over spatial coordinates and then evaluate perturbation sums over intermediate states. In this finite–dimensional numerical approximation to the Green's function approach, we first sum over intermediate states (or whatever the expansion basis is), and then perform a numerical integration over spatial coordinates. Formally, the two approaches are equivalent, but the numerical behaviour of each is quite different from the other.

We may clarify these differences by concentrating on a single radial auxiliary function, $S^0(r_2,r_1;z)$. This algebraic function is typical of the numerical sum which we attempt to construct from a finite–dimensional L–spinor expansion of the radial Dirac–Coulomb Green's function and we may easily simulate the effect of L–spinor basis set truncation error by terminating the sum in the auxiliary function after a finite number of terms have been added.

Table 2. Convergence of the partial sums of the Real (A) and Imaginary (B) parts $S^0(2,1;0.1i)$ with respect to the number of terms added (M). Powers of ten in parentheses.

M	A	B
1	0.201627208857(−01)	0.159991778539(−04)
2	0.206484121252(−01)	0.160869073304(−04)
3	0.193090054168(−01)	0.158933600174(−04)
4	0.177494561132(−01)	0.157055615022(−04)
5	0.171058663843(−01)	0.156391329045(−04)
6	0.170735260581(−01)	0.156362121287(−04)
7	0.169612023026(−01)	0.156271949067(−04)
8	0.165160571688(−01)	0.155950327507(−04)
9	0.159434905136(−01)	0.155574250496(−04)
10	0.155450890638(−01)	0.155334376596(−04)
.	.	.
50	0.167869681063(−01)	0.155688931513(−04)
51	0.167669614074(−01)	0.155686204155(−04)
52	0.167589992268(−01)	0.155685138833(−04)
.	.	.
98	0.163332272973(−01)	0.155638847494(−04)
99	0.163419293884(−01)	0.155639470001(−04)
100	0.163546580874(−01)	0.155640371627(−04)

We see that the pointwise convergence of the Dirac Coulomb Green's function expansion is poor and not readily accessible to any of the usual methods available for the acceleration of convergence of such series. Similar observations are made if we work with a finite–dimensional spectral expansion of the radial auxiliary functions in terms of the L–spinors amplitudes derived from the numerical diagonalization of the Dirac matrix. It is initially puzzling that a function which is so poorly defined may be used to produce electron correlation energies to very high accuracy using many–body perturbation theory, but the numerical evidence supporting the use of the basis set method in many–body problems, both relativistically and non–relativistically, is overwhelming. The above illustration of the poor pointwise convergence of Laguerre polynomial expansions of the electron propagator suggests the following explanation for the success of the finite basis set method in usual applications of relativistic many–body theory; the argument is adapted from an analysis of L^2 approximations in scattering theory (Yamani and Reinhardt, 1975).

The asymptotic behaviour of the multipole expansion of a realistic atomic or molecular potential, such as the Coulomb or Breit potential, is proportional to $1/r$, where r is the distance to the point of expansion. Solution of the generalized matrix eigenvalue problem

$$H X = E S X \qquad [5.48]$$

for either the Schrödinger or Dirac kinetic energy operator and a bounded potential constructed from the nuclear and mean–field potentials involves the solution of a characteristic equation for the roots of a secular determinant. The eigenfunctions which result from this diagonalization form an orthonormal set, the radial parts of which behave like the product of a polynomial and an exponential decay factor. The eigenvalues form the abscissae of a generalized quadrature over the complete set. Heller and Yamani (1974) have demonstrated that, for example, if we discretize the non–relativistic Schrödinger operator, the abscissae form an approximate Gauss–Chebychev quadrature scheme; in general, we form a Gaussian–type quadrature determined by the detailed form of the operator.

Orthogonal polynomial or Fourier series expansions of a general function are optimal representations in a mean (least–square) rather than pointwise sense. The rate of convergence of the expansion averaged over an operator and some arbitrary function is determined both by the nature of the operator and by whether the arbitrary function is well–represented in the finite–dimensional space of the expansion set (Yamani and Reinhardt, 1975). The Coulomb potential and the multipole expansion of the Hartree–Fock potential are rapidly convergent with respect to basis set size if we use S– or L–spinors, because these expansion sets are orthogonal with respect to integration over the operator $1/r$, which is the asymptotic limit of both the Coulomb and Breit potentials. If some other numerical method, such as finite–differences or piecewise polynomials, were chosen as an expansion set then we would expect that a much larger basis would be necessary. Within this larger basis, however, the accuracy of an MBPT calculation would be similar to that which we have found in S– and L–spinor calculations of modest dimension, because all these approaches are forming generalized quadratures of the same type, depending only on the Dirac–Fock operator, for example, and not on the numerical method of discretization. This is consistent with the comments of Johnson *et al* (1988), who remark that one may safely exclude high energy discrete states from MBPT summations without changing significantly the value of numerical calculations: the large basis of primitive functions is necessary only to provide accurate low–energy quadrature points. In the current analytic finite basis set approach, this is equivalent to stating that the spatial integrand implicit in any MBPT diagram is well approximated *in the mean* by a low–order generalized polynomial in each radial coordinate and a fixed exponential decay factor.

6. Calculation of the Coulomb–Field Self–Energy Shift

If we follow directly the relativistic extension of Bethe's calculation, replacing the low–frequency dipole interaction with retarded electromagnetic matrix elements, the strategy for the development of numerical methods is:

a) Develop methods for evaluating integrals of the type $J_\nu(k)$ and $I^\pm_\nu(k)$

b) Construct perturbation sum at fixed k by summing over discrete states

c) Integrate over k in the range $[0,\infty)$ taking account of principal values

d) Renormalize, either after or during b) and c).

In self–energy calculations, the matrix elements of Step (a) have kernels involving spherical Bessel functions. We have investigated both analytic and numerical methods in the evaluation of the L–spinor matrix elements of the retarded Breit interaction. The primitive integral

$$\int_0^\infty r^\gamma \exp(-ar) j_\nu(\omega r) dr = \frac{\pi^{\frac{1}{2}} \omega^n \Gamma(\gamma+n+1)}{2^{n+1}[(a^2+\omega^2)^{\gamma+n+1}]^{\frac{1}{2}} \Gamma(n+\frac{1}{2})} \times {}_2F_1(\frac{\gamma+n+1}{2}, \frac{1-\gamma+n}{2}; n+\frac{3}{2}; \frac{\omega^2}{a^2+\omega^2})$$

[6.1]

appears useful, because the Laguerre polynomials may, in principle, be expanded and integrated term–by–term using the above formula. The function ${}_2F_1(a,b,c;z)$ is the Gauss hypergeometric function

$${}_2F_1(a,b,c;z) = F(a,b,c;z)$$

$$= \frac{\Gamma(c)}{\Gamma(a)\Gamma(b)} \sum_{n=0}^\infty \frac{\Gamma(a+n)\Gamma(b+n)}{\Gamma(c+n)} \frac{z^n}{n!}$$

[6.2]

The argument of the hypergeometric function, $z = \omega^2/(a^2+\omega^2)$, always lies in the range $(0,1)$ and a quadratic transformation exists for the above parameter values which further restricts the range to $(0,1/2)$. All required hypergeometric functions in this analytic approach may be generated by the application of the Gauss contiguous relations (Abramowitz and Stegun, 1970), by which means any function $F(a\pm k, b\pm m, c\pm n; z)$ may be evaluated from two seed values which are known accurately from the series or continued fraction representation of the hypergeometric function, for integer values (k,m,n). This approach is not satisfactory for polynomials of higher degree than about 20, however, because the polynomial coefficients which arise from the product of two Laguerre polynomials are huge, resulting in the severe cancellation of terms and a significant loss of numerical accuracy. The integral [6.1] also arises in determining the expansion coefficients of a bound electron state in a basis of spherical wave states: this result is of use in an alternative approach to mass renormalization, discussed later. The generating function for Laguerre polynomials;

$$(1-z)^{-a-1} \exp(\frac{xz}{z-1}) = \sum_{n=0}^\infty L_n^a(x) z^n,$$

[6.3]

has already been used to determine the L–spinor matrix elements involving simple operators ([5.35]–[5.36]), but becomes unwieldy when the integrand involves spherical Bessel functions. We have been unable to devise a reliable analytic approach to the evaluation of the radial integrals which form the matrix elements of the retarded Breit interaction, so that numerical methods were developed.

The photon parameter, k, is continuous on $(0,\infty)$, and the partial wave expansion of the radiation field is formally exact. For large values of k, the spherical Bessel functions which we have used as an expansion basis for the photon field emphasize deficiencies in the finite L-spinor basis which represents the electron-positron field. For any fixed value of the basis dimension, it is always possible to find a value of the photon parameter, k, where the results are sensitive to the value of the L-spinor exponential parameter, λ. The value of λ defines an approximate scale of length; increasing λ decreases the effective radius of the box in which the system has been quantized, and increases the magnitude of each of the single-particle energies in the discrete spectrum. The rapid oscillations in the continuous spectrum of spherical Bessel functions are lost in the slow oscillations of the discrete L-spinor basis, and the basis set method fails. This is not a failure peculiar to the L-spinor approach: it is to be expected also in the finite difference or piecewise polynomial approaches which differ only in the methods used to construct the discrete amplitudes: if a discrete basis is used for the electron-positron field, the representation of the photon field will eventually oscillate too rapidly in the fixed integration grid. It is not possible to demonstrate that the numerical results are independent of the choice of the arbitrary parameters which define any finite basis set representation of the spectrum, so that the method is unreliable for high-energy interactions.

The apparent failure of this approach is linked to previous comments regarding basis set expansions as representations in a mean sense. We may no longer expect the same behaviour that is encountered in electron correlation studies, because the magnitudes of matrix elements over the discrete basis set and highly oscillatory spherical Bessel functions do not fall off rapidly as we add more basis functions. The calculation is moderately sensitive to the basis set parameters, because the spherical Bessel functions and the orthogonal polynomial functions are oscillatory functions which have many nodes; resonances are encountered. Put another way, the spatial integrand of the second-order self-energy shift at fixed photon energy no longer resembles a low-order polynomial function weighted by an exponential decay factor; the spherical Bessel functions are not well-represented by a low-order polynomial over the effective range of numerical integration, except at very low photon energies.

The brute force scheme for assembling the computational pieces into a finite basis set QED scheme was unsuccessful. The lesson to be learned is that basis set approximations form representations in only in the mean and that they resemble quadrature schemes of Gaussian type. If the spatial operator resembles the multipole expansion of the Coulomb or Breit operators which are encountered in electron correlation studies we can expect the basis set scheme to be successful. When the spatial integrand is highly oscillatory, it seems that the pointwise deficiencies of the spectral expansion of the Green's function are emphasized, and the observed rate of convergence of any calculation is poor: it is then meaningless to perform the subsequent integration over the photon parameter, k. On reflection, it was probably foolish to try to construct the integrand at fixed k in the first place; a more sensible approach in view of previous experience with basis set expansions is to construct an effective spatial operator, decomposed into multipole contributions, by first performing the principal value integration over k.

Consider, for simplicity, the unregulated operator:

$$B^{\Lambda=\infty}(\mathbf{r}_1,\mathbf{r}_2;\omega) = \int_0^\infty dk \frac{\sin(kR)}{R}[\boldsymbol{\alpha}_1\cdot\boldsymbol{\alpha}_2 - I]\left\{\frac{1}{\omega+k}\right\} \qquad [6.4]$$

Instead of working with radial matrix elements [5.23–5.24], the partial wave analysis which was performed earlier [5.12 – 5.22] may be used to define an integrated multipole contribution to the Breit–like operator:

$$I_\nu(r_1,r_2;E) = \int_0^\infty j_\nu(r_1\omega) j_\nu(r_2\omega) \frac{\omega}{E+\omega} d\omega \qquad [6.5]$$

The structure $1/(E + \omega)$ arises from the sum–over–states expressions, where the parameter E takes the form of an energy difference between states. Again for simplicity, we assume that $E > 0$, noting that in the general case we take the principal value of the integral. No analytic expression for the above integral is known and no stable recurrence relation for integrals of this type is known to exist. The integrand resembles that of the discontinuous Weber–Schafheitlin integral (Watson, 1944), and any scheme which we devise should recover known forms in the limit $E \to 0$.

The spherical Bessel functions of the First Kind, $j_n(z)$, may be transformed into spherical Bessel functions of the Third Kind (spherical Hankel functions), $h_n^i(z)$, using the identity

$$j_n(z) = \tfrac{1}{2}[h_n^1(z) + h_n^2(z)] \qquad [6.6]$$

The primary integral is rewritten as

$$I_n(r_1,r_2;E) = I_n^1(r_1,r_2;E) + I_n^2(r_1,r_2;E) \qquad [6.7]$$

where

$$I_n^i(r_1,r_2;E) = \frac{1}{2} \int_0^\infty h_n^i(r_1 k) j_n(r_2 k) \frac{k}{E+k} dk. \qquad [6.8]$$

This integral is analytic everywhere on the complex k–plane, except at isolated poles determined by the sign and magnitude of the real number E. In order to restrict the auxiliary integrals to finite values, only the case $r_1 > r_2$ is considered; it is clear that the original integral is symmetric in r_1 and r_2.

There seems to be no conventional notation for spherical Bessel functions of imaginary argument. It is convenient to define "modified spherical Bessel functions", I_n and K_n:

$$j_n(by\ e^{i\pi/2}) = i^n \sqrt{\frac{\pi}{2by}} I_{n+\frac{1}{2}}(by) := i^n \mathcal{I}_n(by) \qquad [6.9]$$
$$-\pi < \arg(by) \le \pi/2$$

$$h_n^1(ay\ e^{i\pi/2}) = -\frac{2}{\pi} i^{-n} \sqrt{\frac{\pi}{2ay}} K_{n+\frac{1}{2}}(ay) := -\frac{2}{\pi} i^{-n} \mathcal{K}_n(ay) \qquad [6.10]$$
$$-\pi < \arg(ay) \le \pi/2$$

Let

$$f^i(k) = \frac{1}{2} h_n^i(r_1 k) j_n(r_2 k) \frac{k}{E+k}. \qquad [6.11]$$

For $f^1(k)$, we deform the contour of integration along the positive imaginary axis to some value of large magnitude and complete the contour by a quarter circle connecting the positive imaginary and real axes. For $f^2(k)$, the contour is deformed along the negative imaginary axis and completed by a quarter circle to large positive real values of k. It may be shown that the contribution from either arc vanishes as its radius is taken to infinity, so that only the integrals along the positive and negative segments of the imaginary axis need be calculated:

$$\int_0^\infty f_n^1(k)dk = i \int_0^\infty f_n^1(ye^{i\pi/2})dy \qquad [6.12]$$

$$\int_0^\infty f_n^2(k)dk = -i \int_0^\infty f_n^2(ye^{-i\pi/2})dy$$

$$= \int_0^\infty \overline{if_n^1(ye^{i\pi/2})}dy \qquad [6.13]$$

The sum of the two auxiliary integrals is real, as expected, because the integrals are related as a complex conjugate pair. Combining the integrals and eliminating all parts related as complex conjugates:

$$\mathcal{I}_n(r_1, r_2; E) = \frac{1}{\pi} \int_0^\infty \mathcal{K}_n(r_2 y) \mathcal{I}_n(r_2 y) \left[\frac{y}{E+iy} + \frac{y}{E-iy}\right]$$

$$= \frac{2E}{\pi} \int_0^\infty \mathcal{K}_n(r_2 y) \mathcal{I}_n(r_2 y) \left[\frac{y}{E^2 + y^2}\right]. \qquad [6.14]$$

Examining the case n=0 and inserting explicit expressions for the modified spherical Bessel functions:

$$I_0(r_1,r_2;E) = \frac{E}{r_1 r_2} \int_0^\infty \frac{e^{-r_1 y}}{E^2 + y^2} \left[\frac{\sinh(r_2 y)}{y}\right] dy \qquad r_1 \geq r_2 \qquad [6.15]$$

$$I_0(r_1,r_2;E) = \frac{E}{r_1 r_2} \int_0^\infty \frac{e^{-r_2 y}}{E^2 + y^2} \left[\frac{\sinh(r_1 y)}{y}\right] dy \qquad r_1 < r_2 \qquad [6.16]$$

Consider the limit $E \to 0$ of the integral $I_n(r_1,r_2;E)$. Many integrals for small absolute values of E are required in a self–energy calculation, because the scale of E is chosen such that $E = 1$ corresponds to the rest–mass energy of a free electron. Small values of E correspond to sums over low–lying bound and continuum states, and $E > 2$ corresponds to negative– energy terms in the second–order perturbation sum formula. Expanding the function $\omega/(E+\omega)$, we obtain:

$$\lim_{E \to 0} I_\nu(r_1,r_2;E) = \lim_{E \to 0} \int_0^\infty j_\nu(r_1\omega) j_\nu(r_2\omega) \frac{\omega}{E + \omega} d\omega$$

$$= \lim_{E \to 0} \int_0^\infty j_\nu(r_1\omega) j_\nu(r_2\omega) \left[1 - \frac{E}{\omega} + \frac{E^2}{\omega^2} - \frac{E^3}{\omega^2} \cdots \right] d\omega$$

$$= \int_0^\infty j_\nu(r_1\omega) j_\nu(r_2\omega) \, d\omega. \qquad [6.17]$$

Apart from simple factors, this is a special case of the discontinuous Weber–Schafheitlin integral:

$$\int_0^\infty j_\nu(r_1\omega) j_\nu(r_2\omega) \, d\omega = \frac{\pi}{2} (r_1 r_2)^{-\frac{1}{2}} \int_0^\infty \frac{J_{\nu+\frac{1}{2}}(r_1\omega) J_{\nu+\frac{1}{2}}(r_2\omega)}{\omega} d\omega.$$

$$[6.18]$$

Using the result (Watson, 1944):

$$\int_0^\infty \frac{J_\mu(at)J_\mu(bt)}{t} \, dt = \frac{1}{2\mu}\left[\frac{b}{a}\right]^\mu \quad \text{for } b < a$$

$$= \frac{1}{2\mu}\left[\frac{a}{b}\right]^\mu \quad \text{for } b \geq a, \qquad [6.19]$$

the low-energy form of the integrated operator becomes:

$$I_\nu(r_1,r_2;0) = \int_0^\infty j_\nu(r_1\omega) j_\nu(r_2\omega) \, d\omega = \frac{\pi}{2}(2\nu+1)^{-1} \frac{r_<^\nu}{r_>^{\nu+1}}, \qquad [6.20]$$

using an obvious notation for the lesser and greater of the pair $\{r_1,r_2\}$. In the limit of small energy differences between single–particle states, we recover a structure which is familiar from atomic structure studies based on the Gaunt and low–frequency Breit operators. Unfortunately, the expansion which was made in powers of (E/ω) may not be used as a computational scheme for evaluating these radiative integrals, because the Weber–Schafheitlin integrals which appear in the series are divergent for inverse powers $(E/\omega)^k$ whenever $\text{Re}(2\mu-k+1) \leq 0$.

The deformation of the contour of integration along the imaginary axis allows us to define an effective potential for QED calculations which is similar to the low–frequency Breit interaction used in atomic structure theory, but which is valid for all single–particle energy differences, including those corresponding to virtual pair–creation. Since the resulting operator is very similar to the Breit operator, we now expect that the accuracy which is attainable using basis set methods to represent the intermediate electron states will be similar to that found in electron correlation studies. We have not eliminated the divergence problems, but rather transformed the divergent self–energy integral into a divergent partial wave expansion; each partial wave component is, however, finite. In order for the scheme to be successful, renormalization at the partial wave level is highly desirable, since we have abandoned the regulator and the known analytic divergences of the free–particle self–energy. The compact analytic expansion of the low–frequency Breit operator as a sum of products functions in the variables r_1 and r_2 has been replaced by a two–dimensional tabular function of the radial coordinates for each value of E, together with analytic expressions for the angular matrix elements. The integral which defines the radial multipole components of the effective potential , $I_n(r_1,r_2;E)$, may be readily evaluated using Gauss–Rational quadrature for any value of E, although special care is required around the limit $E \to 0$. For small values of E in the range $[0,0.1]$, a combination of algebraic expansions and Newton–Cotes integration formulae makes possible the evaluation of the potential function over the entire energy range.

The expectation value of each partial wave contribution to the self–energy operator is finite if the regulated interaction is adopted, involving only perturbation sums over the matrix elements of non–singular bounded operators. The unrenormalized Coulomb field self–energy shift takes the form of a divergent partial wave expansion. We now outline a complete finite basis set calculation of the self–energy shift of a bound electron in a state labelled $|\psi_a\rangle$. Renormalization is achieved by adapting the non–covariant prescription of French and Weisskopf (1949) to the spherical wave expansion of the radiation field. The unrenormalized part of the relativistic self–energy shift is, from equation [5.2]:

$$\Delta E_a = -\frac{\alpha}{\pi} \int d^3r_1 \int d^3r_2 \left\{ +\sum_m{}^{+} \psi_a^\dagger(r_1)\psi_m^\dagger(r_2) F(r_1,r_2;E_m-E_a)\psi_m(r_1)\psi_a(r_2) \right.$$

$$\left. -\sum_m{}^{-} \psi_a^\dagger(r_1)\psi_m^\dagger(r_2) F(r_1,r_2;E_a-E_m)\psi_m(r_1)\psi_a(r_2) \right\}$$

[6.21]

where the effective Feynman gauge potential, $F(r_1,r_2;\omega)$ is:

$$F(r_1,r_2;\omega) = \int_0^\infty dk \, [\boldsymbol{\alpha}_1 \cdot \boldsymbol{\alpha}_2 - I] \frac{\sin(kR)}{R} \frac{1}{\omega + k} \qquad [6.22]$$

and $R = |r_1 - r_2|$. The external field states, $\{\psi_k\}$, satisfy

$$H \psi_k = E_k \psi_k \qquad [6.23]$$

and the free-particle states, $\{\varphi_k\}$ satisfy

$$H_0 \varphi_k = E_k^0 \varphi_k, \qquad [6.24]$$

where H and H_0 are the Dirac single-particle operators for the external field and free particle systems, respectively. In order to renormalize the mass of the bound state we expand $|\psi_a\rangle$ in the complete basis of spherical wave free-particle states; the expansion is naturally restricted to free-particle states of the same symmetry-type as the bound state.

$$|\psi_a\rangle = \sum_k |\varphi_k\rangle \langle \varphi_k | \psi_a\rangle \qquad [6.25]$$

We subtract from the radiative shift of the bound state [6.21] the corresponding self-energy shift of each free-particle component of the bound state, because we define the self-energy shift of a free-electron to be identically equal to zero. Following French and Weisskopf (1949), the mass renormalization counterterm is

$$\Delta M_a = -\frac{\alpha}{\pi} \int d^3r_1 \int d^3r_2$$

$$\left\{ +\sum_{p\,m}{}^{+} \psi_a^\dagger(r_1)\varphi_m^\dagger(r_2) F(r_1,r_2;E_m-E_p)\varphi_m(r_1)\varphi_p(r_2)\langle\varphi_p|\psi_a\rangle \right.$$

$$\left. -\sum_{p\,m}{}^{-} \psi_a^\dagger(r_1)\varphi_m^\dagger(r_2) F(r_1,r_2;E_p-E_m)\varphi_m(r_1)\varphi_p(r_2)\langle\varphi_p|\psi_a\rangle \right\} \qquad [6.26]$$

The index p runs over the complete set, but the sum over m excludes p, and is further restricted by the different formulae for positive- and negative-energy intermediate states. This prescription ensures that the identity of free electrons and positrons is preserved throughout the calculation. We call this physically motivated approach the Kramer's renormalization prescription, to distinguish it from the covariant renormalization prescription embodied in the Feynman integral cutoff technique. The observable shift, ΔS_a, calculated using finite basis set representations of the external field and free-particle states is:

$$\Delta S_a = \Delta E_a - \Delta M_a. \qquad [6.27]$$

This general prescription is now developed into a computational algorithm, suitable for the finite basis set technique. In practice, the subtractions implied in [6.27] may be matched, so that free-field contribution from the intermediate state labelled m renormalizes the contribution from the external-field state with the corresponding label, m, in order to minimize errors from truncation of the spectrum. It is assumed, for convenience, that the free- and external-field basis set representations have the same dimension.

Starting with the tensor decomposition of the Møller interaction matrix element [5.16], we define the auxiliary amplitude functions:

$$A_{an}(r) = P_a(r)P_n(r) + Q_a(r)Q_n(r) \qquad [6.28]$$

$$B^{\pm}_{an}(r) = P_a(r)Q_n(r) \pm Q_a(r)P_n(r) \qquad [6.29]$$

$$C^{\nu\lambda}_{an}(r) = \frac{(\kappa_a - \kappa_n)B^{+}_{an}(r) + \nu B^{-}_{an}(r)}{[\nu(2\nu+1)]^{\frac{1}{2}}} \quad \text{for } \lambda = \nu-1$$

$$= -\frac{(\kappa_a + \kappa_n)B^{+}_{an}(r)}{[\nu(\nu+1)]^{1/2}} \quad \text{for } \lambda = \nu$$

$$= \frac{(\kappa_a - \kappa_n)B^{+}_{an}(r) + (\nu+1)B^{-}_{an}(r)}{[(\nu+1)(2\nu+1)]^{\frac{1}{2}}} \quad \text{for } \lambda = \nu+1$$

$$[6.30]$$

For the sake of clarity I will consider only the positive-energy sum over external field states implied by, [6.27]; the remaining terms are obtained by changing index labels. The integrations in [6.21] over the angular coordinates are performed analytically using the results summarized in equation [5.26]. A two-dimensional numerical integration over the radial coordinates r_1 and r_2 is required, which can be carried out using the techniques developed for the GRASP package. The integrand is symmetric in the radial coordinates and has a discontinuous derivative along $r_1 = r_2$. The numerical behaviour of the integrand is modified by the use of a finite photon regulator, and may be a useful numerical device if sufficiently large values of the regulator, Λ, may be used and if the limit $\Lambda \to \infty$ of the renormalized and regulated interaction may be found.

The positive–energy unrenormalized term in the self-energy calculation, $T_a{}^1$, is:

$$T_a{}^1 = \frac{2\alpha}{\pi} \sum_{\kappa_t} (2j_t+1) \int_0^\infty dr_1 \int_0^{r_1} dr_2$$

$$\times \sum_{\nu=0}^\infty (2\nu+1) \sum_n{}^+ I_\nu(r_1, r_2; E_n-E_a) \left\{ \begin{bmatrix} j_a & \nu & j_t \\ \tfrac{1}{2} & 0 & -\tfrac{1}{2} \end{bmatrix}^2 \Pi^e(\kappa_a, \kappa_t; \nu) \; [A_{an}(r_1) A_{na}(r_2)] \right.$$

$$\left. - \Pi^o(\kappa_a, \kappa_t; \nu) \sum_{\lambda=\nu-1}^{\nu+1} \begin{bmatrix} j_a & \lambda & j_t \\ \tfrac{1}{2} & 0 & -\tfrac{1}{2} \end{bmatrix}^2 [C_{an}^{\nu\lambda}(r_1) C_{an}^{\nu\lambda}(r_2)] \right\}$$

[6.31]

The sum over κ_t includes all symmetry–types, $\kappa_t = \{\pm 1, \pm 2, \ldots\}$. There are three similar terms, $T_a{}^2$, $T_a{}^3$ and $T_a{}^4$, which are calculated in addition to [6.31]. The terms are combined according to [6.27] so that only renormalized amplitudes appear in the final two–dimensional kernel. One of the remaining terms, $T_a{}^2$, corresponds to the negative–energy external field summation in [6.21], and the remaining two terms arise from the positive– and negative–energy renormalization counterterms in [6.26]. We need calculate only a single, two–dimensional integral over a renormalized kernel to determine the observable shift, but in practice it is probably preferable to monitor the convergence of the series by performing an integration for each value of the symmetry label, κ_t.

As yet I am unable to provide numerical evidence that the resulting series in the parameter κ_t converges, but preliminary tests are encouraging and programs have been developed to evaluate the integrands of [6.21] and [6.26] within an L–spinor basis. I have also been made aware that a similar approach to the self–energy problem is under development at Göteborg using numerical basis functions (Lindgren, private communications, 1990). Although the techniques described in this report represent a formal step backwards to the non–covariant perturbation theory of Kroll and Lamb (1949) and French and Weisskopf (1949), the correct results obtained in those early calculations suggests that it is worth persevering, because the approach seems to be directly applicable to the numerical calculation of self–energy shifts in non–Coulomb fields. The elegant covariant techniques are well–suited to purely algebraic manipulations of QED corrections using well–known integral identities. Similar techniques have been used, with some difficulty, in the Coulomb–field problem, but this is possible only because the an exact representation of the Dirac–Coulomb electron propagator is available, enabling contour integration techniques to simplify the calculation. One approach for many–electron atoms is to expand a mean–field propagator in a Dirac–Coulomb basis, since the self–energy shift is dominated by the bare–nuclear charge and many of the analytic tools required for the Coulomb–field problem can be retained. If, on the other hand, QED calculations using the finite basis set approximation can be devised to give results in agreement with the covariant calculations of Mohr and others, then we will have a scheme in which single–particle, many–body and quantum electrodynamic effects in atoms may be calculated within a unified computational framework. It is this goal which remains the motivation for the current project, which aims to study quantum electrodynamical interactions in many–electron atoms and ions.

Acknowledgments: I wish to thank Ian Grant and Stephen Wilson for many stimulating discussions and for their collaboration and encouragement in all aspects of this project. I wish also to thank Ann–Marie Mårtensson–Pendrill and Ingvar Lindgren for their correspondence on the many–body and QED aspects of this work. This work was supported by the Science and Engineering Research Council.

References

M.Abramowitz and I.Stegun, *Handbook of Mathematical Functions*,
  Dover Publications (1970)
P.J.C.Aerts, *Towards Relativistic Quantum Chemistry*,
  Thesis, University of Groningen, (1986)
M.Baranger, H.A.Bethe and R.P.Feynman, *Phys.Rev.*, **92**, 482 (1953).
H.A.Bethe, *Phys.Rev.*, **72**, 339 (1947).
H.A.Bethe and E.E.Salpeter, *Quantum Mechanics of One– and Two–Electron Systems*, Springer–Verlag (1957)
J.T.Broad, *Phys.Rev.*, **A31**, 1494 (1985).
J–P.Desclaux, *Computer Phys. Commun.* **9**, 31 (1975).
Ю.Ю. Дмитриев, Г.Л.Климчицкая, Л.Н.Лабзовский, "Релятивистские
  Эффекты в Спектрах Атомных Систем", Москва Энергоатомиздат, (1984)
G.W.F.Drake, *Can.J.Phys.*, **66**, 586 (1988).
G.W.F.Drake and S.P.Goldman, *Phys.Rev.*, **A23**, 2093 (1981).
G.W.F.Drake and S.P.Goldman, *Phys.Rev.*, **A25**, 2877 (1982).
G.W.F.Drake and S.P.Goldman, *Adv.At.Mol.Phys*, **25**, 393 (1988).
K.G.Dyall, I.P.Grant, E.P.Plummer, C.T.Johnson and F.A.Parpia,
  Computer Phys. Commun. **55**, 425 (1989).
R.P.Feynman, *Phys.Rev.*, **76**, 749 (1949).
R.P.Feynman, *Phys.Rev.*, **76**, 769 (1949).
R.P.Feynman, *Quantum Electrodynamics*, Benjamin–Cummings (1962).
J.B.French and V.F.Weisskopf, *Phys.Rev.*, **75**, 1240 (1949).
W.H.Furry, *Phys.Rev.*, **81**, 115 (1951).
S.P.Goldman and G.W.F.Drake, *Phys.Rev.*, **A25**, 2877 (1982).
O.Gorceix, P.Indelicato and J–P.Desclaux, *J.Phys.B*, **20**, 639 (1987).
I.P.Grant, *Adv.Phys.*, **19**, 747 (1970).
I.P.Grant, *Methods in Computational Chemistry*,
  (ed. S.Wilson) vol 2, p1, Plenum Press (1988).
I.P.Grant and H.M.Quiney, *Adv.At.Mol.Phys*, **23**, 37 (1988).
I.P.Grant, *Relativistic, Quantum Electrodynamic and Weak Interaction Effects in Atoms*, (eds. W.R.Johson, P.J.Mohr and J.Sucher),
  pp 235–253, AIP Conference Proceedings 189 (1989).
E.J.Heller and H.A.Yamani, *Phys.Rev.*, **A9**, 1201 (1974)
H.Hochstadt, *Integral Equations*, John Wiley and Sons (1973)
L.Hostler, *J.Math.Phys.*, **24**, 2366 (1983).
L.Hostler, *J.Math.Phys.*, **28**, 2984 (1987).
E.A.Hylleraas, *Z.Phys.*, **48**, 469 (1930)
P.Indelicato, O.Gorceix and J–P.Desclaux, *J.Phys.B*, **20**, 651 (1987).
Y.Ishikawa, R.Baretty and K.M.Sando, *Chem.Phys.Lett.*, 117, 444 (1985)
Y.Ishikawa, R.Baretty and R.C.Binning, *Chem.Phys.Lett.*, 121, 130 (1985)
C.Itzykson and J–B.Zuber, *Quantum Field Theory*, McGraw–Hill Inc., (1980).
W.R.Johnson and J. Sapirstein, *Phys.Rev.Lett.*, **57**, 1126 (1986)
W.R.Johnson, S.A.Blundell and J. Sapirstein, *Phys.Rev.*, **A37**, 2764 (1988)
W.R.Johnson, S.A.Blundell and J. Sapirstein, *Phys.Rev.*, **A38**, 2699 (1988)
W.R.Johnson, *Relativistic, Quantum Electrodynamic and Weak Interaction Effects in Atoms*, (eds. W.R.Johson, P.J.Mohr and J.Sucher),
  pp 209–216, AIP Conference Proceedings 189 (1989).
Y–K.Kim, *Phys.Rev.*, **154**, 17 (1967)
N.M.Kroll and W.E.Lamb, *Phys.Rev.*, **75**, 388 (1949).
W.Kutzelnigg, *Int.Journ.Quant.Chem.*, **25**, 107 (1984).
Y.S.Lee and A.D.Maclean, *J.Chem.Phys.*, **76**, 735 (1982)

I.Lindgren and J.Morrison, *Atomic Many–Body Theory*, 2nd Edition
 Springer–Verlag, (1986).
I.Lindgren, *Relativistic, Quantum Electrodynamic and Weak Interaction Effects in Atoms*, (eds. W.R.Johson, P.J.Mohr and J.Sucher),
 pp 371–392, AIP Conference Proceedings 189 (1989).
I.Lindgren, *J.Phys.B.*, **23**, 1085 (1990).
E.Lindroth, *Phys.Rev.*, **A37**, 316 (1988).
E.Lindroth and A–M Mårtensson–Pendrill, *Phys.Rev.*, **A39**, 3794 (1989).
E.Lindroth, A–M Mårtensson–Pendrill, A.Ynnerman and P.Öster,
 J.Phys.B, **22**, 2447 (1989)
M.Lieber, *Relativistic, Quantum Electrodynamic and Weak Interaction Effects in Atoms*, (eds. W.R.Johson, P.J.Mohr and J.Sucher),
 pp 445–459, AIP Conference Proceedings 189 (1989).
N.L.Manakov, L.P.Rapoport and S.A.Zapryagin, *J.Phys.B*, **7**, 1076 (1974).
F.Mandl and G.Shaw, *Quantum Field Theory*, John Wiley and Sons (1984)
P.J.Mohr, *Ann.Phys (N.Y.)*, **88**, 26 (1974).
P.J.Mohr, *Ann.Phys (N.Y.)*, **88**, 52 (1974).
W.Pauli and F.Villars, *Rev.Mod.Phys.*, **21**, 434 (1949).
H.M.Quiney, *Finite Basis Set Studies of the Dirac Equation*,
 D.Phil Thesis, University of Oxford, (1987).
H.M.Quiney, *Methods in Computational Chemistry*,
 (ed. S.Wilson) vol 2, p227, Plenum Press (1988).
H.M.Quiney, I.P.Grant and S.Wilson, *J.Phys.B*, **18**, 2805 (1985).
H.M.Quiney, I.P.Grant and S.Wilson, *J.Phys.B*, **18**, 577 (1985).
H.M.Quiney, I.P.Grant and S.Wilson, *J.Phys.B*, **20**, 1413 (1987).
H.M.Quiney, I.P.Grant and S.Wilson, *Physica Scripta*, **36**, 460 (1987).
H.M.Quiney, I.P.Grant and S.Wilson, *Many–Body Methods in Quantum Chemistry*, (ed U.Kaldor), Lecture Notes in Chemistry **52**,
 p307, Springer–Verlag (1989).
H.M.Quiney, I.P.Grant and S.Wilson, *J.Phys.B*, **23**, L271 (1990)
M.Rotenberg, *Adv.At.Mol.Phys.*, **6**, 233 (1970)
S.Salomonson and P.Öster, *Phys.Rev.*, **A40**, 5559 (1989)
J.Schwinger, *Phys.Rev.*, **76**, 790 (1949).
J.Schwinger, *J.Math.Phys.*, **5**, 1606 (1964).
R.E.Stanton and S.Havriliak, *J.Chem.Phys.*, **81**, 1910 (1984).
J.Sucher, *Relativistic, Quantum Electrodynamic and Weak Interaction Effects in Atoms*, (eds. W.R.Johson, P.J.Mohr and J.Sucher),
 pp 28–46 and pp 337–370, AIP Conference Proceedings 189 (1989).
B.Swirles, *Proc.Roy.Soc. (London)*, **A152**, 625 (1935).
G.N.Watson, *A Treatise on the Theory of Bessel Functions*,
 Cambridge University Press (1944)
V.F.Weisskopf, *Phys.Rev.*, **56**, 72 (1939).
J.Weitsman and P.L.Hagelstein, *J.Phys.B*, **19**, L59 (1986).
E.Wichmann and N.M.Kroll, *Phys.Rev.*, **101**, 843 (1956)
S.Wilson, *Electron Correlation in Molecules*, Clarendon Press, Oxford (1984)
H.A.Yamani and W.P.Reinhardt, *Phys.Rev.*, **A11**, 1144 (1975)

RELATIVISTIC CALCULATIONS OF ELECTRON IMPACT IONISATION CROSS-SECTIONS OF HIGHLY CHARGED IONS

D.L. Moores

Department of Physics and Astronomy
University College London
Gower Street
London WC1E 6BT

THEORY

In this paper we report calculations of electron-impact ionisation cross sections of highly charged positive ions, using a fully relativistic distorted wave method. In the processes we consider, a target N-electron ion interacts with an incident electron to give a (N-1)-electron ionised ion plus an ejected and a scattered electron. A relativistic expression for the cross section for this process has been given by Pindzola et al. [1] One starts from lowest-order QED theory, including the two Feynman diagrams depicted in figure 1, representing direct and exchange ionisation. The double line represents a bound orbital of the target ion, while the continuum electrons are represented by single lines. Let the subscripts b, i, e and f signify the bound, incident, ejected and scattered electrons respectively. Using standard notation, [2] the ionisation amplitude may be expressed as

$$S = -\frac{i}{2} \delta(\varepsilon_f + \varepsilon_e - \varepsilon_i - \varepsilon_b) M, \quad (1)$$

where
$$M = M_d - M_x, \quad (2)$$

$$M_d = \int d^3x \int d^3y \left\{ \bar{\psi}_f(\underline{x}) \gamma_u \psi_i(\underline{x}) \right\} e^{i|x-y|} \frac{(\varepsilon_f - \varepsilon_i)/c}{|x-y|} \left\{ \bar{\psi}_e(\underline{y}) \gamma^u \psi_b(\underline{y}) \right\}, \quad (3)$$

$$M_x = \int d^3x \int d^3y \left\{ \bar{\psi}_e(\underline{x}) \gamma_u \psi_i(\underline{x}) \right\} e^{i|x-y|} \frac{(\varepsilon_e - \varepsilon_i)/c}{|x-y|} \left\{ \bar{\psi}_f(\underline{y}) \gamma^u \psi_b(\underline{y}) \right\} \quad (4)$$

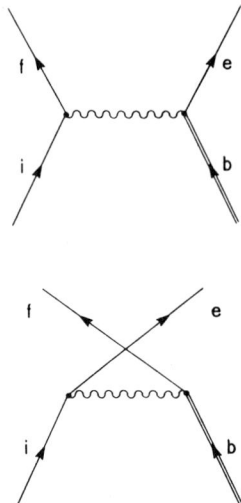

Figure 1. Lowest-order Feynman diagrams for electron scattering with a with a bound-state electron (double line).

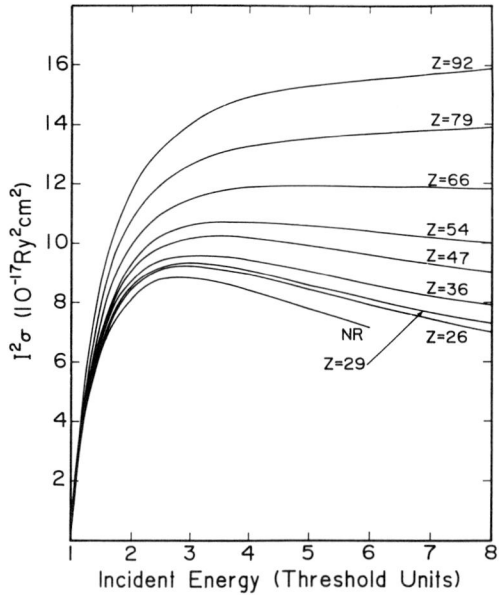

Figure 2. I^2 for the 1s subshell ionisation of hydrogenic ions, where I is the ionisation potential and is the cross section. The non-relativistic result (NR) is independent of Z. The relativistic results are labelled by the nuclear charge Z.

In (3) and (4), is the electron energy minus the rest energy, the $\psi(x)$ are Dirac bispinors and the γ_u are 4x4 matrices. The Lorentz gauge has been chosen for the photon propagator and integration over time and momentum transfer has been carried out. In this work we assume a MCDF approximation [3] to describe both the target and ionised ion structure, and run the GRASP [4] package to obtain the bound orbitals. The wave functions for the target ion (ionised ion) are taken to be a superposition of configurations α_m (β_n) with mixing coefficients C_m (d_n). The continuum waves are taken to be in the form of relativistic partial wave expansions in which the radial functions satisfy the Dirac equations

$$[\frac{d}{dr} + \frac{\kappa}{r}] P_{\varepsilon l j}(r) + \frac{1}{c}[V_n(r) - \varepsilon - 2c^2 + V_{DF}] Q_{\varepsilon l j}(r) = 0, \tag{5}$$

$$[\frac{d}{dr} - \frac{\kappa}{r}] Q_{\varepsilon l j}(r) - \frac{1}{c}[V_n(r) - \varepsilon + V_{DF}] P_{\varepsilon l j}(r) = 0, \tag{6}$$

where $V_n(r)$ is the nuclear potential, and the distorting potential operator V_{DF} is constructed from Dirac-Fock (DF) target orbitals, generated by GRASP.

The quantum number κ is defined by

$$\kappa = -2(j-l)(j + 1/2) \tag{7}$$

The ionisation cross section may then be expressed in the form

$$= \int_0^{E/2} d\varepsilon_e \frac{16}{P_i{}^3 P_e P_f} \sum_{l_i l_f l_e} \sum_{j_i j_f j_e} \sum_{m n m' n'} \frac{(2j_i + 1)(2j_f + 1)(2j_e + 1)}{(2j_b + 1)}$$

$$G(\alpha_m, \beta_n) G(\alpha_m', \beta_n') C_m d_n C_m' d_n' |Vav|^2 \tag{8}$$

where $G(\alpha_m, \beta_n)$ are angular coefficients and $|Vav|^2$ is defined by Pindzola et al [1] the momenta p are related to the energy by the usual relativistic expression

$$p^2 = 2\varepsilon + \varepsilon^2 \tag{9}$$

and the total energy

$$E = \varepsilon_e + \varepsilon_f \tag{10}$$

RESULTS

In figure 2 we show some results for ionisation of the 1s state

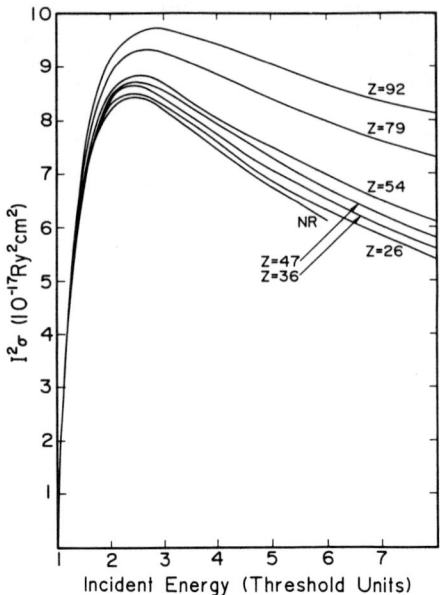

Figure 3. I^2 for the 2s subshell ionisation of hydrogenic ions, where I is the ionisation potential and is the cross section. The non-relativistic result (NR) is independent of Z. The relativistic results are labelled by the nuclear charge Z.

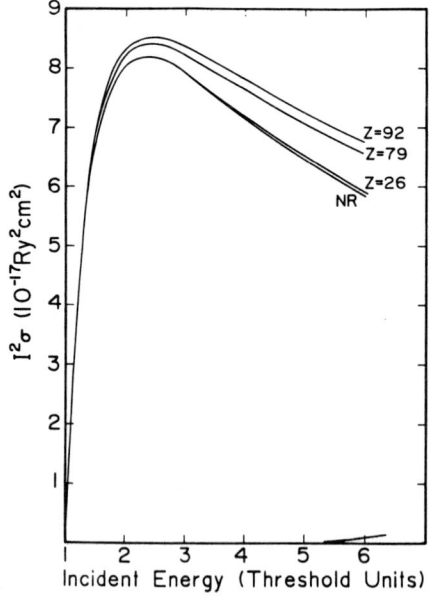

Figure 4. I^2 for the 3s subshell ionisation of hydrogenic ions, where I is the ionisation potential and is the cross section. The non-relativistic result (NR) is independent of Z. The relativistic results are labelled by the nuclear charge Z.

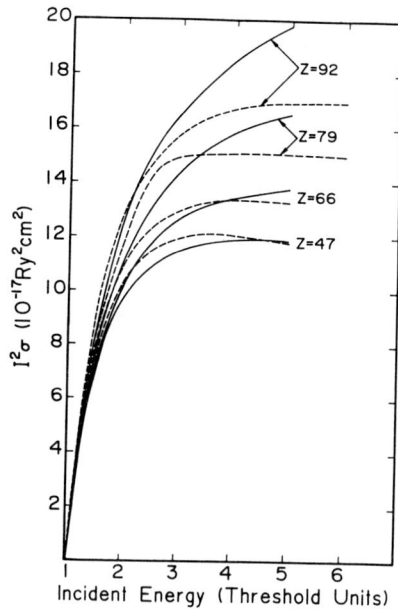

Figure 5. $I^2\sigma$ for the 1s subshell ionisation of hydrogenic ions, where I is the ionisation potential and σ is the cross section. Solid curves - fully relativistic results with static, magnetic and retardation interactions; dashed curves - relativistic results with static interactions only.

Table 1. Electron-impact ionisation cross sections of Ne-like Xe $2p^5$ 3s (J=1)→$2p^4$ 3s (J=1/2) in cm^2. The numbers in parentheses indicate the power of 10 by which the number should be multiplied.

Transition (see text)	X = 1.25	X = 1.5
1 → 3	7.5 (- 23)	1.4 (- 22)
4	1.2 (- 22)	1.6 (- 22)
5	3.1 (- 25)	4.4 (- 25)
2 → 3	4.8 (- 25)	6.9 (- 25)
4	1.3 (- 22)	1.8 (- 22)
5	1.7 (- 22)	2.5 (- 22)

of hydrogenic ions. The quantity $I^2\sigma$, where I is the ionisation energy, varies slowly with the nuclear charge Z and is plotted for $26 \geqslant Z \geqslant 92$ against X, the incident energy divided by the ionisation energy. The curves are labelled by the value of the nuclear charge. The curve marked NR is a non-relativistic calculation, [5] ignoring screening, which is independent of Z. Figure 2 clearly illustrates the increasing importance of relativity as both Z and the incident energy are increased. The effect is progressively less marked for 2s ionisation (figures 3) and 3s ionisation (figure 4). The results shown in figures 2—4 do not include magnetic or retardation effects: only the Coulomb part of the interaction was retained. The effects of including these relativistic corrections to the electron-electron interaction are shown in figure 5, which shows a comparison between no-exchange results with and without them, for 1s ionisation, $47 \geqslant Z \geqslant 92$. For each ion, the curves tend to diverge from each other when the incident energy exceeds about 250 kev. For H-like U, Z=92, the cross section has been measured [6] at an incident electron energy of 222 kev, which corresponds to X=1.7. The experimental cross section is a factor of about 3.5 times higher than theory. A similar comparison for He-like U gives a result a factor of 5 higher. The reasons for these discrepancies is not understood. It is unlikely that the calculations are in error by this amount.

As an example of a multi-configuration case, we consider the transitions

$$2p^5 3s \ (J = 1) \rightarrow 2p^4 3s \ (J = 1/2)$$

in Ne-like Xe (Z = 54)

The initial state is taken to be an admixture of the two configurations (1) $2\bar{p}^2 \ 2p^3 \ 3s$ and (2) $2\bar{p} \ 2p^4 \ 3s$, each outside a closed shell Core $1s^2 \ 2s^2$. The final state is an admixture of (3) $2\bar{p}^2 \ 2p^2 \ 3s$ (4) $2\bar{p} \ 2p^3 \ 3s$ and (5) $2p^4 \ 3s$.
Here $2\bar{p}$ represents 2p (j=1/2) and 2p represents 2p(j=3/2).

The cross sections for two different values of the incident energy are shown in table 1. The transitions (1) → (5) and (2) → (3) have small cross sections since they involve double electron jumps and are only possible through configuration interaction.

It is hoped to extend these calculations to the case of electron impact excitation in future work.

REFERENCES

1. M.S. Pindzola, D.L. Moores and D.C. Griffin, Electron-impact ionization of highly-charged ions in lowest-order QED theory, Phy. Rev A. 40:4941 (1989)

2. J.D. Bjorken and S.D. Drell, "Relativistic Quantum Mechanics," McGraw-Hill, New York (1964).

3. I.P. Grant, Relativistic Calculation of Atomic Structures, Advan Phys 19: 747 (1970).

4. K.G. Dyall, I.P. Grant, C.T. Johnson, F.A. Parpia and E.P. Plummer, Computer Physics Communications, 55:425 (1989).

5. D.L. Moores, L.B. Golden and D.H. Sampson, Ionisation from the 3p and 3d sublevels of highly charged ions, J.Phys B 13:385 (1980).

6. N. Claytor, B. Feinberg, H. Gould, C.E. Bemis, Jr, J. Gomez del Campo, C.A. Ludemann and C.R. Vane, Electron Impact Ionization of U^{88+} U^{91+}, Phys.Rev Lett. 61:2081 (1988).

MOLECULES

Nonsingular Relativistic Perturbation Theory and Relativistic Changes of Molecular Structure

W.H.E. Schwarz, A. Rutkowski, and G. Collignon

Theoretical Chemistry, The University
POB 10 12 40, D–5900 Siegen, West Germany, and
Division of Physics, Higher Pedagogical School, Zolnierska 14,
PL–10561 Olsztyn, Poland

Abstract

The new relativistic perturbation approach for the Dirac equation, which does not suffer from divergences in any order, is presented. The corresponding double perturbation theory, where the second perturbation is an additional potential, is applied to the potential curve of H_2^+. Relativistic changes of bond length and force constant are calculated and discussed. The equivalence of different paradoxical calculation– and explanation–schemes are numerically verified with high accuracy. Au_2 is also discussed using, however, finite double perturbation theory within the effective core potential approach. Finally the small relativistic bond angle changes, as compared to the large bond length changes, are explained.

1. Introduction

In order to systematize, rationalize and finally understand the mass of experimental and theoretical data on matter, a hierarchy of simplifying models is needed. Concerning the basic chemical data such as energies, geometries and interatomic forces in molecules and crystals, the nonrelativistic Schrödinger–Coulomb and the relativistic Projected–Dirac-Breit Hamiltonians within the framework of the Born–Oppenheimer approximation are useful starting points.

The differences between the rather well understood nonrelativistic model, and the more realistic relativistic model, are most clearly revealed with the help of perturbation theory. Traditionally, the Schrödinger model is used as the unperturbed equation in one component form. As is well known (compare e.g. the review by Kutzelnigg 1989), the transition from the four–component Dirac model to the one (or two) component Schrödinger (or Pauli, resp.) model is severely hampered by singularities, irrespective of whether the elimination or the transformation method is applied (Moss 1970). There, the physical constant $\alpha = c^{-1} \to 0$ is used as perturbation parameter, although the relativistic Hamiltonian contains terms of order c and c^2, and although the wavefunction is nonanalytic in c^{-1}.

Therefore, one of us (Rutkowski 1986) has recently developed a new perturbation approach within a completely four–component framework, which is free from singularities and divergences. The essentials of this approach are described in sec. 2.

The above mentioned basic chemical parameters are defined by derivatives of the energy with respect to the nuclear positions. Relativistic changes of structural parameters and force constants are therefore given by double perturbation energies, where the first perturbation is relativity (ρ) and the second one is nuclear displacement (D). The corresponding double perturbation energy expressions are presented and discussed in sec. 3.

The so-called double perturbation interchange relations are of fundamental importance concerning the interpretation of relativistic changes of chemical parameters. The present approach is applied with high accuracy to the smallest molecular system, H_2^+, in sec. 4. The numerical equivalence and physical meaning of the complementary schemes is substantiated.

In the case of heavy molecules with very many electrons, *ab initio* approaches become unwieldy, and pseudopotential valence-only approaches are more recommendable. The interchange relation, and the equivalence of different paradoxical explanations of the relativistic bond length contraction are verified for the Au_2 molecule by a finite perturbation pseudopotential approach in sec. 5.

Bending force constants of molecular angles are typically an order of magnitude weaker than the stretching force constants of bond lengths. Nevertheless, and although the angular behavior of relativistic atomic orbitals is strongly influenced by spin-orbit coupling, relativistic bond angle changes are in general negligibly small, whereas relativistic bond length changes are of the order of $(Z/c)^2$ (Pyykkö 1988). This empirical paradoxon is rationalized within the relativistic Hückel approximation in sec. 6.

Our conclusions are presented in sec. 7.

2. Nonsingular Relativistic Pertubation Theory

We write the Dirac equation for a single electron or positron in an external field (V,A) in the following form:

(2.1) $$[mc^2]\Psi = [\beta(\hbar\omega - eV) + \gamma(c\mathbf{p} - e\mathbf{A})]\Psi$$

where ω is the frequency factor, Ψ is a four-component spinor, and β and γ read in the standard representation as

(2.2) $$\beta = \begin{bmatrix} 1_2 & 0_2 \\ 0_2 & -1_2 \end{bmatrix} = \beta^0 - \beta^1 = \begin{bmatrix} 1_2 & 0_2 \\ 0_2 & 0_2 \end{bmatrix} - \begin{bmatrix} 0_2 & 0_2 \\ 0_2 & 1_2 \end{bmatrix}, \quad \gamma = \begin{bmatrix} 0_2 & -\sigma \\ \sigma & 0_2 \end{bmatrix}$$

with the index $_2$ indicating 2x2-matrices. The energy of the electron or positron is given by

(2.3) $$E = mc^2 + \epsilon = +\hbar\omega \text{ or } -\hbar\omega ,$$

respectively. In order to allow for a nonproblematic transition between the relativistic and nonrelativistic models, also the nonrelativistic Schrödinger equation is written in four-component form (Levy-Leblond 1967). In the case of electrons it reads (compare eq. 2.1)

(2.4) $$[mc^2]\Psi^0 = [\beta mc^2 + \beta^0(\epsilon^0 - eV) + \gamma(c\mathbf{p} - e\mathbf{A})]\Psi^0$$

The superscript ^{0} indicates the nonrelativistic model, ϵ^0 and Ψ^0 are the nonrelativistic reduced energy and the nonrelativistic Dirac spinor. Its upper two components Ψ_+^0 are just the Schrödinger–Pauli nonrelativistic functions for α and β spin, and the two lower components are given by $\Psi^0_- = (\sigma p/2mc)\Psi^0_+$. The difference between eqs. (2.1,2.4),

(2.5) $$\Delta H = \beta^1(\epsilon - eV) ,$$

is the nonpathological operator of relativistic corrections for electrons.

Introducing the relativistic pertubation parameter ρ with $\rho=0$ for the nonrelativistic case and $\rho=1$ for the relativistic case, eqs. (2.1) and (2.4) may be combined. (In later sections, ρ will denote the electron density, however.) In appropriate (2x2)–block form, the unified eq. reads

(2.6) $$\begin{bmatrix} \epsilon^\rho - eV & \sigma(cp-eA) \\ \sigma(cp-eA) & 2mc^2 + \rho(\epsilon^\rho - eV) \end{bmatrix} \cdot \begin{bmatrix} +\Psi^\rho_+ \\ -\Psi^\rho_- \end{bmatrix} = 0$$

Contrary to the traditional approaches, relativistic, nonrelativistic and perturbation operators act here on the same Hilbert space.

Expanding Ψ and ϵ in a power series of ρ, we define the perturbation approach. It differs somewhat from the conventional Rayleigh–Schrödinger approach because of the peculiar appearance of ϵ in the perturbation ΔH, eq. (2.5). The system of perturbation equations for the one–electron case was given by Rutkowski (1986a). By the standard procedure of perturbation theory, matrix elements of different perturbed functions can be substituted by each other, so that only the m'th order function $\Psi^m = (\Psi^m_+, \Psi^m_-)$ is needed to determine the (2m+1)–st order energy. We here display just the first three relativistic perturbation energies for the common case of vanishing **A**

(2.7) $\epsilon^1 = <\Psi^0_-|eV - \epsilon^0|\Psi^0_-> = <\Psi^0_+|\sigma p(V-\epsilon^0)\sigma p|\Psi^0_+>/4c^2 \quad = \eta^1.c^{-2}$

(2.8) $\epsilon^2 = <\Psi^1_-|eV-\epsilon^0|\Psi^0_-> + <\Psi^1_+|-\epsilon^1|\Psi^0_+> + <\Psi^0_-|-\epsilon^1|\Psi^0_-> = \eta^2.c^{-4}$

(2.9) $\epsilon^3 = <\Psi^1_-|eV-\epsilon^0|\Psi^1_-> + <\Psi^1_+|-\epsilon^1|\Psi^1_+> + <\Psi^0_-|-\epsilon^1|\Psi^0_->$

$\qquad + 2<\Psi^1_-|-\epsilon^1|\Psi^0_-> + 2<\Psi^1_+|-\epsilon^2|\Psi^0_+> \qquad = \eta^3.c^{-6}$

In the most serious case of a singular point–charge Coulomb potential, Ψ behaves as $r^\zeta.e^{-\lambda r}$, and the perturbation functions Ψ^i behave as $(\ln r)^i.\Psi$ near the singularity. Therefore all integrals remain finite. Especially we note that only overlap and simple potential integrals are needed, and that no high–order derivatives such as p^{2i} appear in the present approach. Still, it must be admitted that the higher order perturbation functions need extended basis sets because of the weak logarithmic singularity, what is less desirable in numerical variation–perturbation calculations (Rutkowski 1986b). Finally we note that the ϵ^i are exactly i! times the i'th Taylor coefficient of the power expansion of the relativistic energy in terms of c^{-2}. So this perturbation approach converges provided the Taylor–expansion of the exact relativistic energy converges.

Finally we mention that the first order perturbation Hamiltonian in eq. (2.7) may be split up into

(2.10) $\sigma p(eV-\epsilon^0)\sigma p = p(eV-\epsilon^0)p + 4c^2.h_{so} = [T,eV-\epsilon^0]_+ + (T-V) + \nabla V \times p.\sigma$

where the first, second and third terms on the rhs., respectively, are the mass–velocity, Darwin and spin–orbit terms of the conventional Pauli Hamiltonian. These terms have the well–known singularities at the origin of Coulomb potentials, which cancel each other if treated correctly.

3. Double Perturbation Theory

Rutkowski (1986c) and Jankowski and Rutkowski (1987) have generalized the one–electron approach to many–electron systems. The electron interaction was treated as a second perturbation. Quite successfull numerical applications to many–electron atoms have been published recently (Jankowski and Rutkowski 1989). Up to second order, relativistic and correlation effects are additive in this approach, except for the Breit correction.

This approach has now been extended to a general perturbing potential (Rutkowski and Schwarz 1990) where the potential is expanded in a power series of the parameter D. Examples are homogeneous external electric or magnetic fields, or changes of the electron–nuclear interaction due to changes of the form or of the position of the nuclei. Expanding the energy as

$$(3.1) \qquad E = \Sigma_{r,p} \, \rho^r . D^p . E^{rp}$$

the perturbation energies E^{rp} of relativistic order r and potential order p can be symbolically written as

$$(3.2) \qquad E^{10} = <00|10|00>$$

$$(3.3) \qquad E^{20} = <00|20|00> + <10|10|00>$$

$$(3.4) \qquad E^{11} = <00|11|00> + 2<10|01|00>$$
$$\text{or} \quad = <00|11|00> + 2<01|10|00>$$

$$(3.5) \qquad E^{30} = <00|30|00> + 2<10|20|00> + <10|10|10>$$

$$(3.6) \qquad E^{21} = <00|21|00> + 2<10|11|00> + <10|01|10>$$
$$+ 2<01|20|00> + 2<10|10|01>$$

$$(3.7) \qquad E^{40} = <00|40|00> + 2<10|30|00> + <10|20|10>$$
$$+ <20|20|00> + <20|10|10>$$

$$(3.8) \qquad E^{31} = <00|31|00> + 2<10|21|00> + <10|11|10>$$
$$+ 2<01|30|00> + 2<10|20|01>$$
$$+ 2\{<11|20|00> + <11|10|10>\}$$
$$\text{or} \quad + 2\{<20|11|00> + <20|01|10> + <20|10|01>\}$$

$$(3.9) \qquad E^{22} = <00|22|00> + 2<10|12|00> + 2<01|21|00>$$
$$+ <10|02|10> + <01|20|01> + 2<10|11|01> + <11|00|11>$$
$$+ 2\{<20|01|01> + <20|02|00>\}$$
$$\text{or} \quad + 2\{<02|10|10> + <02|20|00>\}$$

The perturbation energies with inverted indices have the same symbolic form. $|rp\rangle$ is the perturbation function of r–th order in relativity and of p–th order in the potential. $|ij|$ is a perturbation operator of orders i and j, respectively. The explicit forms are somewhat involved; they contain terms of lower order operators and lower order perturbation energies, multiplied by projection operators β_k^1 onto the lower components of the many–electron wavefunction. For details see Jankowski and Rutkowski (1987,1989), and Rutkowski and Schwarz (1990).

We now comment on the different perturbation energies. The <u>first order</u> energies consist of a single symbolic term. According to eq. (2.10) the relativistic first order energy contains $\mathbf{p}(V-E)\mathbf{p}$ and the conventional spin–orbit term h_{so}. For many electron systems, also two–electron terms such as h_{soo}, h_{ss} etc. appear, also terms of the type $(V_i+T_i)T_j$, but no terms of the type $T_iT_i=p_i^4/4m^2$. That is, <u>sums</u> of the velocity–mass and Darwin terms are replaced by more well–behaved expressions. Similarly, no divergent expressions appear in higher orders.

The <u>second order</u> energies E^{20}, E^{11}, E^{02} consist of two terms. The original expression for the mixed perturbation energy,

$$(3.10) \qquad E^{11} = \langle 00|11|00\rangle + \langle 10|01|00\rangle + \langle 01|10|00\rangle$$

where both first order functions $|10\rangle$ and $|01\rangle$ appear, can be simplified with the help of the double perturbation relation (Dalgarno and Stewart 1956)

$$(3.11) \qquad \langle 10|01|00\rangle = \langle 01|10|00\rangle \;;$$

see the two different alternative forms presented in eq. (3.4).

The <u>third order</u> energies at least need the first order functions: E^{30} needs $|10\rangle$, and there are three terms (see eq. 3.5); E^{21} (and E^{12}) at least needs both $|10\rangle$ and $|01\rangle$ and consists of 5 terms (see eq. 3.6). The original expression for E^{21} (containing also the second order functions $|20\rangle$ and $|11\rangle$) consists of 5 terms, too. In general there are rp+r+p terms for E^{rp}. Thus, in order to "understand" the values of these perturbation energies in a physical manner, one must at first develop an intuitive understanding of the sign and magnitude of these different terms. This will turn out not too difficult. However, one must also intuitively understand the sign and order of magnitude of the corresponding <u>sum</u> of terms. In how far a physical understanding of the numerical values of the total perturbation energies is achievable, will be discussed in the subsequent sections.

4. Relativistic Changes of Molecular Geometries. The Bond Length of H_2^+

In the following, Δ shall always denote relativistic-nonrelativistic differences. For instance

$$(4.1) \qquad \Delta\epsilon = \epsilon - \epsilon^0$$

is the relativistic correction to the energy.

The equilibrium geometry D_e of a molecule is defined by $(dE/dD)_{D_e}=0$, where D are the internuclear geometric parameters like bond lengths or bond angles. From eq. (3.1) we obtain for the relativistic (ρ=1) equilibrium position

$$(4.2) \qquad \Sigma_{p=1}\left[p \cdot D_e^{p-1} \cdot (\Sigma_{r=0} E^{rp}) \right] = 0 \;.$$

Defining the nonrelativistic equilibrium geometry as $D_e^0 = 0$ so that $E^{01} = 0$, one obtains for the relativistic D_e the following recursion relation

(4.3) $$\Delta D_e = D_e = - \left[\sum_{r=1} E^{r1} \right] / \left[\sum_{p=2} p \cdot D_e^{p-2} \cdot (\sum_{r=0} E^{rp}) \right]$$

which yields

(4.4) $-2 \cdot \Delta D_e = (E^{11}/E^{02}) + (E^{02}E^{21} - E^{11}E^{12} + 3/4 \cdot (E^{11})^2 E^{03}/E^{02})/(E^{02})^2 + O(c^{-6})$

(4.4a) $-2 \cdot \Delta D_e \approx 2 E'_{D=0} / \bar{k}$

where $E'_{D=0}$ is the slope of the relativistic energy curve at the nonrelativistic equilibrium position, while the slope of the nonrelativistic curve vanishes at that point, $(E^0)'_{D=0} = 0$. k is the relativistic force constant, $k = d^2E/dD^2$, and $\bar{k} = (k_{D=0} + k_{D_e})/2$.

For stable molecules, $k > 0$. Therefore, the sign of ΔD_e, i.e. whether there is relativistic bond length (or bond angle) increase or decrease, is determined by E', i.e. to lowest relativistic order by E^{11}.

4.1 First contribution to ΔD_e

According to eq. (3.4), E^{11} consists of two terms. The **first** contribution is

(4.5) $$E^{11}(1) = <00|11|00> = <\Psi_+^0 | \sigma p(\partial V/\partial D) \sigma p | \Psi_+^0 > / 4c^2$$

i.e. the expectation value of the geometric change $\partial/\partial D$ of the relativistic correction operator $\sigma p V \sigma p / 4c^2$, i.e. $<\partial h_{so}/\partial D> + <p|\partial V/\partial D|p>$. For a σ-bond, the spin–orbit coupling is often rather unimportant, as in the present case of a s–s bond. $<p\Psi_+^0 | p\Psi_+^0>$ is the local kinetic energy density, which is positive everywhere; for a bonding orbital it is very small between the nuclei, so it mainly weights the region behind the nuclei. $\partial V/\partial D$ is positive between the nuclei and negative behind them. Consequently, $E^{11}(1)$ is negative or bond expanding.

Another chain of reasoning is the following: the analogue of the present term $\sigma p(V-E^{00})\sigma p$ in the conventional Pauli approach is $h_{mv} + h_d + h_{so}$. The mass–velocity term $h_{mv} \sim p^4$ does not contribute because $\partial(h_{mv})/\partial D = 0$. As already mentioned, the spin–orbit interaction is not important in the underlying case. The Darwin–term h_d is strongly localized at the Coulomb singularities of the nuclei, and $\partial(h_d)/\partial D$ measures the change of slope of the density on the nuclei between the bond and the back side:

(4.6) $$Z\pi/8 \, (|\rho'_{bond}| - |\rho'_{back}|) \, \alpha^2 \, .$$

Since the slope is steep on the back side and small on the bond side of the nuclei for a bonding σ orbital, we again arrive at the conclusion, that $E^{11}(1)$ should be negative. Indeed, for the H_2^+ molecule, very accurate calculations (Rutkowski and Schwarz 1990) yield

(4.7) $$E^{11}(1) = -0.09225 \, \alpha^2$$

4.2 Second contribution to ΔD_e

For the **second** contribution to E^{11}, there exist, as mentioned above, two alternative expressions. The one is

(4.8) $\quad E^{11}(2a) = 2<10|01|00> = 2<\Psi_+^{10}| \partial V/\partial D |\Psi_+^{00}>/4c^2$

$\quad\quad\quad\quad = \int F^{hf}.\Delta^1\rho.dv^3$

$\partial V/\partial D = F^{hf}$ is the electrostatic Hellmann–Feyman Coulomb force, which acts on the relativistic first order change of the electron density, $2\Psi_+^{10}\Psi_+^{00}/4c^2 = \Delta^1\rho$.

Covalent bonds are usually due to overlapping, partially occupied s– and p–type atomic valence orbitals. In general, these are relativistically contracted (Rose et al. 1978, Schwarz et al. 1989). So we may assume with some justification that the corresponding bonding molecuar orbitals are relativistically contracted, too. Consequently $E^{11}(2)$ should be positive and bond–contracting (Schwarz 1988). An accurate calculation on H_2^+ yields

(4.9) $\quad\quad\quad\quad E^{11}(2a) = + 0.16402\, \alpha^2$

So we may say: the σ–bond in H_2^+ is shortened relativistically, because the bond–contracting <u>Hellmann–Feynman force, resulting from the nonrelativistic Coulomb interaction</u> between the nuclei and the <u>relativistically contracted valence electron charge cloud</u> is more important than the relativistic Darwin reduction of the interaction between the nuclei and the nonrelativistic charge distribution in the spatial vicinity of the nuclei.

A completely different explanation of the relativistic bond contraction can be based on the alternative expression

(4.10) $\quad E^{11}(2b) = 2<01|10|00> = 2<\Psi_+^{01}| \sigma p(V-E^{00})\sigma p/4c^2 - E^{10}|\Psi_+^{00}>$

$\quad\quad\quad\quad = \int [\sigma V\sigma p/4c^2 - E^{00}.p^2/4c^2 - E^{10}].\partial\rho^{00}/\partial D.dv^3$

The terms in the brackets under the integral represent the relativistic first order perturbation, $|10|$, which acts on the nonrelativistic change of electronic density with bond length, $\partial\rho^{00}/\partial D = 2\Psi_+^{01}\Psi_+^{00}$.

The relativistic perturbation is energy lowering in the vicinity of the nuclei (see Schwarz et al. 1989). Concerning $\partial\rho/\partial D$, we should distinguish between two cases, hydrogen atoms without occupied core shells, and "normal" atoms with occupied cores. In the case of hydrogen, the density around the nucleus increases significantly upon bond formation (Kutzelnigg and Schwarz 1982). In the case of "normal" atoms, the Pauli principle, or alternatively the orthogonality constraint of orbitals, results in an <u>overall</u> keeping the valence electrons of atom A out of the core volume of atom B. But the orthogonality constraint results nevertheless in <u>some</u> local density increases, especially in the nuclear vicinity. Consequently $E^{11}(2b)$ is expected to be positive.

Again we may also use the Pauli form of the perturbation operator for an alternative reasoning. The most important term in the underlying case is the mass–velocity operator, $h_{mv} = -T^2/2c^2$. In the case of bonded hydrogen atoms, electronic density and curvature of wavefunction are in general strongly increased near the nuclei, so that $E^{11}(2b)$ should be positive and bond contracting. In the case of bigger atoms with occupied core shells, the overlap of an atomic valence shell with the occupied core of an adjacent atom leads to the so–called Pauli repulsion. This happens because of the (nonrelativistic) kinetic energy increase accompanying the increase of curvature which is caused by orthogonalization of the valence orbital on the adjacent core. Relativity (h_{mv}) will therefore cause a reduction of the Pauli repulsion. Those parts of space contribute most to this relativistic bond contraction where T is large, and where $-T^2/2c^2$ is most important. It is interesting to note that (within the framework of low order perturbation theory) the relativistic bond contraction originates in the very tails of the valence orbitals near the nuclei, while the outer core shells contribute much to the nonrelativistic Pauli repulsion. So we arrive at a slightly different rationalization than Pyykkö et al. (1981).

The accurate evaluation $E^{11}(2b)$ yields

(4.11) $$E^{11}(2b) = +0.16397\, \alpha^2$$

which differs by only $5.10^{-5}.\alpha^2$ from the result (4.9). Here we may say: the σ-bond in H_2^+ is shortened relativistically, because the bond-contracting action of the relativistic correction of the Hamilton on the increased nonrelativistic electronic charge density near the nuclei (an increase which happens in most cases, when σ-bonds are formed) is more important than the bond expanding action of $\partial(\Delta H)/\partial D$ on the unperturbed density.

4.3 Further remarks on relativistic changes of bond lengths

Summarizing, if we denote relativistic changes by Δ and bond length changes by δ, we may write

(4.12) $$E^{11} = \int \{\delta(\Delta H).\rho + (\Delta H).\delta\rho\}.dv^3 = \delta\int \Delta H.\rho.dv^3$$

or

(4.13) $$E^{11} = \int \{\Delta(\delta H).\rho + (\delta H).\Delta\rho\}.dv^3 = \Delta\int \delta H.\rho.dv^3$$

with

(4.14) $$\Delta H = H^{rel} - H^{nrel}, \quad \Delta\rho = \rho^{rel} - \rho^{nrel}$$
$$\delta H = \partial H/\partial D, \quad \delta\rho = \partial\rho/\partial D$$

Thus, relativistic bond length changes may be either rationalized as the variation with bond length (δ) of the expectation value of the relativistic perturbation hamiltonion, $\delta<\Delta H|\rho>$, or alternatively as the relativistic correction (Δ) of the expectation value of the Hellmann–Feynman force, , $\Delta<\delta H|\rho>$.

Concerning the next higher order terms in eq. (4.4), numerical calculations on H_2^+ (where the small spin–orbit contribution to E^{21} has been neglected) yielded

(4.15) $$E^{21}/E^{02} = -0.047\,\lambda^2$$
$$-E^{11}E^{12}/(E^{02})^2 = +1.216\,\lambda^2$$
$$0.75(E^{12})^2 E^{03}/(E^{02})^3 = -1.237\,\lambda^2.$$

As a rule of thumb, relativistic effects in the valence shell of heavy atoms scale with $\lambda = (Z\alpha)^2$ (Schwarz 1990, Pyykö 1988), where Z is the unscreened nuclear charge. Accordingly, we expect significant contributions to ΔD_e from E^{12} and E^{03}, i.e. from the relativistic and the bond length changes of the force constant, Δk and anharmonicity, respectively. We note that Δk of many heavy molecules is not just a small correction, but is a drastic change. For instance, Schwerdtfeger et al. (1989) found for Au_2 at the

SCF level $k^0 = 0.57$, $k^{rel} = 1.37$ N/cm and $\Delta k/k = 1.74$ λ^2 (the values are increased by electron correlation, the experimental value being $k = 2.11$ N/cm).

5. Relativistic Bond Length Changes in the Effective Core Potential Approach. Au_2

Using an effective Hamiltonian means that both the operators and the wavefunctions have been subject to some transformation. Accordingly, physically meaningful quantities like E^{11} should remain invariant, whereas different contributions to it need not be invariant individually. In the pseudopotential or effective core potential approaches, for instance, the *ab initio* relativistic correction operator mainly acting in the vicinity of nuclei is simulated by a centrosymmetric potential for each atom which is attractive in the outer–core/inner–valence–shell region. Consequently H^{11} will be dominantly positive in this region where we have, in general, electronic density accumulation for σ–bonded systems. Therefore, the first term in eq. (3.4) will no longer be negative as in the *ab initio* approach, but will be significantly positive, while the second term will be significantly smaller in magnitude than its *ab initio* counterpart.

In order to present a numerical example, we have performed accurate pseudopotential calculations on Au_2 at the relativistic CI level. The Au atom is treated as a one–electron system with 78 electrons in the polarizable core. The parameters for the effective potential of the core including relaxation and Pauli repulsion were taken from Preuss et al. (1990). The different perturbation energies were calculated by the finite perturbation method at different internuclear distances at the relativistic and nonrelativistic levels. We obtained

$$\left.\begin{array}{l} E^{11}(1) = +0.018 \\ E^{11}(2a) = +0.001 \\ E^{11}(2b) = +0.000 \end{array}\right\} E^{11} \approx 0.018 \text{ a.u.} \quad (5.1)$$

First, $E^{11}(2a)$ and $E^{11}(2b)$ are approximately equal. Second, the numerical results corroborate our intuitive expectations concerning the sign and relative magnitudes of the E^{11}–terms. Third, these results refute the reservations of Christiansen and Ermler (1985) against the double perturbation scheme.

Our calculated relativistic bond length contraction for Au_2 is $-.31$Å the breakup of which into contributions is shown in the following table:

Contributions to the relativistic bond length contraction ΔD_e of Au_2 [Å]

	This work	Christiansen & Ermler (1985)
from $E^{11}(1)$	− 0.30	− 0.47
from $E^{11}(2a)$	− 0.00	+ 0.13
from $E^{11}(2b)$	− 0.01	÷
total	− 0.31	− 0.34

In Christiansen and Ermler's effective core approach, Au is treated as an eleven–electron system with 68 electrons in a frozen core. We note that different effective core potential approaches yield quite different individual contributions, but rather similar results for the final values of E^{11} or ΔD_e.

6. Relativistic Changes of Bond Angles

Nonrelativistic p–AOs are triply degenerate. Those linear combinations which are orthogonal in Hilbert space, and which allow for maximum overlap with neighbouring atoms, are the p_x, p_y, p_z ones. They are orthogonal in real space. Consequently atoms from the right side of the periodic system, which form $p\sigma$–bonds, tend to form 90^0 bond angles.

Spin–orbit coupling breaks the orbital degeneracy. $p3/2$ spinors have less pronounced directional properties, their quadrupole moments being only half as large as those of the p_x, p_y, p_z–AOs. The $p1/2$ spinor even has a spherically symmetric density. Therefore the question arises, whether spin–orbit coupling will modify the bond angles, which are comparatively soft with small bending force constants $k(\theta)$. It is well known that spin–orbit coupling yields significantly weaker and longer bonds in many cases (Hafner et al. 1981, Schwarz 1990). Since bond lenghts D are in general stiffer than bond angles θ, with $k(D) >> k(\theta)$, one may expect quite large relativistic bond angle changes $\Delta\theta$.

The numerical findings in the literature (see the review by Pyykkö 1988) as well as unpublished results from this laboratory seem to indicate, however, that in most cases relativistic changes of bond angles are small, and in general smaller then 2^0 even for the heaviest molecules. Recently, Balasubramanian (1988) found a relativistic bond angle change of 10^0 for an excited triplet state of PbH_2. However, we could not reproduce such a large value, our biggest $\Delta\theta$ for excited states of PbH_2 being smaller than 4^0 (Collignon 1989).

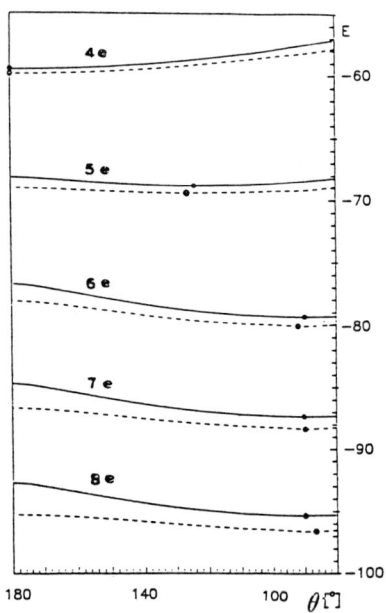

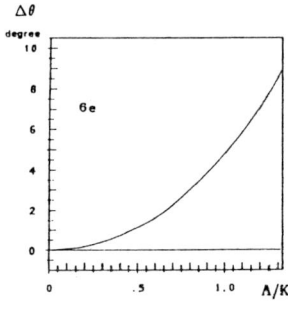

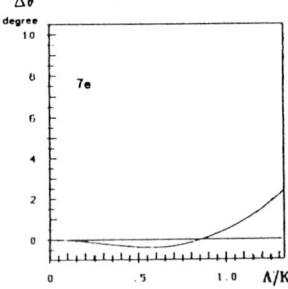

Fig.1 Hückel orbital energy sums for AL_2 systems with 4 to 8 electrons as a function of bond angle with (- - - -) and without (———) spin–orbit coupling. • indicates the potential energy curve minima.

Fig.2 Change of bond angle as a function of spin–orbit coupling over resonance interaction for AL_2 molecules with 6 and 7 electrons

In order to better understand the low sensitivity of bond angles on relativity, especially on spin–orbit coupling, we will use a simple model of the LCAO–MO–Hückel–Walsh type for an AL_2–molecule. We assume a minimal basis of s, p_x, p_y, p_z AOs on the central atom A, and a single σ AO on each ligand L. The two most important parameters are 1) the resonance parameter K for the interaction between the s/p–AOs of A and the σ–AOs on L, and 2) the spin orbit coupling constant Λ of A. For more details see Collignon (1989).

Total energies with and without spin–orbit coupling as a function of the bond angle for systems with 4 to 8 valence electrons are shown in Fig. 1. The relativistic bond angle changes $\Delta\theta$ for 6– and 7–electron systems are plotted as a function of Λ/K in Fig. 2.

For small spin–orbit parameters the interaction is obviously quenched by the molecular field so that the bond angle change is only of second order in Λ. In a basis of real functions p_x, p_y, p_z the matrix-elements of the spin–orbit operator are purely off–diagonal. There is no linear diagonal spin–dependent contribution. Except for nonrelativistic molecular orbital (near–) degeneracy, where first–order <u>degenerate</u> perturbation theory applies, Λ can appear in the energy only in the second order. If the total energy of the molecule is represented as a function of bond angle θ and spin–orbital coupling Λ of the form, as obtained from nondegenerate second order pertubation theory,

(6.1) $\quad E = f(\theta) + g(\theta) \cdot \Lambda^2$

we obtain for the change of bond angle with Λ

(6.2) $\quad d\theta_e/d\Lambda = -(\partial^2 E/\partial\theta\partial\Lambda)/(\partial^2 E/\partial\theta^2) = O(\Lambda^1)$

so that

(6.3) for small Λ: $\theta_e \approx \theta_e^0 + \text{const} \cdot \Lambda^2$, i.e. $\Delta\theta_e = O(\Lambda^2)$

Only if Λ is larger than the covalent resonance interaction K, the corresponding bond angle change will become appreciable and behave as a linear or first order relativistic effect:

(6.4) for large Λ: $\theta_e \approx \tilde{\theta} + \text{const} \cdot \Lambda^1$, i.e. $\Delta\theta_e \approx O(\Lambda^1)$

However, even for the heaviest atoms of the periodic systems, Λ is not much bigger than typical K values. In contrast, the bond length changes which are mainly due to the velocity–mass and Darwin, or **pVp**–terms, are a relativistic first order effect and are detectable already for much smaller relativistic corrections.

7. Summary and Conclusions

We have presented a relativistic pertubation approach which does not lead to singular terms even in higher orders. The corresponding double perturbation approach yields the energy coefficients for a description of relativistic changes of many properties of atoms and molecules. Second order energy coefficients consist of two terms, third order pure, and mixed, perturbation energies consist of 3, and 5 terms, respectively, and fourth order pure and mixed energies, resp., consist of 5 and 7–9 terms. For the m–th order energy, at least perturbation functions of the $[m/2]$–th order are needed. In general, there exist several different equivalent formulae.

Because of the increasing number of terms with increasing order, only second order effects seem to offer the possibility for an intuitive, direct understanding in basic physical terms. Relativistic changes of molecular structure belong to this class of effects. The relevant parameter is the second order mixed pertubation energy, first order in relativity and first order in the change of molecular structure.

The first contribution $E^{11}(1)$ to this E^{11} is the expectation value of the relativistic correction to the Hellmann Feynman force operator for electron–<u>nuclear</u> interaction. In *ab initio* approaches this relativistic correction to the force is mainly due to the Darwin effect. In special cases, also the spin–orbit coupling may become important. Since the Darwin effect is destabilizing, it will in most cases <u>lengthen</u> the nonrelativistic bond distance. Since the Darwin opperator acts only at the nuclear positions (or in the vicinity of the nuclei, if the equivalent expression of the present approach is used) $E^{11}(1)$, will be important for those bonding orbitals which have considerable density at the nuclei. Because of the core orthogonalization tails, this is the case for all σ-bonds, irrespective of whether they originate from s, p, or d valence AOs.

In semiempirical or pseudopotential approaches, however, where one uses an effective or a model Hamiltonian (Durand and Malrieu 1987), the situation is quite different. Here the relativistic correction to the effective Hellmann–Feynman force between atomic <u>cores</u> and valence electrons is predominantly stabilizing and most important in the inner valence-shell region. Therefore the first contribution to E^{11} is positive and bond <u>shortening</u>. This is no contradiction to the *ab initio* result: only the physically relevant total value of E^{11} is uniquely defined, but not the individual formal contribution $E^{11}(1)$.

The individual contributions $E^{11}(1)$ and $E^{11}(2)$ obtained using different effective core potential approaches do not only differ from their *ab initio* counterparts, but also differ from each other. The changes of specific values of $E^{11}(2)$ are compensated by opposite changes of $E^{11}(1)$, and the values of the total E^{11}–correction are (almost) independent of the calculational approaches used.

The second formal contribution $E^{11}(2)$ to relativistic changes of molecular structure can be represented by two equivalent, but very different expressions. Of course, both expressions must yield the same value, $E^{11}(2a) = E^{11}(2b)$, and numerical calculations of sufficiently high accuracy are fully consistent with theory. In the present case, $E^{11}(2a)$ describes the structural change as due to the nonrelativistic Coulomb-Hellmann–Feynman force acting on the relativistic modification of the molecular charge distribution. In many cases, both the atomic and molecular charge distributions contract relativistically. However, it is an unjustified oversimplification to assume that the relativistic <u>molecular</u> charge correctian is given by the sum of relativistic <u>atomic</u> charge corrections. Anyhow, in *ab initio* approaches $E^{11}(2a)$ is strongly bond contractiong for many molecules. It overcompensates the bond expanding term $E^{11}(1)$. In effective core approaches, on the other hand, $E^{11}(2a)$ is smaller in value than $E^{11}(1)$ and in general only modifies the bond contracting $E^{11}(1)$–term slightly.

The alternative is to consider $E^{11}(2b)$, thereby relating the relativistic change of molecular structure to the relativistic energy correction due to the "bond charge", that is the change of electronic charge distribution accompanying the change of internuclear distances. It must be kept in mind that bonding does not only affect the "valence electronic charge distribution" in the outer spatial "valence region" but also in the "inner core region" near the nuclei. These small inner portions of the "bond charge" become decisive for operators which heavily weight the inner core regions. This is the case, for instance, for the nonrelativistic Hellmann–Feynman force (as demonstrated by Spackman and Maslen in 1985), and also for the *ab initio* form of the relativistic correction to the Hamiltionian. Since the stabilizing velocity–mass contribution dominates over the destabilizing Darwin contribution, and because of the factor 2 in eqs.(3.4), $E^{11}(2b)$ turns out to be bond–contracting in the majority of cases, and to overcompensate $E^{11}(1)$. On the other hand, in the effective core potential framework, the relativistic correction of the Hamiltonian is energy lowering in the core/valence border region, where the bond density is already small. Consequently, in the latter frame work, $E^{11}(2b)$ is small or even slightly negative, i.e. bond–expanding.

Summarizing, different theoretical and computational schemes lead to agreeing numerical results, and they lead to logically consistent and equivalent, but quite differently looking — even paradoxical, complementary ways of explanation in physically evident terms. Some care is needed, however, in order not to stumble into conceptual inconsistencies.

Our discussions have focussed upon changes of bond lengths. Although bond lengths are in general stiffer than bond angles, the latter ones are in general less sensitive to relativistic corrections. The qualitative explanation for this fact in the LCAO—MO framework is that in the standard real AO basis the spin—orbit matrix elements are angle dependent but off-diagonal, while the velocity—mass & Darwin matrix elements do not vanish on the diagonal and are in general more bond—length sensitive than bond—angle sensitive.

Acknowledgement. A.R. thanks for hospitality at the University of Siegen; W.H.E.S. thanks for hospitality at the Higher Pedagogical School of Olsztyn. We thank for financial support by Deutsche Forschungsgemeinschaft, by Fonds der Chemischen Industrie, and by the Polish Academy of Sciences.

References

Balasubramanian, K., 1988, J. Chem. Phys. 89:5731
Christiansen, P.A., and Ermler, W.C., 1985, Mol. Phys. 55:1109
Collignon, G., 1989, Thesis, University of Siegen; Collignon, G., and Schwarz, W.H.E., to be submitted to Chem. Phys. Let.
Dalgarno, A., and Stewart, A.L., 1956, Proc. Roy. Soc. A 238:269
Durand, P., and Malrieu, J.P., 1987, Adv. Chem. Phys. 67:321
Hafner, P., Habitz, P., Ishikawa, Y., Wechsel—Trakowski, E., and Schwarz, W.H.E., 1981, Chem. Phys. Let. 80:311
Jankowski, K., and Rutkowski, A., 1987, Phys.Scripta 36: 464
Jankowski, K., and Rutkowski, A., 1989, J. Phys. B 22: 2669
Kutzelnigg, W., 1989, Perturbation theory of relativistic effects, in "Quantum Chemistry: Basic Aspects, Current Trends", Elsevier, Amsterdam; Z. Phys. D 11:15; 1990, Z. Phys. D 12:27
Kutzelnigg, W., and Schwarz, W.H.E., 1982, Phys. Rev. A 26:2361
Levy—Leblond, J.M., 1967, Commun. Math. Phys. 6:288
Moss, R.E., 1973, Advanced Molecular Quantum Mechanics, Chapman and Hall, London
Preuss, H., et al., 1990, Arbeitbericht 27:25, Institut für Theoretische Chemie, Universität Stuttgart; see also Schwerdtfeger et al. 1989
Pyykkö, P., 1988, Chem. Rev. 88:563
Pyykkö, P., Snijders, J.G., and Baerends, E.J., 1981, Chem. Phys. Let. 83:432
Rose, S.J., Grant, I. and Pyper, N.C., 1978, J. Phys. B 11:1171
Rutkowski, A., 1986a, J. Phys. B 19:149
Rutkowski, A., 1986b, J. Phys. B 19:3431
Rutkowski, A., 1986c, J. Phys. B 19:3443
Rutkowski, A., and Schwarz, W.H.E., 1990, Theor. Chim. Acta 76:391
Schwarz, W.H.E., 1987, Phys. Scripta 36:403
Schwarz, W.H.E., 1990, in "Theoretical Models of Chemical Bonding", Ed. Maksic, Z., Springer Berlin
Schwarz, W.H.E., van Wezenbeek, E.M., Baerends, E.J., and Snijders, J.G., 1989, J. Phys. B 22:1515
Schwerdtfeger, P., Dolg, M., Schwarz, W.H.E., Bowmaker, G.A., and Boyd, P.D.W., 1989, J. Chem. Phys. 91:1762
Spackman, M.A., and Maslen, E.N., 1985, Acta Cryst. A 41:347

BASIS SET EXPANSION DIRAC-FOCK SCF CALCULATIONS AND MBPT REFINEMENT

Yasuyuki Ishikawa

Department of Chemistry
The University of Puerto Rico, Rio Piedras
Puerto Rico 00931 U.S.A.

INTRODUCTION

The majority of relativistic electronic structure calculations on atoms and ions have been done using finite difference numerical methods (Desclaux, 1973; Grant et al., 1980). They will remain important, especially in atomic calculations, because effective means have been devised for augmenting them to account for correlation effects by, viz., multiconfiguration expansion and many-body perturbation theory (MBPT) techniques.

Recently the basis set expansion Dirac-Fock (DF) self-consistent field (SCF) method has been developed sufficiently to approach numerical methods in accuracy (Quiney, Grant and Wilson, 1989a; Ishikawa et al., 1985, 1988; Goldman, 1989). The DF basis set expansion method provides a simple alternative to direct numerical approaches to atomic structure calculations, and it can be extended in a straightforward way to molecular calculations.

The following section outlines our research on the Gaussian basis set expansion DF method (Ishikawa and Quiney, 1987; Ishikawa and Sekino, 1988, 1990) for efficiently determining relativistic SCF wavefunctions. The final section outlines the relativistic MBPT method for improving DF wavefunctions to account for Coulomb correlation effects.

DIRAC-FOCK GAUSSIAN BASIS SET EXPANSION CALCULATIONS

Gaussian-type functions (GTF) are employed in most multicenter Hartree-Fock (HF) basis set expansion calculations for reasons of computational economy, and a number of efficient Gaussian integral evaluation packages are available. However, it is also well-known that the behavior of GTF near the origin is inappropriate to properly represent a cusp in the familiar point nucleus approximation. Consequently, an expansion in GTF, to represent that behavior, must employ a relatively large number of functions (Wilson, 1987).

The use of GTF in DF SCF calculations (Malli, 1979) on atoms and molecules is likely to prove more difficult than in the nonrelativistic case if nuclei are modelled as particle points. The cusp at a point nucleus is infinite in relativistic wavefunctions. For heavy atoms and heavy-atom-containing molecules, the high nuclear charge creates a more pronounced cusp (Schwarz and Wallmeier, 1982). Therefore a number of GTF of high exponent must be included in a basis set to mimic, in the least-squares sense, the wavefunction near the origin (Ishikawa et al., 1985).

In the DF scheme, the behavior of an electron in a central field potential, V, is described by a radial DF equation of the form (Kim, 1967),

$$H_r \phi_{nk}(r) = \epsilon_{nk} \phi_{nk}(r) \qquad (1)$$

where

$$H_r = \begin{bmatrix} V & c\pi \\ c\pi^+ & V-2c^2 \end{bmatrix}$$

with $\quad \pi = -d/dr + k/r \quad$ and $\quad \pi^+ = d/dr + k/r \qquad (2)$

Here

$$\phi_{nk}(r) = \begin{pmatrix} P_{nk}(r) \\ Q_{nk}(r) \end{pmatrix} \qquad (3)$$

The radial functions $P_{nk}(r)$ and $Q_{nk}(r)$ are referred to as the upper and lower components, respectively. c is the speed of light.

In calculations on $s_{1/2}$ states, P(r) is expanded in 1s GTF, $\{x^L_i(r)\} = \{r \exp(-\alpha_i r^2)\}$,

$$P(r) = \sum_i r \exp(-\alpha_i r^2) \xi_i \qquad (4a)$$

where N is the number of GTF. In previous studies of hydrogenic systems, P(r)/r and Q(r)/r have been expanded in the same sets of from 1 to 14 1s GTF. With each expansion, the exact $1s_{1/2}$ energy was obtained as a minimum (Ishikawa et al., 1985). Such results are obviously spurious; a single GTF is a grossly incorrect form for the solution of either the Dirac or the Schrödinger equation.

A criterion of basis set "balance"

$$x^S_i(r) = \pi_k^+ x^L_i(r) \qquad (4b)$$

has been proposed (Lee and McLean, 1982) to avoid spurious solutions as well as blatant "bounds failure." This "balance" criterion suggests that the lower component be expanded in $\{x^S_i(r)\} = \{\pi_k^+ r \exp(-\alpha_i r^2)\} = \{r^2 \exp(-\alpha_i r^2)\}$,

$$Q(r) = \sum_i r^2 \exp(-\alpha_i r^2) \eta_i \qquad (5)$$

where in equations (4a) and (4b), $\{\xi_i\}$ and $\{\eta_i\}$ are linear variation parameters. The balanced GTF expansions, (4a) and (4b), have been used in the point nucleus approximation (Aerts and Nieuwpoort, 1985; Ishikawa et al., 1985).

With the point representation of the nucleus, both the exact $P(r)/r$ and $Q(r)/r$ have an infinite cusp at the origin. Variational solutions approximated by (4a) and (4b) fail to satisfy this condition because $P(r)/r$ approximated by (4a) has a finite value and zero slope at the origin while $Q(r)/r$ as represented by (4b) vanishes (Ishikawa and Quiney, 1987). Expansions (4a,b) therefore require some high exponent GTF to approximate the behavior of the exact wavefunction near the origin, and thereby to reproduce the relativistic kinematics. In calculations on $p_{1/2}$ states, the intuitive choice for the expansion of $P(r)/r$ is the 2p-radial GTF, $\{r \exp(-\alpha_i r^2)\}$,

$$P(r) = \sum_i r^2 \exp(-\alpha_i r^2) \xi_i \qquad (6a)$$

$Q(r)/r$ is expanded in "extended" kinetically balanced GTF (Aerts and Nieuwpoort, 1985), $\{\exp(-\alpha_i r^2)\}$ and $\{r^2 \exp(-\alpha_i r^2)\}$,

$$Q(r) = \sum_i r \exp(-\alpha_i r^2) \eta_i + \sum_j r^3 \exp(-\alpha_j r^2) \omega_j \qquad (6b)$$

If $P(r)$ is expanded in N GTF, (6b) requires that $Q(r)$ be expanded in 2N GTF. The balanced GTF expansion, (6a,b), has been used in atomic and molecular DF calculations in the point nucleus approximation (Aerts and Nieuwpoort, 1985). The variational solution approximated by (6a,b), however, fails to satisfy the infinite cusp condition. $P(r)/r$ approximated by the expansion (6a) vanishes at the origin, whereas $Q(r)/r$ represented by (6b) has a finite value and zero slope there (Ishikawa and Quiney, 1987).

The balanced basis expansions, (4a,b) and (6a,b), are well defined only as c approaches infinity (Ishikawa and Quiney, 1987; Quiney, 1988) because they are derived from criterion (5), valid only in the limit. Recent work has focussed on choosing basis sets for upper and lower components of the wavefunction which are balanced in the sense that the relativistic wavefunction approaches the wavefunction obtained with the Schrödinger Hamiltonian as the speed of light approaches infinity (Lee and McLean, 1982; Mark and Schwarz, 1982; Dyall et al., 1984; Aerts and Nieuwpoort, 1985; Stanton and Havriliak, 1984).

It is not necessarily true that the use of a basis set which is balanced in this sense must lead to a solution of the DF equations, nor that spurious solutions are precluded for finite c. The following example illustrates the point. In calculations on the $1s_{1/2}$ state of the hydrogenic systems, use of a single Slater-type basis function of integer power, $r \exp(-\xi r)$, in the expansions of both the $P(r)$ and $Q(r)$ gives a balanced basis expansion, and produces the exact $1s_{1/2}$ energy as a minimum when $\xi = Z/[1-(Z/c)^2]^{1/2}$. Although the infinite cusp is absent, the variational solution is a reasonably accurate one for small nuclear charge, Z. However, the solution is an incorrect form for the wavefunction for the large Z case in which the infinite cusp is more pronounced (Ishikawa and Quiney, 1987). This example illustrates how failing to impose correct boundary conditions on basis functions can lead to spurious solutions. Achieving the limiting energy alone is no guarantee that the variational solution has converged to the true wavefunction. Attention must be given to satisfying the relativistic boundary conditions for finite c (Quiney, 1988; Ishikawa and Sekino, 1990).

Ishikawa et al. (1987, 1990) and Quiney (1988) have stressed that the criterion for basis set balance, (5), is well defined only as c approaches infinity; it can lead to a relationship between upper and lower component basis sets which need not produce the correct wavefunction at finite c. For an electron in an external field V, the radial upper and lower components of the Dirac four-spinor are related by

$$Q(r) = [2c^2 - (V - \epsilon)]^{-1} c\pi_k^+ P(r) \qquad (7)$$

Basis set balance criterion (5) emerges from Eq. (7) as $c \to \infty$, and thus is valid at the nonrelativistic limit; the wavefunction is not thereby constrained to satisfy the relativistic boundary when the speed of light is finite (Ishikawa and Quiney, 1987).

Ishikawa et al. (1987, 1990) and Quiney (1988) have examined the implications which imposition of correct boundary conditions on DF solutions may have on basis set selection. With a point representation of the nucleus, basis sets of Slater- or Gaussian-type functions with integer power of r are not suitable for DF calculations because they are unable to satisfy the relativistic constraint (Eq.(7)), and thus, cannot reproduce the relativistic boundary.

With a point representation of the nucleus, a form of basis function which reproduces the relativistic boundary for finite c and is "kinetically balanced" as c approaches infinity, is the noninteger "S-spinor" (Quiney et al., 1989a; Grant and Quiney, 1988),

$$x^T(r) = r^p \exp(-\zeta r)[A^T + B^T \zeta r], \quad p = +(k^2 - Z^2/c^2)^{1/2}$$

where T = L or S, for the radial upper or lower components, respectively. Here A^T and B^T are functions of k, nuclear charge Z, and c.

The precise conditions under which the "kinetically balanced" GTF of (4a,b) and (6a,b) can be made to reproduce the relativistic boundary associated with the point nucleus model for finite c is not known. The infinite cusp, however, is an artifact of the point nucleus model, and the boundary condition on the wavefunction at the origin can be changed by abandoning that model. In the change the difficulty of approximating the boundary with GTF is eliminated. In recent work, we have emphasized alteration of the boundary conditions such that "kinetically balanced" GTF become the best form for basis functions (Ishikawa et al., 1985; Ishikawa and Quiney, 1987). Representing the nucleus as a finite body of uniform proton charge does that. In the finite nuclear model "kinetically balanced" GTF of integer power of r are appropriate basis functions because imposition of the finite nuclear boundary results in a solution which is Gaussian at the origin.

In the uniform charge model of the finite nucleus, the potential inside the nucleus is

$$V_{in} = -(Z/2R)(3 - r^2/R^2)$$

Here R, the nuclear radius, is given by

$$R = 2.2677 \times 10^{-5} A^{1/3}$$

with A the atomic mass number. The potential outside the nucleus is Coulombic, $V_{out} = -Z/r$. With this representation of the potential, the exact $s_{1/2}$ solutions near the origin are (Bohr and Weisskopf, 1950)

$$P(r)/r = 1 + g_2 r^2 + g_4 r^4 + \ldots \tag{8a}$$

$$Q(r)/r = f_1 r + f_3 r^3 + \ldots \tag{8b}$$

so that, for α arbitrary parameters (Ishikawa and Quiney, 1987),

$$P(r) = r + g_2 r^3 + \ldots \approx r \exp(-\alpha r^2)$$

$$Q(r) = f_1 r^2 + f_3 r^4 + \ldots \approx r^2 \exp(-\alpha r^2)$$

In (8a,b), $P(r)/r$ has a finite value and zero slope at the origin, whereas $Q(r)/r$ vanishes there.

The $p_{1/2}$ solutions near the origin may be represented as,

$$P(r)/r = r + g_3 r^3 + \ldots \tag{9a}$$

$$Q(r)/r = f_0 + f_2 r^2 + \ldots \tag{9b}$$

Then the upper and lower components may be written

$$P(r) = r^2 + g_3 r^4 + \ldots \approx r^2 \exp(-\alpha r^2)$$

$$Q(r) = f_0 r + f_2 r^3 + \ldots \approx r \exp(-\alpha r^2) + r^3 \exp(-\alpha r^2)$$

In Eqs. (9a,b), $P(r)/r$ vanishes at the origin, whereas $Q(r)/r$ has finite value and zero slope there. The behaviors of the exact solutions, (8a,b) and (9a,b), near the origin are precisely what the balanced GTF expansions, (4a,b) and (6a,b), provide. Thus a Gaussian basis, balanced in the sense we have suggested, provides a natural description of the wavefunction within the nucleus (Ishikawa et al., 1985).

Klahn and Morgan (1984) have postulated that the rate of convergence of a variational calculation is determined by how quickly the basis functions replicate the analytic structure of the unknown function one is trying to approximate. A GTF expansion mimics exactly the behavior of the wavefunction near the origin of a uniformly charged spherical nucleus. This is the very reason why DF GTF expansions, with a finite nucleus, exhibit fast and smooth convergence.

Using the expansion schemes, (4a,b) and (6a,b) together with the uniform charge model of the finite nucleus, we have performed calculations on neon, systematically enlarging the GTF basis sets (Ishikawa and Sekino, 1988). The DF spinors have been expanded in GTF basis sets of van Duijneveldt, optimized in nonrelativistic calculations. Table 1 shows the orbital and total energies. The total energy in each instance is an upperbound to the numerical limit. It decreases steadily toward the numerical limit as expansion size increases. The largest basis set, (14s9p), gives a total energy of -128.69166 a.u., near the numerical limit (-128.6919 a.u.).

Matsuoka and Huzinaga (1987) have proposed an expansion scheme for k > 0 states,

$$Q_k(r) = \sum_i [(n+1)r^{n-1} - 2\alpha_i r^{n+1}] \exp(-\alpha_i r^2) \eta_i, \quad n = k+1 \quad (6c)$$

This form of expansion was proposed by Stanton (1984), who observed bounds failures in calculations on several $p_{1/2}$ states in which the point nuclear approximation was employed. The extent to which variational failures occur is greater for larger Z, where the cusp is pronounced; for small Z the collapse is not noticeable. The Q(r) as expanded in (6c) cannot satisfy the relativistic boundary associated with the point nucleus for a finite value of c because $Q(r)/r \rightarrow$ (nonzero) constant as $r \rightarrow 0$, whereas the exact $Q(r)/r \rightarrow \infty$ as $r \rightarrow 0$, in the point nucleus approximation. Failure to satisfy the relativistic boundary is a consequence of using expansion (6c), derived from the kinetic balance criterion (5), and valid only for $c \rightarrow \infty$ (Ishikawa et al., 1990).

When the point nucleus approximation is forfeited in favor of the finite uniform charge representation, expansion (6c) becomes more appropriate. However, it is not flexible enough to reproduce the correct behavior near the origin. The "contraction" coefficients in (6c), derived in the nonrelativistic limit, are unsuitable at finite c. Consequently, DF calculations which employ expansion (6c) are susceptible to variational failure.

Table 1. Orbital and total energies of Ne (a.u.).

Basis set	$1s_{1/2}$	$2s_{1/2}$	$2p_{1/2}$	$2p_{3/2}$	T.E.[b]
7s3p	-32.8017	-1.90669	-0.78821	-0.78470	-128.42597
9s4p	-32.8107	-1.92431	-0.83422	-0.83011	-128.63241
10s5p	-32.8139	-1.93232	-0.84754	-0.84317	-128.67608
11s6p	-32.8163	-1.93507	-0.85157	-0.84757	-128.68928
12s7p	-32.8174	-1.93572	-0.85255	-0.84802	-128.69043
13s8p	-32.8175	-1.93584	-0.85276	-0.84822	-128.69133
14s9p	-32.8175	-1.93585	-0.85280	-0.84825	-128.69166
Limit[a]	-32.8175	-1.93585	-0.85283	-0.84827	-128.69194

[a]Computed using Desclaux's finite difference DF program.
[b]Total energy.

In contrast, the uncontracted expansion (6b) can represent the exact wavefunction near the origin because of the flexibility conferred by parameters $\{\eta_i\}$ and $\{\omega_j\}$ in the variational process (Quiney, 1988).

In order to examine the bounding property associated with the two expansion schemes, we have performed two sets of DF calculations on highly ionized and neutral atom species using either (6a,b) or (6a,c) for $k > 0$ symmetries along with (4a,b) for $k < 0$ symmetries. The method which employs "uncontracted" expansions, (4a,b) and (6a,b) is referred to as the method 1, and that which employs "contracted" expansions, (4a,b) and (6a,c) as method 2.

Table 2 displays the results of calculations on the six-electron neutral C atom and Hg^{+74} ion in their closed-shell configuration, $(1s_{1/2})^2(2s_{1/2})^2(2p_{1/2})^2$. The total energies of these systems computed by method 1 agree with the numerical DF limits. The energies of neutral C atom computed using method 2 agree exactly with those computed with method 1, showing no sign of variational failure. However, the energies of Hg^{+74} ion computed with method 2 exhibit variational failure; the calculated total energy lies 0.048 a.u. below the numerical limit.

For both methods 1 and 2, there is a clean separation of negative and positive energy branches of the spectra of the six-electron systems (Ishikawa et al., 1990). The results indicate that the basis functions for the large and small components need not be matched in pairs to obtain a clean separation of the spectrum.

Matsuoka and Huzinaga (1987) and Mohanty and Clementi (1989) have performed DF GTF basis set calculations on a series of neutral atoms using expansion method 2. Their results show no sign of variational failure for neutral atoms up to Rn, whereas, in the present study, the highly ionized, high-Z species, Hg^{+74}, shows failure clearly. To determine whether calculations on neutral Hg exhibit variational failure, we have used method 2 in DF calculations, enlarging the basis set systematically (Ishikawa et al., 1990). As expansion length increases, the total energy converges monotonically to the numerical limit. In each instance, the computed total energy is an upperbound to the numerical limit. Our results on the neutral Hg atom, taken with the results of Mohanty and Clementi (1989) on the neutral heavy elements indicate that, even when Z is large, neutral species do not exhibit a tendency to variational collapse. The shielding effect of a large number of electrons seems to reduce the magnitude of variational collapse to negligible size.

In summary, expansion method 1 does not show any sign of variational failure in any application. It can be applied to highly ionized, high-Z species, for which high precision calculations are often essential. A disadvantage is that, as the basis set becomes large, for spinors of $k > 0$, calculations using the method become unwieldy. An additional disadvantage lies in the fact that the basis functions for the upper and lower components are not matched in pairs, and therefore half the negative energy solutions do not have a positive energy counterpart.

Table 2. Results for neutral C atom and Hg^{+74} (a.u.).

Expansion method[a]	C	Hg^{+74}
1	-37.6572	-10369.3180
2	-37.6572	-10369.3695
Numerical limit[b]	-37.6574	-10369.3211

[a] Expansion lengths for the neutral carbon and mercury ion are respectively, 16 and 18.
[b] Computed using Desclaux's finite difference DF program.

Slow convergence in perturbation theory calculations of properties which involve summation of the negative energies could result. Method 2 appears to be free of significant variational failure when applied to neutral species. It is computationally simpler than method 1, and hence suited to calculations on neutral heavy atom species. A disadvantage is that it may show variational failure in studies of highly ionized, high-Z species. The basis functions for the upper and lower components are, however, matched in pairs, and a clean separation of the spectrum is obtained.

RELATIVISTIC MANY-BODY PERTURBATION THEORY CALCULATIONS

Dirac-Fock wavefunctions provide results which may be directly compared to those obtained with the most popular nonrelativistic computational method, the Hartree-Fock method. But accurate treatment of atomic and molecular systems requires that DF treatments be refined to include the effects of electron correlation.

Accurate treatment of many-electron atoms requires techniques which are computationally tractable and capable of recovering nearly all of the correlation energy. Two methods proven successful in treating relativistic many-body problems are the variational and perturbative types, configuration interaction (CI) (Beck and Cai, 1988) and many-body perturbation theory (MBPT) (Johnson et al., 1987; Quiney et al., 1987). MBPT developed from field theory and QED and thus provides a natural framework within which to treat relativistic effects. It was introduced into atomic physics by Kelly (1963). MBPT is employed to calculate electronic properties because it is computationally efficient, and because the formalism is well developed (Kelly, 1969). It is size consistent (Bartlett, 1989), but not variational. An advantage is that perturbation theoretical formulations are most often applied to account for fine effects; MBPT is therefore a natural form with which to approach the correlation problem. MBPT is probably best suited for the analysis of correlation effects in many-electron systems because it permits a decomposition of the correlation energy of a system into component contributions.

MBPT is less studied with basis sets than with the finite-difference approach. Basis sets, however, have some advantage over finite-difference methods in MBPT calculations in that the infinite summations over bound and continuum states which appear in finite difference calculations are replaced in basis set calculations with summations over a finite pseudospectrum (Johnson and Sapirstein, 1986; Quiney et al., 1987). MBPT up to third order has been applied to relativistic many-body calculations using basis set expansion DF wavefunctions as reference states (Johnson et al., 1987; Quiney et al., 1987; Ishikawa et al., 1989). It is a tractable technique for recovering correlation energy. In highly ionized systems the perturbation expansion converges rapidly; a substantial fraction of the correlation energy in these systems is recovered with the second order term (Ishikawa et al., 1989). Thus, a finite basis set expansion approach to relativistic DF and MBPT calculations may be regarded as an accurate and versatile approximation technique.

The usual starting point for the development of relativistic MBPT calculations is the time-honored Dirac-Coulomb Hamiltonian (H_{DC}). The leading effects of transverse photon exchange, however, may be included in the zero-order Hamiltonian by adding the Breit operator to the instantaneous Coulomb operator (Quiney et al., 1989b). This approach has the advantage that all effects through order α^2 are included in the Hamiltonian. The use of such a zero-order Hamiltonian in variational calculations naturally leads to DF equations which include the Breit interaction. As Quiney, Grant and Wilson (1987) have pointed out, incorporating the Breit term in the SCF process results in the Coulomb and Breit interactions being included in the SCF potential to the same order. Thus the instantaneous Coulomb and Breit terms are treated on an equal basis in the two-electron interaction for relativistic many-body perturbation theory calculations.

The starting point for our development of relativistic MBPT calculations in c-number theory is the relativistic "no-pair" Dirac-Coulomb Hamiltonian (H_{DC}^+), originally introduced to avoid the "continuum dissolution" problem associated with relativistic many-body calculations (Sucher, 1980).

$$H_{DC}^+ = \sum_i h_D(i) \quad + \quad L_+ (\sum_{i<j} 1/r_{ij}) L_+$$

where $L_+ = L_+(1)...L_+(N)$. H_{DC} is the sum of Dirac one-electron Hamiltonians

$$h_D = c\alpha \cdot p + \beta c^2 + V_{nuc}$$

and the instantaneous Coulomb interaction between electrons is

$$V_{e-e} = \sum_{i<j} 1/r_{ij}$$

with $L_+(i)$ the projection operator onto the space spanned by the positive energy eigenfunctions of the DF operator. The no-pair Dirac-Coulomb Hamiltonian, then, treats one-electron effects relativistically, two-electron effects being "nonrelativistic". The correlation energy induced by the Breit interaction is significant for inner-shell spinors of heavier systems. In the present work, however, we neglect that interaction.

Negative energy states, as part of the complete set of states, play a role in many-body calculations. Contributions from the negative energy states due to creation of virtual electron-positron pairs, though, are small, of order α^3, and are neglected here.

In q-number theory, the normally ordered second-quantized "no-pair" Hamiltonian

$$H_N = \sum_{r,s} f_{rs}[r^+s] + 1/4 \sum_{\substack{r,s \\ t,u}} <rs \| tu> [r^+s^+ut]$$

may be employed in relativistic MBPT calculations (Quiney, 1988; Sucher, 1987). Here f_{rs} and $<rs \| tu>$ are, respectively, one-electron DF and antisymmetrized two-electron Coulomb interaction matrices over the Dirac four-component spinors, r, s, t, and u. Normal ordering implies that, in the vacuum state, annihilation operators are moved to the right of creation operators as if all anticommutators vanish. The Fermi level is shifted to the highest occupied positive-energy state. The creation operator then appears to the right of a normally ordered set when it refers to an occupied positive-energy state, while the annihilation operator remains on the right for a positive-energy virtual state. In this form the no-pair Hamiltonian is restricted to contributions from the positive-energy branch of the spectrum.

The no-pair approximation is a natural extension of nonrelativistic many-body theory (Sucher, 1987). Diagrammatic summation is done within the subspace of the positive energy branch of the spectrum. In linked-cluster MBPT, both Brandow and Hugenholtz diagrams utilize antisymmetry, and thus provide compact diagrammatic notation. With Goldstone diagrams, however, there is no explicit use of antisymmetry. The diagrams involve only ordinary one- and two-electron matrix elements which connect one-particle states. They are, therefore, most closely related to the original Feynman diagrams. In recent studies, Goldstone diagrams have been summed to compute Dirac-Coulomb correlation corrections up to third order (Quiney et al., 1988; Ishikawa et al., 1989).

There is numerical evidence that, for all but the lightest systems, the second order MBPT correction represents the bulk (> 95%) of the correlation energy. In Ne, for instance, the second order correction recovers more than 99% of the "experimental" total correlation energy. In the present study on the neutral He, Ne, Ar, and Kr atoms, Goldstone diagrams are summed to second order.

In nonrelativistic MBPT, the use of single configuration Hartree-Fock wavefunctions as a reference for perturbation theory calculations is well justified by the fact that the most commonly used definition of correlation energy refers to the difference between the Hartree-Fock and exact nonrelativistic energies. In our relativistic MBPT calculations, single configuration Dirac-Fock ground state reference wavefunctions are employed as a starting point. Virtual spinors in our studies are calculated in the field of the nucleus and all electrons (V^N potential). Our relativistic MBPT calculations may, then, be directly compared to those obtained with the most popular nonrelativistic MBPT calculations, facilitating an understanding of relativistic effects, correlation effects, and their interaction in heavier systems.

We have performed relativistic MBPT calculations on rare gas atoms in which the basis sets employed have been systematically expanded. Representative results of DF and second order MBPT calculations on He, Ne, Ar, and Kr atoms with even- or well-tempered GTF basis sets are shown in Table 3. The speed of light used to obtain the relativistic results was c = 137.0370 a.u. To simulate the non-relativistic limit, a value of c of 10^4 a.u. was used. An even-tempered (14s10p8d7f6g5h4i) GTF basis set was employed in the He calculations. For the Ne, Ar and Kr calculations basis sets, respectively, of (14s10p8d6f5g4h4i), (16s11p9d7f6g5h) and (16s13p8d7f6g5h) well-tempered GTF were used.

Lindroth (1988) has used relativistic pair equations to compute the second order Coulomb correlation correction (E_2) of He using a partial wave expansion with L_{max} = 4. E_2 thus obtained was -0.036965 a.u. Our second order energy, -0.36911 a.u. obtained with L_{max} = 4 (a 14s10p8d7f6g basis set) agrees with the value reported by Lindroth. The error in the second order energies due to basis set truncation is about 0.15%.

Using finite difference pair equations, Lindgren and Salomonson (1980) have computed the nonrelativistic second order correlation energy of Ne with a partial-wave expansion having L_{max} = 6. The second order energy obtained, -0.38355 a.u., may be compared to our nonrelativistic result, -0.38018 a.u. Das et al. have estimated the "experimental" correlation energy of Ne to be -0.3890 a.u.

Quiney et al. (1989b) have computed the second order Dirac-Coulomb correlation energy of Ar with a large STF basis set consisting of 17 s-spinors of each symmetry type, up to L_{max} = 3. The point nucleus approximation was employed in the calculations. A second order energy of -0.639424 a.u. was reported. This value may be compared with our second order energy, -0.63383 a.u. obtained with L_{max} = 3. Assuming the effect on E_2 of treating the nucleus differently is small, the discrepancy of 0.0056 a.u. between the results is best attributed to basis set truncation error in our calculation.

For every system listed in Table 3, the Coulomb correlation energy computed using the no-pair Dirac-Coulomb Hamiltonian is different from that computed at the nonrelativistic limit. The difference represents interference between relativistic and correlation effects. In helium this nonadditive contribution is about 10^{-6} a.u.; it increases in Kr to 10^{-2} a.u.

SUMMARY AND CONCLUSIONS

Theoretical methods developed for electronic structure studies of truly many-electron atoms must be able to account for relativistic and electron correlation effects. They must be computationally efficient because they will eventually be applied to high-Z neutral atoms. And they should be capable of being extended in a straightforward manner to the study of molecules. The present study has employed one such approach, the solution of the Dirac-Fock equations by expansion in Gaussian-type functions and many-body perturbation refinement of the solutions.

Table 3. Second order Coulomb correlation energies of He, Ne, Ar and Kr

	Relativistic		Nonrelativistic limit	
	E_{DF}	E_2	E_{HF}	E_2
He	-2.861812	-0.037132	-2.861679	-0.037135
Ne	-128.6916	-0.38041	-128.5468	-0.38018
Ar	-528.6815	-0.68226	-526.8157	-0.68128
Kr	-2788.809	-1.71957	-2752.033	-1.70683

Dirac-Fock SCF calculations by conventional expansion in a basis exhibit variational failure. The major barrier to the development of <u>ab initio</u> relativistic quantum chemistry in the past decade has been the lack of a basis set expansion scheme which prevents variational failure and the appearance of spurious solutions. The work we report here should convey the idea that no substantial problems remain; our Dirac-Fock GTF basis expansion method together with its MBPT refinement is a reliable, accurate and versatile approximation in relativistic quantum mechanics.

References

Aerts, P.J.C. and Nieuwpoort, W.C., 1985, On the use of Gaussian basis sets to solve the Hartree-Fock-Dirac equations, <u>Int. J. Quantum Chem.</u>, S19:267.

Bartlett, R.J., 1989, Coupled-cluster approach to molecular structure and spectra: A step toward predictive quantum chemistry, <u>J. Phys. Chem.</u>, 93:1697.

Beck, D.R. and Cai, Z., 1988, A relativistic many body theory of electric dipole oscillator strengths with application to Tl^+ $6s^2$ -- 6s6p, <u>Phys. Rev.</u>, A37:4481.

Beck, D.R., 1988, Relativistic and many-body effects in first-row transition-metal negative ions, <u>Phys. Rev.</u>, A37:1847.

Bohr, A. and Weisskopf, V.F., 1950, The influence of nuclear structure on the hyperfine structure of heavy elements, <u>Phys. Rev.</u>, 77:94.

Desclaux, J.P., 1973, Relativistic Dirac-Fock expectation values for atoms with Z = 1 to Z = 120, <u>At. Data Nucl. Data Tables</u>, 12:311.

Dyall, K.G., Grant, I.P., and Wilson, S., 1984, The Dirac equation in the algebraic approximation II: Extended basis set calculations for hydrogenic ions, <u>J. Phys.</u>, B17:1201.

Goldman, S.P., 1989, Generalized Laguerre representation: Application to relativistic two-photon decay rates, Phys. Rev., A40:1185.

Grant, I.P., McKenzie, B.J., Norrington, P.H., Mayers, D.F., and Pyper, N.C., 1980, An atomic multiconfiguration Dirac-Fock package, Comput. Phys. Commun., 21:207.

Ishikawa, Y., Baretty, R., and Binning, R.C., 1985, Relativistic Gaussian basis set calculations on one-electron ions with a nucleus of finite extent, Chem. Phys. Lett., 121:130

Ishikawa, Y., Baretty, R., and Binning, R.C., 1985, Gaussian basis for the Dirac-Fock discrete basis expansion calculations, Int. J. Quantum Chem., S19:285.

Ishikawa, Y. and Quiney, H.M., 1987, On the use of an extended nucleus in Dirac-Fock Gaussian basis set calculations, Int. J. Quantum Chem., S21:523.

Ishikawa, Y. and Sekino, H., 1988, On the use of Gaussian-type functions in Dirac-Fock basis set expansion calculations, Int. J. Quantum Chem., S22:457

Ishikawa, Y., Sekino, H., and Binning, R.C., 1989, Relativistic many-body perturbation theory calculations on Be, Ne^{+6}, Ar^{+14} and Ne, Chem. Phys. Lett., 160:206.

Ishikawa, Y. and Sekino, H., 1990, Variational bounds in Dirac-Fock basis set expansion calculations, Chem. Phys. Lett., 165:243.

Johnson, W.R. and Sapirstein, J., 1986, Computation of second order many-body corrections in relativistic atomic systems, Phys. Rev. Lett., 57:1126.

Johnson, W.R., Blundell, S.A., and Sapirstein, J., 1987, Many-body perturbation-theory calculations of energy levels along the sodium isoelectronic sequence, Phys. Rev., A35:3218.

Kelly, H.P., 1963, Correlation effects in atoms, Phys. Rev., 131:684

Kelly, H.P., 1969, Applications of many-body diagram techniques in atomic physics, Adv. Chem. Phys., 14:129.

Kim, Y.-K., 1967, Relativistic self-consistent field theory for closed-shell atoms, Phys. Rev., 154:17

Klahn, B. and Morgan III, J.D., 1984, Rate of convergence of variational calculations and of expectation values, J. Chem. Phys., 81:410.

Lee, Y.S. and McLean, A.D., 1982, Relativistic effects on R_e and D_e in AgH and AuH from all-electron Dirac-Hartree-Fock calculations, J. Chem. Phys., 76:735.

Lindgren, I. and Salomonson, 1980, A numerical coupled-cluster procedure applied to the closed-shell atoms Be and Ne, Phys. Scr., 21:335.

Lindroth, E., 1988, Numerical solution of the relativistic pair equation, Phys. Rev., A37:316.

Malli, G., 1979, Spherical Gaussian basis sets in relativistic quantum chemistry, Chem. Phys. Lett., 68:529.

Mark, F. and Schwarz, W.H.E., 1982, New representation of the $\alpha \cdot p$ operator in the solution of Dirac-type equations by the linear-expansion method, Phys. Rev. Lett., 48:673.

Matsuoka, O. and Huzinaga, S., 1987, Relativistic well-tempered Gaussian basis sets, Chem. Phys. Lett., 140:567.

Mohanty, A. and Clementi, E., 1989, Kinetically balanced geometric Gaussian basis set calculations for relativistic many-electron atoms with finite nuclear size, Chem. Phys. Lett., 157:348.

Quiney, H.M., 1988, Relativistic many-body perturbation theory, in: "Methods in Computational Chemistry, Volume 2: Relativistic Effects in Atoms and Molecules," S. Wilson, ed., Plenum, New York.

Quiney, H.M., Grant, I.P., and Wilson, S., 1987, The Dirac equation in the algebraic approximation, Phys. Scr., 36:460.

Quiney, H.M., Grant, I.P., and Wilson, S., 1988, On the relativistic many-body perturbation theory of atomic and molecular electronic structure in: "Many-body methods in quantum chemistry," U. Kaldor, ed., Springer, Berlin.

Quiney, H.M., Grant, I.P., and Wilson, S., 1989a, On the accuracy of Dirac-Hartree-Fock calculations using analytic basis sets, J. Phys., B22:L15

Quiney, H.M., Grant, I.P., and Wilson, S., 1989b, Relativistic many-body perturbation theory using analytic basis functions, preprint.

Schwarz, W.H.E. and Wallmeier, H., 1982, Basis set expansions of relativistic molecular wave equations, Mol. Phys., 46:1045.

Stanton, R. and Havriliak, S., 1984, Kinetic balance: A partial solution to the problem of variational safety in Dirac calculations, J. Chem. Phys., 81:1910.

Sucher, J., 1980, Foundations of the relativistic theory of many-electron atoms, Phys. Rev., A22:348

Sucher, J., 1987, Relativistic many-electron Hamiltonians, Phys. Scr., 36:271

Wilson, S., 1987, Basis sets, Adv. Chem. Phys., Part I: Ab initio Methods in Quantum Chemistry, 67:439.

COMMENT ON "BASIS SET EXPANSION DIRAC–FOCK SCF CALCULATIONS AND MBPT REFINEMENT" BY Y. ISHIKAWA

Pekka Pyykkö

Department of Chemistry, University of Helsinki
Et. Hesperiankatu 4, 00100 Helsinki, Finland

Fock (1930) derived, for systems with Coulomb ($\pm r^{-1}$) potentials, the equations

$$c\langle \alpha \cdot p \rangle + \langle V \rangle = 0 \qquad (1)$$

$$E = mc^2 + \beta \qquad (2)$$

Later derivations and discussions are listed in Table 2.5 of Pyykkö (1986) who, however, missed the papers by Bethe (1948), and Epstein and Epstein (1962), and failed to quote in the table Breit and Brown (1948) and Brown (1950) giving

$$(Z\alpha/2) \langle r^{-2} \rangle = k \int fg\, dr \qquad (3)$$

and Epstein (1976), giving several relations involving α, β and r^n. All these papers are related to hydrogen–like atoms.

The equations (1–2) were mentioned in the basis–set calculations of Kim (1967). Since then, this possibility for further testing has hardly been used. For instance, one could ask whether the variational energy minimum and equation (1–2) are simultaneously satisfied in parameter space. the necessary conditions were discussed by Data and Devaiah (1988).

References

H.A. Bethe, 1948, Bemerkungen über die Wasserstoff–Eigenfunktionen in der Diracschen Theorie, Z. Naturforschung **3a**, 470–477.

G. Breit and G.E. Brown, 1948, Effect of nuclear motion on the fine structure of hydrogen, Phys. Rev. **74**, 1278–1284.

G.E. Brown, 1950, Note on a relation in Dirac's theory of the electron, Proc. Natl. Acad. Sci. USA **36**, 15–17.

S.N. Datta and G. Devaiah, 1988, the minimax techniques in relativistic Hartree–Fock calculations, Pramana **30**, 387–405.

J.H. Epstein and S.T. Epstein, 1962, Some applications of hypervirial theorems to the calculation of average values, Am. J. Phys. **30**, 266–268.

S.T. Epstein, 1976, A differential equation for the energy eigenvalues of relativistic hydrogen atoms, and its solution, Am. J. Phys. **44**, 251–252.

V.A. Fock, 1930, Bemerkung zum Virialsatz, Z. Physik **63**, 855–858.

Y.–K. Kim, 1967, Relativistic self–consistent–field theory for closed–shell atoms, Phys. Rev. **154**, 17–39.

P. Pyykkö, 1986, *Relativistic theory of atoms and molecules. A bibliography 1916–1985*, Lecture Notes in Chemistry **41**, Springer, Berlin.

REPLY TO "COMMENT ON "BASIS SET EXPANSION DIRAC–FOCK SCF CALCULATIONS AND MBPT REFINEMENT" BY P. PYYKKÖ

Yasuyuki Ishikawa

Department of Chemistry, The University of Puerto Rico
Rio Piedras, Puerto Rico 00931 USA

Satisfying the virial theorem is a necessary, but not a sufficient condition for the variational minimum. In general, there are more than one set of nonlinear parameters (i.e., basis function exponents) for which the virial theorem is satisfied. The Hartree–Fock energy is optimal only for one of these sets. For this reason, the virial theorem has rarely been used in the Hartree–Fock energy minimization processes.

In his pioneering work on Dirac–Fock basis set expansion calculations, Kim (1967) introduced the idea of using the relativistic virial theorem to **identify** the best Dirac–Fock energies. This was most likely necessary because, in earlier work, relativistic self–consistent–field calculations which used an analytic basis were plagued by the problem of variational failure, a phenomenon that the computed total energy falls below the numerical limit. In our case, the Dirac–Fock basis set expansion method (method 1) provides an upper bound to the numerical limit, and thus, the relativistic virial theorem has not been explicitly used to minimize the total energy. We do, however, evaluate the relativistic virial to assess the quality of a given basis set used in the Dirac–Fock GTF basis expansion calculations.

In our Dirac–Fock calculations, we have been using both even– and well– tempered Gaussian basis sets that are moderately large. The number of non–linear parameters involved in the even– and well– tempered basis sets are two and four, respectively. And they can easily be energy optimized without the explicit use of the virial theorem. For the optimal nonlinear parameters, the relativistic viral theorem is usually satisfied to 7 figures of accuracy.

In a recent work, Goldman (1987) exploited the relativistic virial theorem to construct a new optimization technique for basis set exponents in Dirac–Fock expansion calculations. This procedure may become essential if one tries to optimize each basis set exponent individually.

References

Y.–K. Kim, 1967, Relativistic self–consistent field theory for closed–shell atoms, Phys. Rev. **154**, 17.

S.P. Goldman, 1987, Variational Dirac–Hartree–Fock energy optimization by the virial–theorem method, Phys. Rev. A**36**, 3054.

COMMENT ON "BASIS SET EXPANSION DIRAC–FOCK SCF CALCULATIONS AND MBPT REFINEMENT BY Y. ISHIKAWA"

K. G. Dyall

Eloret Institute, 3788 Fabian Way Palo Alto
California 94303, USA[1]

We have so far observed no bounds failure for $2p_{1/2}$ states using your method 2, with a Gaussian nuclear charge distribution. Could the failure be due to the uniform sphere model of the nucleus that you are using? This would arise because of the change of the potential from harmonic to Coulombic at the nuclear boundary, making Gaussians inappropriate functions, whereas for a Gaussian nuclear model, the potential remains an even function of r, no such change occurs, and Gaussians remain appropriate functions for all r.

REPLY TO "COMMENT ON "BASIS SET EXPANSION DIRAC–FOCK SCF CALCULATIONS AND MBPT REFINEMENT" BY K.G. DYALL

Yasuyuki Ishikawa

Department of Chemistry, The University of Puerto Rico
Rio Piedras, Puerto Rico 00931 USA

We have introduced the finite nucleus of uniform proton charge distribution in our Dirac–Fock GTF basis expansion calculations primarily because the boundary conditions associated with the uniform charge model are most suitable for the kinetically balanced Gaussian–type functions.

In the uniform charge model, the Dirac–Fock wavefunctions expanded in terms of the balanced GTF are the most accurate representation inside the nucleus. However, the solutions of the Dirac equation with a uniform charge model have discontinuities in the second derivative and higher at the nuclear boundary. I do not know if this is the reason why method 2 causes the bounds failure for $2p_{1/2}$ states. We will look into this possibility in our future study.

---

[1] Mailing address: NASA Ames Research Center, MS RTC 230–3, Moffett Field, CA 94035–1000, U.S.A.

POLYATOMIC MOLECULAR DIRAC-HARTREE-FOCK CALCULATIONS WITH GAUSSIAN BASIS SETS

Kenneth G. Dyall[†], Knut Fægri, Jr.[‡] and Peter R. Taylor[†]

[†]Eloret Institute, 3788 Fabian Way, Palo Alto
California 94303, U.S.A. *
[‡]Department of Chemistry, University of Oslo
P.O. Box 1033, Blindern, N-0315 Oslo 3, Norway

INTRODUCTION

The majority of research in *ab initio* quantum chemistry is performed on molecules containing light atoms, and indeed a large part of chemical research generally is concerned with such systems. *Ab initio* quantum chemical calculations are limited by the performance of computer hardware and software, and advances in these areas have increased the scope of quantum chemistry to the point where it is now possible to perform accurate calculations on systems containing the lighter transition metals. However, there are many important chemical processes involving heavy elements, and here another obstacle to accurate quantum chemical calculations presents itself — the increasing importance of relativistic effects with increasing atomic number.

The chemical effects of relativity have been extensively documented (for exhaustive reviews see Pyykkö 1978, 1986, 1988, Pitzer 1979, Pyykkö and Desclaux 1979). The traditional methods of computational chemistry are clearly inadequate for the description of important phenomena such as relativistic orbital contractions and spin-orbit splitting — effects that may decisively influence structure as well as reactivity of heavy-atom molecules. Yet molecules containing heavy atoms are important in a number of areas of chemistry, such as catalysis and surface chemistry. The need for accurate quantum chemical calculations in these areas has motivated

---

* Mailing address: NASA Ames Research Center, MS RTC 230-3, Moffett Field, CA 94035-1000, U.S.A.

recent theoretical and methodological developments which are aimed at overcoming the obstacles to a relativistic treatment of systems containing heavy atoms.

One widely-used approach to calculations on molecules containing heavy atoms is the relativistic effective core potential (RECP) method. While this method has considerable advantages, there are situations in which it is inappropriate, and in most applications spin-orbit splitting is not included. Effective core potential methods also need calibration against *ab initio* calculations for validation. An all-electron relativistic method is clearly required, both for rigour and understanding, and in evaluating alternatives.

Numerical methods have been used successfully in atomic Dirac-Hartree-Fock (DHF) calculations for many years (Desclaux 1975, Grant et al. 1980). Some DHF calculations using numerical methods have been done on diatomic molecules (Laaksonen and Grant, 1984, Sundholm et al. 1987, Sundholm 1988), but while these serve a useful purpose for calibration, the computational effort in extending this approach to polyatomic molecules is prohibitive. An alternative more in line with traditional quantum chemistry is to use an analytical basis set expansion of the wave function. This approach fell into disrepute in the early 1980s due to problems with variational collapse and intruder states, but has recently been put on firm theoretical foundations (Grant 1986, Grant and Quiney, 1988, Quiney 1988). In particular, the problems of variational collapse are well understood, and prescriptions for avoiding the most serious failures have been developed. Consequently, it is now possible to develop reliable molecular programs using basis set methods. We describe such a program in this paper, and report results of test calculations to demonstrate the convergence and stability of the method.

THEORY

With a single determinant many-electron wave function constructed from 4-spinors $|j\rangle$, we may write the (unrestricted) Dirac-Fock energy as

$$E = \sum_{j=1}^{n} \langle j | \hat{h}_D | j \rangle + \frac{1}{2} \sum_{j,k=1}^{n} [\langle jk | \hat{g} | jk \rangle - \langle jk | \hat{g} | kj \rangle]. \qquad (1)$$

The one-electron operator in the field of the nuclei is

$$h_D = -ic\,\alpha.\nabla + (\beta - 1)c^2 + V^{nuc}, \qquad (2)$$

where $\alpha = (\alpha_x, \alpha_y, \alpha_z)$; $\alpha_x, \alpha_y, \alpha_z$ and β are 4×4 matrices,

$$\alpha_x = \begin{pmatrix} 0 & \sigma_x \\ \sigma_x & 0 \end{pmatrix}, \ \alpha_y = \begin{pmatrix} 0 & \sigma_y \\ \sigma_y & 0 \end{pmatrix}, \ \alpha_z = \begin{pmatrix} 0 & \sigma_z \\ \sigma_z & 0 \end{pmatrix}, \ \beta = \begin{pmatrix} I_2 & 0 \\ 0 & -I_2 \end{pmatrix}.$$

σ_x, σ_y and σ_z are the Pauli spin matrices, and I_2 is the 2×2 unit matrix. The fully covariant electron-electron interaction can be expanded in a power series in c^{-2}. The lowest order term, which is $\mathcal{O}(c^0)$, is the Coulomb interaction,

$$\hat{g} \equiv \hat{g}(1,2) = \frac{1}{r_{12}}. \qquad (3)$$

The term which contributes at the next order, $\mathcal{O}(c^{-2})$, is the Breit interaction, whose contribution to the energy comes mainly from the region near the nuclei. For present purposes, the Coulomb interaction is an adequate description of the electron-electron interaction. Writing the 4-spinors in terms of large and small component 2-spinors $\mathbf{j}^L$ and $\mathbf{j}^S$,

$$|j\rangle = \begin{pmatrix} \mathbf{j}^L \\ i\mathbf{j}^S \end{pmatrix} \tag{4}$$

where the superscripts L and S indicate large and small components respectively, we obtain for the matrix elements of the one- and two-electron operators

$$\begin{aligned}\langle j|\hat{h}_D|j\rangle &= c\left[\langle \mathbf{j}^L|(\sigma.\nabla)|\mathbf{j}^S\rangle - \langle \mathbf{j}^S|(\sigma.\nabla)|\mathbf{j}^L\rangle\right] \\ &+ \langle \mathbf{j}^L|V^{nuc}|\mathbf{j}^L\rangle + \langle \mathbf{j}^S|V^{nuc}-2c^2|\mathbf{j}^S\rangle,\end{aligned} \tag{5}$$

$$\begin{aligned}\langle jk|\hat{g}|jk\rangle &= (\mathbf{j}^L\mathbf{j}^L|\mathbf{k}^L\mathbf{k}^L) + (\mathbf{j}^L\mathbf{j}^L|\mathbf{k}^S\mathbf{k}^S) \\ &+ (\mathbf{j}^S\mathbf{j}^S|\mathbf{k}^L\mathbf{k}^L) + (\mathbf{j}^S\mathbf{j}^S|\mathbf{k}^S\mathbf{k}^S)\end{aligned} \tag{6a}$$

$$\begin{aligned}\langle jk|\hat{g}|kj\rangle &= (\mathbf{j}^L\mathbf{k}^L|\mathbf{k}^L\mathbf{j}^L) + (\mathbf{j}^L\mathbf{k}^L|\mathbf{k}^S\mathbf{j}^S) \\ &+ (\mathbf{j}^S\mathbf{k}^S|\mathbf{k}^L\mathbf{j}^L) + (\mathbf{j}^S\mathbf{k}^S|\mathbf{k}^S\mathbf{j}^S).\end{aligned} \tag{6b}$$

We expand the large and small components in a basis of 2-spinors $\{\mu^L\}$ and $\{\mu^S\}$

$$|\mathbf{j}^L\rangle = \sum_{\mu=1}^{N} c_{\mu j}^L |\mu^L\rangle; \quad |\mathbf{j}^S\rangle = \sum_{\mu=1}^{N} c_{\mu j}^S |\mu^S\rangle, \tag{7}$$

and define the nuclear potential energy, overlap, kinetic energy and density matrix elements by

$$V_{\mu\nu}^{XX} = \langle \mu^X|V^{nuc}|\nu^X\rangle, \tag{8}$$

$$S_{\mu\nu}^{XX} = \langle \mu^X|\nu^X\rangle, \tag{9}$$

$$\Pi_{\mu\nu}^{XY} = \langle \mu^X|\sigma.\nabla|\nu^Y\rangle, \tag{10}$$

$$D_{\mu\nu}^{XY} = \sum_{j=1}^{n} c_{\mu j}^{X\dagger} c_{\nu j}^{Y}, \tag{11}$$

respectively, where X and Y can be L or S, with the restriction that for the kinetic energy matrix elements, $X \neq Y$. The Dirac-Fock energy can then be written

$$E = \sum_{\mu\nu}^{N} \left[c(D_{\mu\nu}^{LS} \Pi_{\mu\nu}^{LS} - D_{\mu\nu}^{SL} \Pi_{\mu\nu}^{SL}) + D_{\mu\nu}^{LL} V_{\mu\nu}^{LL} + D_{\mu\nu}^{SS} (V_{\mu\nu}^{SS} - 2c^2 S_{\mu\nu}^{SS}) \right]$$

$$+ \frac{1}{2} \sum_{\mu\nu\kappa\lambda}^{N} \left[D_{\mu\nu}^{LL} D_{\kappa\lambda}^{LL} \{ (\mu^L \nu^L | \kappa^L \lambda^L) - (\mu^L \lambda^L | \kappa^L \nu^L) \} \right. \tag{12}$$

$$+ D_{\mu\nu}^{SS} D_{\kappa\lambda}^{SS} \{ (\mu^S \nu^S | \kappa^S \lambda^S) - (\mu^S \lambda^S | \kappa^S \nu^S) \}$$

$$\left. + 2 D_{\mu\nu}^{LL} D_{\kappa\lambda}^{SS} (\mu^L \nu^L | \kappa^S \lambda^S) - 2 D_{\mu\nu}^{LS} D_{\kappa\lambda}^{SL} (\mu^L \lambda^L | \kappa^S \nu^S) \right].$$

Differentiating with respect to the large and small component coefficients, we obtain the following matrix representation of the Dirac-Fock equations,

$$\begin{pmatrix} \mathbf{F}^{LL} - \epsilon \mathbf{S}^{LL} & \mathbf{F}^{LS} \\ \mathbf{F}^{SL} & \mathbf{F}^{SS} - \epsilon \mathbf{S}^{SS} \end{pmatrix} \begin{pmatrix} \mathbf{c}^L \\ \mathbf{c}^S \end{pmatrix} = 0. \tag{13}$$

The elements of the various blocks of the Fock matrix are defined by

$$F_{\mu\nu}^{LL} = V_{\mu\nu}^{LL} + \sum_{\kappa\lambda}^{N} \left[D_{\kappa\lambda}^{LL} \{ (\mu^L \nu^L | \kappa^L \lambda^L) - (\mu^L \lambda^L | \kappa^L \nu^L) \} \right. \tag{14a}$$

$$\left. + D_{\kappa\lambda}^{SS} (\mu^L \nu^L | \kappa^S \lambda^S) \right],$$

$$F_{\mu\nu}^{SS} = V_{\mu\nu}^{SS} - 2c^2 S_{\mu\nu}^{SS} + \sum_{\kappa\lambda}^{N} \left[D_{\kappa\lambda}^{SS} \{ (\mu^S \nu^S | \kappa^S \lambda^S) - (\mu^S \lambda^S | \kappa^S \nu^S) \} \right. \tag{14b}$$

$$\left. + D_{\kappa\lambda}^{LL} (\mu^S \nu^S | \kappa^L \lambda^L) \right],$$

$$F_{\mu\nu}^{LS} = c \Pi_{\mu\nu}^{LS} - \sum_{\kappa\lambda}^{N} D_{\kappa\lambda}^{SL} (\mu^L \lambda^L | \kappa^S \nu^S) = F_{\mu\nu}^{SL\dagger}. \tag{14c}$$

The 2-spinors may be written as a combination of scalar functions $|a\rangle$ and $|b\rangle$ with spin functions:

$$|\mu^X\rangle = T_{\mu a}^{X\alpha} |a\rangle \begin{pmatrix} 1 \\ 0 \end{pmatrix} + T_{\mu b}^{X\beta} |b\rangle \begin{pmatrix} 0 \\ 1 \end{pmatrix}. \tag{15}$$

where $\begin{pmatrix} 1 \\ 0 \end{pmatrix}$ and $\begin{pmatrix} 0 \\ 1 \end{pmatrix}$ are unit vectors in spin space corresponding to $m_s = \frac{1}{2}$ and $m_s = -\frac{1}{2}$, conventionally labelled α and β. These may be used to further reduce the Fock matrix expressions:

$$F_{\mu\nu}^{XY} = \sum_{\sigma\tau} \sum_{ab} T_{\mu a}^{X\sigma\dagger} F_{ab}^{\sigma\tau} T_{\nu b}^{Y\tau}, \tag{16}$$

where σ and τ run over both spin indices. The Fock matrix elements in the scalar basis are given by the following expressions. If functions a and b belong to the same component (L or S),

$$F_{ab}^{\sigma\sigma} = V_{ab}^{\sigma\sigma} + \sum_{cd}(ab|cd)\left[D_{cd}^{\sigma\sigma} + D_{cd}^{\tau\tau}\right] - \sum_{cd}(ad|cb)D_{cd}^{\sigma\sigma}, \qquad (17a)$$

$$F_{ab}^{\sigma\tau} = -\sum_{cd}(ad|cb)D_{cd}^{\tau\sigma}. \qquad (17b)$$

The sums over c and d for the direct integrals ($ab|cd$) extend over both components, while those for exchange integrals ($ad|cb$) extend only over the same component as a and b. For the blocks connecting the large and small components,

$$F_{ab}^{\sigma\sigma} = (2\sigma)c\Pi_{ab}^{0} - \sum_{cd}(ad|cb)D_{cd}^{\sigma\sigma} \qquad (17c)$$

$$F_{ab}^{\sigma\tau} = c\Pi_{ab}^{2\tau} - \sum_{cd}(ad|cb)D_{cd}^{\tau\sigma} \qquad (17d)$$

where a and d are large component functions, and b and c are small component functions. The kinetic energy matrix elements are defined by

$$\Pi_{ab}^{0} = \langle a|\frac{\partial}{\partial z}|b\rangle, \quad \Pi_{ab}^{\pm 1} = \langle a|\frac{\partial}{\partial x}\pm i\frac{\partial}{\partial y}|b\rangle. \qquad (18)$$

CONSIDERATIONS FOR IMPLEMENTATION

There are a number of alternatives to be considered in the implementation of the Dirac-Fock method in a basis set, such as choice of basis function type and nuclear model, relations between small and large component basis functions, form of spinor expansions, transformation from the scalar to the 2-spinor basis, inclusion of double group symmetry (including time-reversal symmetry), details of integral storage and transformation and SCF method. We discuss these alternatives in the following subsections.

Basis function type and nuclear model

The principal causes of variational collapse in the attempt to solve the Dirac equation are the failure to satisfy the boundary conditions at the nucleus (Grant and Quiney 1988), and the failure to ensure the proper relations between the large and small component basis functions (Ishikawa et al. 1983, Dyall et al. 1984, Stanton and Havriliak 1984). The second point will be addressed in the next section.

Numerical solutions of the Dirac-Fock equations have the boundary conditions built in, so that any spherically symmetric model of the nuclear charge distribution may be employed. For finite basis set approximations, these boundary conditions will determine the form of the basis functions. Thus, the choice of basis function type and nuclear model are interrelated.

The traditional model used in electronic structure calculations for the nuclear charge distribution is the point nuclear model, which gives rise to the cusp in the non-relativistic electronic wave function at the nucleus. In the solution of the electronic Dirac equation, this cusp is replaced by a singularity, which has to be modelled by the basis functions. For atoms and diatomics, Slater-type functions with *non*-integral exponents of r should be used, such as the S-spinors or L-spinors advocated by Quiney et al. (1989). While these are convenient for atomic calculations, and may also be useful for diatomic molecules, their use for polyatomic systems would be computationally intractable.

The alternative to the point charge model for the nucleus is to use some kind of charge distribution with a finite radius, for which the wave function is no longer singular at the origin. For the purposes of electronic structure calculations, the details of the model for the nuclear charge distribution are not critical, provided they approximately represent the real distribution *. The most popular models in atomic structure calculations are the uniformly charged sphere model,

$$\rho_{nuc}(r) = \rho_0, \ r \leq r_0$$
$$= 0, \ r > r_0$$

and the Fermi distribution,

$$\rho_{nuc}(r) = \rho_0[1 + \exp((r-a)/c)]^{-1}.$$

Visser et al. (1987) have investigated the use of a single Gaussian function for the nuclear charge distribution in basis set DHF calculations,

$$\rho_{nuc}(r) = \rho_0 \exp(\eta_{nuc} r^2).$$

None of these models is anything but a crude representation of the nuclear charge distribution, but they are adequate for electronic structure calculations. The justification for the use of the uniform and Fermi models is that they have been used in the fitting of nuclear scattering data to obtain gross nuclear dimensions. The parameters r_0 for the uniform distribution, a and c for the Fermi distribution, and η_{nuc} for the Gaussian distribution are determined from these fits to nuclear scattering data. The effect of the choice of finite nuclear model on the energy is not large, as shown by some numerical calculations for Hg^{79+} using a modification of the GRASP program (Dyall et al. 1989), given in table 1. The nuclear size effect is of the order of 2 E_h, but the effect of the shape of the nuclear boundary is only of the order of 20 mE_h.

---

* This will not necessarily be true, of course, for properties such as nuclear hyperfine structure and parity non-conservation effects, which may be sensitive to the nuclear model.

Table 1. The effect of the nuclear model on the 1s eigenvalue of Hg^{79+}, given in E_h.

Nuclear model	Eigenvalue E	E − E(point)
Point	−3532.191 849	
Uniform	−3530.174 275	2.017 574
Fermi	−3530.182 156	2.009 693
Gaussian	−3530.193 999	1.997 850

The relation between the choice of basis functions and the nuclear charge model may be clarified by the following analysis. The nuclear potential for any finite nuclear charge distribution may be expanded in a power series about the origin,

$$V^{nuc}(r) = v_0 + v_2 r^2 + v_3 r^3 + \dots.$$

Note that there is no term linear in r, so that for small r, regardless of the details of the nuclear model, the potential is approximately harmonic. The solutions of the Schrödinger equation for a harmonic potential are Hermite Gaussian functions; the solutions of the Dirac equation are not, but may be represented by Gaussian functions. At very short range, then, Gaussian functions will be appropriate basis functions. If *all* terms of odd order in the series expansion of the nuclear potential have zero coefficients, then the solutions of the electronic Dirac equation for a many-electron atom are either pure even or pure odd functions of r. Both the uniform and the Gaussian nuclear charge distribution have such an expansion inside the nuclear radius. The series for the uniform model is *only* valid inside the nuclear radius, but for the Gaussian nuclear model the series is valid for all r. Solutions of the Dirac equation with a uniform nuclear model will have discontinuities in the higher derivatives at the nuclear boundary, but for a Gaussian nuclear model, the solutions will be continuous and differentiable to all orders for all r. The Fermi distribution, on the other hand, has terms of odd order in the series expansion, which will introduce elements in the solutions which have cusps in the higher derivatives at the origin. Use of the Gaussian nuclear model thus leads to solutions which are mathematically more well-behaved than the other two models, and as a result, expansion of the solutions in a finite Gaussian basis set is likely to have superior convergence properties.

One further, practical consideration in the choice of nuclear model needs mention. The calculation of many-centre nuclear attraction integrals would have to be done numerically for the uniform and Fermi models, whereas for the Gaussian model, the integrals may be evaluated using existing technology, since they require only straightforward changes to the expressions for point nuclear integrals.

Kinetic balance

Many of the problems with variational collapse and intruder states disappeared, once it was established that the small component basis functions must at least be related to the large component functions (Ishikawa et al. 1983, Dyall et al. 1984, Stanton and Havriliak 1984) by the condition now known as kinetic balance,

$$\{\mu^S\} \supseteq \{\sigma \cdot \nabla \mu^L\}. \tag{19}$$

Most of the remaining problems relate to the nuclear boundary conditions discussed above. Kinetic balance is, however, only a zeroth-order or non-relativistic approximation, which is useful when the potential V^{nuc} is much smaller than $2c^2$. A relation which more accurately represents the small component where the potential is large (Ishikawa et al. 1983, Dyall et al. 1984), derived from the one-electron Dirac equation, is

$$\{\mu^S\} \supseteq \{[2c^2 + \epsilon - V^{nuc}]^{-1} \sigma \cdot \nabla \mu^L\}. \tag{20}$$

With a Gaussian nucleus, kinetically balanced Gaussian basis functions do not satisfy this relation, but they still provide a good representation of the small component.

Applying $\sigma \cdot \nabla$ to jj-coupled 2-spinors composed of spherical harmonic Gaussian functions of the form

$$\chi_{n,\ell,m}(r,\theta,\phi) = \mathcal{N} r^{n-1} \exp(-\zeta r^2) Y_{\ell m}(\theta,\phi),$$

with $n = \ell + 1$, yields the following results. For the $j = \ell + \frac{1}{2}$ spin-orbit components of the large component 2-spinors, the small component 2-spinors are composed of functions $\chi_{n+1,\ell+1,m'}$, with appropriate values of m'. The small component basis functions in this case have one *more* unit of angular momentum than the large component basis functions. For the $j = \ell - \frac{1}{2}$ spin-orbit components of the large component 2-spinors, the small component 2-spinors are composed of functions $\chi_{n+1,\ell-1,m'}$ and $\chi_{n-1,\ell-1,m'}$. In this case, the small component basis functions have one *less* unit of angular momentum than the large component basis functions, but they are also composed of *two* radial functions rather than just one. Thus, the $1s_{1/2}, 2p_{3/2}, 3d_{5/2}, \ldots$ large component 2-spinors generate $2p_{1/2}, 3d_{3/2}, 4f_{5/2}, \ldots$ small component 2-spinors; but the $2p_{1/2}, 3d_{3/2}, 4f_{5/2}, \ldots$ large component 2-spinors generate $1s_{1/2}, 2p_{3/2}, 3d_{5/2}, \ldots$ and $3s_{1/2}, 4p_{3/2}, 5d_{5/2}, \ldots$ 2-spinors. Kinetic balance also requires equal exponents for corresponding large and small component Gaussian functions.

If we consider the scalar Gaussian basis functions of which the 2-spinors are composed, there are $2\ell + 1$ large component basis functions for each ℓ value, if the two spin-orbit components share the same basis functions. By the above rules, these generate $6\ell + 1$ small component basis functions, except for s, which generates 3 basis functions. The total number of basis functions for a given ℓ shell is then $8\ell + 2$, which represents an approximately *fourfold* increase in the basis set size over the corresponding nonrelativistic basis set size of $2\ell + 1$ spherical harmonic Gaussians. If the two spin-orbit components do *not* share the same basis functions, we must add $2\ell + 1$ functions to the large component basis, making a total of $10\ell + 3$ basis functions, which is an almost *fivefold* increase.

For example, consider a non-relativistic basis set consisting of 15 s, 11 p, 6 d and 3 f functions, which has a total of 99 basis functions. With common basis functions for the spin-orbit components in the large component set, the corresponding relativistic basis set would have these 99 functions for the large component set, and the small component set would consist of 11 s, 21 p, 14 d, 6 f and 3 g pure spherical functions, and 11 $3s$, 6 $4p$ and 3 $5d$ "contaminants" — a total of 257 basis functions for the small component. Including both components, the relativistic basis set would consist of 356 basis functions. Without the use of common basis functions for the spin-orbit components, the relativistic basis set would consist of 440 basis functions in total.

In a contracted basis set, it would generally be necessary to have different contracted functions for the two spin-orbit components of a given shell. Both the large and small component functions would have different contraction coefficients, including the contaminants (which belong to one spin-orbit component) and the pure spherical functions (which belong to the other). If we contracted the above non-relativistic basis to $6s5p3d2f$, for example, the relativistic contracted basis would have 80 large component basis functions and 130 small component basis functions, a total of 210 basis functions, compared with 43 basis functions for the nonrelativistic contracted basis. Thus, contraction as a space-saving mechanism is not quite as advantageous in a relativistic calculation as compared with a nonrelativistic calculation, owing to the need to duplicate the large component set for all basis function types except s.

<u>Choice of spinor expansion</u>

The expansion of the components of the 4-spinors $|j\rangle$ in a basis set can be done in three ways: independent expansions for each of the four components, expansion in terms of 2-spinors, as outlined in this paper, and expansion in 4-spinors, where each component is fixed in relation to the others. Since the last of these has not been much used, we will focus on the first two. Atomic finite basis calculations (see Pyykkö 1986 for references) have usually exploited atomic symmetry, and expanded only the radial parts of the large and small components in a basis. These use equal length expansions for the large and small components. Some molecular calculations have also used a 2-spinor basis (Hegarty and Aerts 1987).

Early molecular calculations (Mark et al. 1980, Lee and McLean 1982) used separate expansions for each of the four spinor components. This corresponds to carrying out the entire calculation in the scalar basis. It inevitably means that the basis set expansions for the large and small components will have different lengths. For example, consider the case of a single s function for the large component. This requires a p-type function in the small component, which consists of three functions: p_1, p_0 and p_{-1}. If these are coupled to the spin to form jj-coupled 2-spinors, we get two functions for the $p_{1/2}$ and four for the $p_{3/2}$ spin-orbit components of the p shell. Only the $p_{1/2}$ set are used to represent the small component for the large component s function. The $p_{3/2}$ functions form a basis for a "negative-energy" (or positron) state of symmetry $j = 3/2$, which will also appear in the spectrum along with the desired states. If the basis is ill-chosen, these extra states may not even be confined to the negative continuum, but appear instead as intruder states in the bound state region or in the positive continuum. A classic example of intruder or spurious states is the calculation of the hydrogenic states for $j = \ell - \frac{1}{2}$ by Drake and Goldman (1982), who observe a state degenerate with the state of the same j, but different ℓ value — for instance, a $p_{1/2}$ state degenerate with the $1s_{1/2}$ state. The features of this approach, then, are that it gives an unbalanced representation of the two branches of the spectrum, with a better description of the negative continuum, which is of no interest for molecular properties, than of the bound states and positive continuum. More than that, the extra functions will give rise to extra eigenvalues, which may occur as intruder states in the parts of the spectrum which are of interest, as Grant and Quiney (1988) have warned.

Some more recent 2-spinor based calculations, both on atoms (Ishikawa and Sekino 1990) and molecules (Visscher et al. 1990), have also employed different expansion lengths for the large and small components. Their expansions have a better representation of the negative energy states than the positive energy states. Experience has generally shown that, provided the basis functions are appropriate for the boundary conditions, the intruder states are to be found in the negative continuum. However, the physical states of interest may be distorted by the higher density of states in the negative continuum (see Stanton and Havriliak 1984), and mask problems of variational collapse or bounds failure. A more serious problem arises in the treatment of electron correlation if the negative continuum is to be included, for example in perturbation sums. Calculations of this type go beyond the no-pair approximation, i.e. they include contributions from creation of electron-positron pairs (see Grant and Quiney 1988 for a discussion). While the inclusion of the negative continuum is not likely to be of importance for molecular properties, the distortion of the positive spectrum, which results from the higher density of states in the negative continuum, may even affect calculations that do not include it.

Though the seriousness of the problems that may arise from the use of different length expansions will depend on the nature of the calculations, we believe it is wise to avoid any possible problems by ensuring that the large and small component spinor basis sets are of the same length, which means using jj-coupled atomic 2-spinors as a basis for both large and small components in matched sets.

Transformation to 2-spinors

The considerations of the previous section affect only the final representation of the eigenstates of the system under study. In the process of the construction of these eigenstates, it is not necessary to keep all quantities in the 2-spinor basis. Indeed, there may be some advantages in using the scalar basis at various stages in the calculations. Computationally, the transformation to 2-spinors is not expensive, as the transformation matrix is very sparse. The transformation will scale only as the number of indices to be transformed. The point at which the transformation is done will then be determined by the practical considerations of storage and efficiency.

In a direct SCF procedure, the transformation can be done while the Fock matrix is being constructed; in conventional SCF, there are two points at which the transformation to 2-spinors may be done. The first is at the stage of two-electron integral generation; the second is after construction of the Fock matrix.

The decision of whether or not to transform the integrals will be influenced by a number of factors, but the overriding consideration will be that of disk space, given the much larger basis sets needed in a relativistic calculation. Most molecular integral codes generate integrals over real basis functions, which may be Cartesian or spherical functions, atom-centred or symmetry-adapted functions. Since the jj-coupled 2-spinors are complex, the permutational symmetry of the integrals will be reduced after the transformation, and hence the number of values to be stored will increase. Time-reversal symmetry regains one degree of permutational freedom. If the integrals are generated over symmetry functions, then the fact that the 2-spinor functions are generally distributed between fewer irreducible representations in the double group than the scalar functions are in the single group will, in general, mean an increase in the number of values stored. These increases are offset by the reduction in size of the small component basis upon transformation. It is not immediately clear that transformation will inevitably increase the total storage requirements, but it may do so. The transformation would be implemented either at the same stage as the spherical harmonic transformation, since it is of a similar nature, or after the symmetry transformation (if the integral code includes this).

If the integrals are kept in the scalar basis, then the Fock matrix must be transformed instead. Since the construction of the Fock matrix is the most time-consuming step in a conventional SCF calculation, it is important to make it efficient. Use of unordered integrals in a non-relativistic calculation does not permit vectorization; to achieve this, ordered integrals or supermatrices are used. Due to the large basis sets needed for relativistic calculations, the extra disk space required for integral ordering may not be available. When constructing the Fock matrix in the scalar basis, however, each $(ab|cd)$ integral with distinct indices contributes in 36 unique places. It is then possible to vectorize the construction of the Fock matrix using sparse vector operations. While the gain in speed from vectorization is smaller than that obtained from the use of ordered integrals or supermatrices, it is nevertheless substantial.

IMPLEMENTATION

Our program uses for the scalar basis spherical harmonic Gaussian functions, which are symmetry-adapted for D_{2h} and its subgroups. The jj-coupled 2-spinors constructed from these are symmetry functions for the corresponding double group. Linear molecules are treated as a special case. The nuclear charge distribution is a single Gaussian with an exponent chosen to match the rms radii of the Gaussian and the nucleus, given by

$$\eta_{nuc} = 3/(2r^2_{rms}). \tag{21}$$

The nuclear rms radius is fitted to a function of the nuclear mass:

$$r_{rms} = \langle r^2 \rangle^{1/2}_{nuc} = 0.836 A^{1/3} + 0.57 \tag{22}$$

where A is in amu and r_{rms} is in fm. This formula is appropriate for the Fermi 2-parameter distribution. The one-and two-electron integrals, which are generated by an adaption of MOLECULE (Almlöf and Taylor, unpublished), are kept in the scalar basis. The Fock matrix is constructed in the scalar basis, then transformed to the 2-spinor basis. Each (double group) symmetry block of the Fock matrix is symmetrically orthonormalized, and diagonalized using standard EISPACK routines. Density damping has been implemented to accelerate convergence.

RESULTS

Testing of any new program requires comparison of a number of calculations of various types with known results. For atoms, the adapted GRASP program (Dyall et al. 1989) was used to obtain numerically accurate Dirac-Fock-Coulomb energies with a Gaussian nuclear charge distribution for comparison with results from the present program. For molecules, fewer results exist for comparison. The numerical diatomic calculations of Laaksonen et al. (1984) and Sundholm (1988) were used for comparison. All calculations were done with uncontracted basis sets.

One-electron calculations

The first sets of calculations were performed on one-electron systems. Tables 2 and 3 give results for the 1s orbital of H and Hg^{79+}. The basis sets for H were taken from Partridge (1989), while those for Hg^{79+} were obtained by scaling the H sets. The results converge from above on the exact solution in both cases. For H, the error in the total energy, due to basis set truncation, converges more slowly than the error in the relativistic correction, as the basis set size is increased. Clearly, non-relativistic basis sets for H do not need modification for use in the Dirac equation. For Hg^{79+}, however, the error in the total energy is almost entirely due to the error in the relativistic correction. The basis set clearly does not describe the relativistic contraction of the 1s orbital well. We optimized the scale factor for each basis, with results as given in Table 4. Some 25% of the missing energy is recovered by this procedure, but the error in the relativistic correction still dominates the error in the

Table 2. Convergence of energies for hydrogen. Values given in μE_h.

Basis	ΔE	$\Delta(\Delta E_{rel})$	% error
7s	16.724	0.021	0.32
8s	5.448	0.010	0.15
9s	1.869	0.005	0.07
10s	0.670	0.002	0.03
11s	0.250	0.000	0.01
12s	0.097	0.000	0.00
13s	0.039	0.000	0.00
14s	0.016	0.000	0.00
15s	0.007	0.000	0.00

Numerical results: $E = -0.500\,006\,656\ E_h$, $\Delta E_{rel} = -6.656\,59\ \mu E_h$.

Table 3. Convergence of energies for Hg^{79+}. Values given in mE_h.

Basis	ΔE	$\Delta(\Delta E_{rel})$	% error
7s	2938.266	2839.618	0.86
8s	1229.427	1199.215	0.36
9s	458.401	448.863	0.14
10s	138.589	135.432	0.041
11s	32.612	31.475	0.010
12s	12.081	11.635	0.004
13s	9.370	9.197	0.003
14s	4.058	3.996	0.001
15s	0.540	0.504	0.0002

Numerical results: $E = -3530.193\,999\ E_h$, $\Delta E_{rel} = -330.482\,066\ E_h$.

Table 4. Convergence of energies for Hg^{79+} with optimized scale factors. Values given in mE_h.

Basis	ΔE	$\Delta(\Delta E_{rel})$	% error
7s	2337.469	2238.815	0.71
8s	937.087	906.869	0.27
9s	329.167	319.623	0.097
10s	94.710	91.547	0.028
11s	23.785	22.642	0.007
12s	11.674	11.205	0.003
13s	8.477	8.299	0.002
14s	2.673	2.605	0.0008
15s	0.432	0.414	0.0001

Table 5. Results for 2p states of Hg^{79+}.

$2p_{1/2}$ orbital energies and differences in E$_h$.

Basis	ϵ	ΔE	$\Delta E(3s)^a$
9p	−904.269 615	0.550 647	0.122 584
10p	−904.653 964	0.166 299	0.034 308
11p	−904.784 643	0.035 620	
11p^b	−904.802 234	0.018 029	0.014 425
numerical	−904.820 263		

$2p_{3/2}$ orbital energies and differences in E$_h$.

Basis	ϵ	ΔE
9p	−817.805 038	0.002 446
10p	−817.806 484	0.001 000
11p	−817.806 640	0.000 844
11p^b	−817.806 318	0.001 166
numerical	−817.807 484	

^{a} Extra energy difference when 3s functions omitted from small component spinors.
^{b} Previous basis with most diffuse function replaced by tight function.

total energy. This indicates that nonrelativistic basis sets need to be re-optimized, at least for the core orbitals, if they are to be used in DHF calculations.

Results of calculations on the $2p_{1/2}$ and $2p_{3/2}$ orbitals of Hg^{79+} with a few different basis sets are given in table 5. The basis sets were taken from atomic calculations on Pb (Fægri 1987). Here, the effect of spin-orbit splitting on the basis sets may be seen. The error for the $2p_{1/2}$ is much larger than that for the $2p_{3/2}$, even when the most diffuse function is replaced by a tighter function. Matsuoka and Okada (1989) found it necessary to add two tight p functions to the non-relativistic basis sets of Fægri (1987) in order to sufficiently reduce the eigenvalue errors for the neutral 6p block elements. Clearly, the basis set requirements for the two spin-orbit components are different, which implies that, in a contracted basis, it will be necessary to use different contractions for the $2p_{1/2}$ and $2p_{3/2}$ functions.

Strict kinetic balance requires the $2p_{1/2}$ small component basis functions to be composed of a 1s and a 3s function in a fixed ratio. It may be argued that the 3s could be represented as a linear combination of 1s functions, and therefore may be omitted. We have investigated the effect of omitting the 3s part of the small component spinor for each basis set size. The results are also given in Table 5. The importance of the 3s decreases with increasing basis set size, as is expected. With large enough basis sets, we conclude that the 3s may be omitted, but for

Table 6. Eigenvalues in E_h for H_2^+ at R = 2 bohr.

Basis	$\epsilon_{1/2g}$	$\epsilon_{1/2u}$
7s	−1.090 734	−0.667 302
15s	−1.090 954	−0.667 337
7s1p	−1.101 289	−0.667 438
7s3p	−1.102 384	−0.667 516
7s4p	−1.102 405	−0.667 517
numerical	−1.102 642	−0.667 553

Table 7. Lowest eigenvalue in E_h for HeH^{2+} at R = 2 bohr.

Basis	$\epsilon_{1/2}$
7s4p	−2.511 977
numerical	−2.512 296

smaller basis sets, which may be necessary for molecular calculations, it may still be beneficial to retain the 3s. It is important also to note that omission of the 3s does *not* cause bounds failure: to the contrary, the energy increases when it is left out. We should emphasize that the coefficients of the 1s and 3s functions in our calculations are *not* independently varied, as in the recent work by Ishikawa and Sekino (1990). Our approach is equivalent to their method 2.

Finally, some calculations were done on H_2^+ and HeH^+. The s basis sets were taken from Partridge (1989), and the p basis sets were chosen in the range normally used for polarization functions on H. The results are given in Tables 6 and 7, along with the numerical results of Laaksonen and Grant (1984). Though these calculations do not approach the basis set limit, they at least demonstrate that there are no problems with variational collapse. Even when the difference between the 7s and the 15s basis results is added to the 7s4p result, as an estimate of the truncation error in the s set, the energy is still above the numerical value.

Many-electron systems

We have done several series of calculations on many-electron atoms, to investigate the trends in basis set errors with basis set size and with atomic number. Results of a representative set of calculations, for 10-electron systems with Z = 10, 20, 40 and 80, are given in Table 8. The Ne basis sets were taken from Partridge (1989). The basis sets for the ions were energy optimized in non-relativistic finite nucleus atomic calculations. Our results for Ne are consistent with those of Hegarty and Aerts (1987). The results are presented in terms of the basis set truncation error and the error in the predicted relativistic correction for each basis set. The trends with atomic number demonstrate again that for the heavy atoms, the er-

Table 8. Comparisons of results for Ne, Ca^{10+}, Zr^{30+} and Hg^{70+}

Percentage error in relativistic correction.

Basis	Ne	Ca^{10+}	Zr^{30+}	Hg^{70+}
$7s3p$	2.50	2.54	2.80	5.12
$9s5p$	0.53	0.41	0.46	1.12
$10s6p$	0.29	0.20	0.22	0.58
$12s7p$	0.11	0.07	0.08	0.24
$13s8p$	0.06	0.03	0.04	0.13

Error in relativistic correction as a percentage of error in total energy.

Basis	Ne	Ca^{10+}	Zr^{30+}	Hg^{70+}
$7s3p$	1.4	8.7	32.2	81.5
$9s5p$	3.9	32.3	61.8	95.5
$10s6p$	6.7	29.9	72.3	97.5
$12s7p$	10.3	43.4	82.7	98.9
$13s8p$	13.6	55.6	88.2	99.4

ror in the total energy comes mainly from the non-optimal nature of the basis for relativistic calculations. However, the fractional error in the relativistic correction increases slowly with atomic number: for example, the error for Hg is only twice that for Ne.

Most of the testing has been on atomic systems. We present also a few calculations on molecular systems, H_2 and H_2O, in Tables 9 and 10. For H_2, the numerical results were taken from Sundholm (1988). The basis set for H_2, which is the same as that used for H_2^+ is modest in size — the basis set truncation error is 5.5 mE_h, but the relativistic correction is accurate to 0.4 μE_h. Using the previous 7s basis results for H atom, we obtain a result for the relativistic contribution to the binding energy of 1.1 μE_h, in agreement with the numerical result to better than 0.1 microhartree. Thus, while the absolute value of the relativistic correction may not be accurate due to an inadequate description of the region near the nucleus, the relativistic contribution to the binding energy is well predicted. Of course, relativistic effects are very small for H_2, but nevertheless, this result may carry over to other molecules. It may be that the re-optimization of the core basis sets, found to be necessary in the calculations on Hg^{79+} in order to obtain accurate core energies, may not be as critical for relativistic contributions to chemical binding. Indeed, Schwarz et al. (1989) comment that the principal differential screening effects that cause expansion or contraction of atomic valence orbitals due to relativity come from the valence and sub-valence shell, and not from orthogonality tails. This implies that a relatively poorer description of the core may not significantly affect calculated chemical properties, and that non-relativistic basis sets may be used

Table 9. Results for H_2 at $R = 1.4\ a_0$. Energies in E_h, differences in μE_h.

Basis	E_{rel}	E_{NR}	ΔE_{rel}
$7s3p$	−1.133 095 73	−1.133 081 64	−14.0
numerical[a]	−1.133 643 97	−1.133 629 57	−14.4

[a] Sundholm (1988)

Table 10. Results for H_2O at 0.96Å, 104.5°. Energies in E_h, differences in mE_h.

Basis	E_{rel}	E_{NR}	ΔE_{rel}
$4s2p, 2s$	−75.168 415	−75.120 506	−47.909

in relativistic calculations without re-optimization. The results of Matsuoka and Okada (1989) on heavy atoms show that the valence eigenvalue errors are of the order of a few mE_h when a large nonrelativistic basis is used without modification, and the improvement on correction of the major deficiency in the core basis gives at most $1mE_h$ improvement. Moreover, the eigenvalue errors are all in the same direction, so that some cancellation of basis set truncation errors occurs in the relative energies. Further investigation of the use of non-relativistic basis sets is necessary, particularly in molecular calculations.

While there are no calculations for H_2O to which we can compare directly, the computed relativistic correction is reasonable when compared with the sum of the relativistic corrections for the atoms, which is dominated by the contribution of 56 mE_h from oxygen. The correction is consistent with the results obtained for atoms, that the relativistic correction to the energy is underestimated in basis set calculations, and converges from above on the true value as the basis set is enlarged.

CONCLUSIONS

In this article we have discussed some of the principles underlying finite basis Dirac-Fock calculations, and described how these have been implemented in a computer program. Although dealing mostly with atoms, our test calculations, as well as results obtained by others using similar approaches, clearly demonstrate the feasibility of carrying out high-quality Dirac-Fock calculations for molecules. The main obstacle to such calculations today appears to be the lengthy basis set expansions required. At the Dirac-Fock level the storage problem this creates can be largely overcome by resorting to direct SCF methods. We feel confident that, with further development, routine quantum chemical calculations of high accuracy for molecules containing heavy atoms will soon be a reality.

ACKNOWLEDGEMENTS

KGD was supported by NASA grant NCC2-552 and PRT by NASA grant NCC2-371. KGD would like to thank Harry Quiney for valuable discussions. KF gratefully acknowledges the hospitality of NASA Ames Research Center and financial support from the Norwegian Research Council for Science and the Humanities (NAVF) during a 1-year sabbatical.

REFERENCES

Desclaux, J.-P., 1975, Comput. Phys. Commun. 9: 31.
Drake, G. W. F., and Goldman, S. P., 1982, Phys. Rev. A 25: 1230.
Dyall, K. G., Grant, I. P., Johnson, C. T., Parpia, F. A., and Plummer, E. P., 1989 Comput. Phys. Commun. 55: 425.
Dyall, K. G., Grant, I. P. and Wilson, S. 1984 J. Phys. B 17: L45, 493, 1201.
Fægri, K., 1987, Theor. Chim. Acta 72: 297.
Grant, I. P., McKenzie, B. J., Norrington, P. H., Mayers, D. F., and Pyper, N. C., 1980, Comput. Phys. Commun., 21: 207.
Grant 1986, J. Phys. B, 19: 3187.
Grant, I. P., and Quiney, H. M., 1988, Adv. At. Mol. Phys., 23: 37.
Hegarty, D., and Aerts, P. J. C., 1987, Physica Scripta, 36: 432.
Ishikawa, Y., Binning, R. C., and Sando, K. M., 1983, Chem. Phys. Lett., 101: 111.
Ishikawa, Y., and Sekino, H., 1990, Chem. Phys. Lett., 165: 243.
Laaksonen, L., and Grant, I. P., 1984, Chem. Phys. Lett., 109: 485, 112: 157.
Laaksonen, L., Grant, I. P., and Wilson, S., 1988, J. Phys. B, 21: 1969.
Lee, Y. S., and McLean, A. D., 1982, J. Chem. Phys., 76: 735.
Mark, F., Lischka, H., and Rosicky, F., 1980, Chem. Phys. Lett., 71: 507.
Matsuoka, O., and Okada, S., 1989, Chem. Phys. Lett., 155: 547.
Partridge, H., 1989, J. Chem. Phys., 90: 1043.
Pitzer, K. S., 1979, Acc. Chem. Res., 12: 271.
Pyykkö, P., 1978, Adv. Quantum Chem., 11: 353.
Pyykkö, P., and Desclaux, J.-P., 1979, Acc. Chem. Res., 12: 276.
Pyykkö, P., 1986, Lect. Notes Chem. 41.
Pyykkö, P., 1988, Chem. Rev., 88: 563.
Quiney, H. M., 1988, in: "Methods in Computational Chemistry, vol. 2," S. Wilson, ed., Plenum, New York.
Quiney, H. M., Grant, I. P. and Wilson, S., 1989, J. Phys. B, 22: L19.
Schwarz, W. H. E., van Wezenbeek, E. M., Baerends, E. J., and Snijders, J. G., 1989, J. Phys. B, 22: 1515.
Stanton, R. E., and Havriliak, S., 1984, J. Chem. Phys., 81: 1910.
Sundholm, D., Pyykkö, P., and Laaksonen, L., 1987, Physica Scripta, 36: 400.
Sundholm, D., 1988, Chem. Phys. Lett., 149: 251.
Visscher, L., Aerts, P. J. C., and Visser, O., 1990, this volume.
Visser, O., Aerts, P. J. C., Hegarty, D., and Nieuwpoort, W. C., 1987, Chem. Phys. Lett., 134: 34.

OPEN SHELL RELATIVISTIC MOLECULAR DIRAC-HARTREE-FOCK SCF-PROGRAM

O. Visser, P.J.C. Aerts and L. Visscher

Laboratory for Chemical Physics
State University of Groningen
Nijenborgh 16
9747 AG Groningen
The Netherlands

ABSTRACT

An open shell version has been developed for the MOLecular Hartree-Fock-DIRac program package MOLFDIR©. This program, originally developed as a closed shell program in 1985, can be used to calculate all-electron four-component Hartree-Fock-Dirac Self Consistent Field solutions for - in principle - molecules of general shape and uses separate gaussian basis sets to expand the large and small component functions. In practice basis sets are chosen to be kinetically balanced; technically, the program is not restricted to such basis sets. Implementation, advantages and limitations due to the Dacre and Elder algorithm for integral evaluation and Fock matrix set-up are discussed and results of test calculations are presented.

INTRODUCTION

In general there are several methods to study molecular relativistic effects (Pyykkö, 1986). We want to use an ab initio method with as few approximations as possible. In addition to the scientific desirability of staying close to the underlying physical model, it is often useful to carry out reference calculations in order to evaluate more approximate methods. In this article we will describe the open shell extension to the MOLFDIR program package, which has been designed as a research program to perform MOLecular Fock DIRac calculations (Aerts, 1986). The Fock Dirac method can be considered as the relativistic analogon of the non-relativistic Hartree Fock method. The package is being extended to include correlation effects.

The relativistic effects are included by using the Dirac-Coulomb hamiltonian (in atomic units):

$$H_D = \sum_i (c\alpha_i \cdot p_i + \beta_i mc^2 + V_i) + \sum_{i<j} r_{ij}^{-1}$$

In this equation, c is the velocity of light in vacuum, m is the mass of an electron, V_i is the electrostatic potential felt by electron i due to the nuclei, α_i and β_i are the Dirac matrices for electron i, and r_{ij}^{-1} is the Coulomb interaction between electrons i and j.

The two-electron part of this hamiltonian is not relativistically invariant. As a correction, the Breit interaction term may be included in the hamiltonian.

The program package has been applied to a number of different systems. At this moment we are very much interested in the series CH_4 -- PbH_4 since the influence of relativity grows systematically important in otherwise almost identical molecules; also, for these systems data are available from other calculations (Aguilar-Ancono et al., 1983; Desclaux and Pyykkö, 1974; Fernandez et al., 1985) which make these systems useful for comparison.

Next planned are Fock-Dirac calculations on the f^n states of EuO_6^{9-} relevant for the optical properties of the Ba_2GdNbO_6:Eu system. Extensive non-relativistic model calculations have been reported by Van Piggelen (1978). Our main interest is getting information on the effects of relativity on his results.

Evidently, the program must be able to handle open shells such as the open f shell of Eu. In practice, such an open shell can give rise to a very large number of configurations and states, which in turn makes it tedious to perform SCF calculations on all those states separately.

Since a multiconfiguration approach will be necessary anyway, it is also not a fruitful way to go.

Instead we have decided to use an average of configuration open shell formalism for the SCF part of the package; the resulting average orbitals can be used as a basis for further CI or MCSCF studies to incorporate the correlation effects.

The use of an average of configuration formalism has several advantages:

- from our experience with non-relativistic calculations on d^n and f^n shells it appears that the use of average orbitals offers a balanced description for the manifold of many-electron states arising from these configurations;

- the computational complexity of the use of different orbitals for different states in further calculations (CI, transition moments) is avoided.

- with some restrictions, it is possible to use the efficient Dacre & Elder algorithm.

The average of configuration open shell method has been implemented in the MOLFDIR program package. Molecular CI or MCSCF studies using the Dirac-Coulomb (and Breit) hamiltonian are scheduled.

BASIS SETS

The program package works with two distinct sets of scalar (contracted) cartesian gaussian functions: a large component set $\{g_i^L\}$ and a small component set $\{g_i^S\}$. It is only possible to contract functions centred on the same atom with the same angular momentum value. General contraction, however, can be applied and has - especially in the case of relativistic basis set calculations - significant advantages (Visscher, 1990). The two-electron repulsion integrals are calculated over the functions belonging to these sets.

From these two scalar sets two new sets of symmetry adapted basis spinors are constructed using the Dirac double group. The spinors can be divided in two sets with the following structure:

$$\{\chi_a\} = \{\chi_l\} + \{\chi_s\}$$

$$\{\chi_l\} = \left\{ \begin{pmatrix} \chi_1^{L\alpha} \\ \chi_1^{L\beta} \\ 0 \\ 0 \end{pmatrix} \right\}$$

$$\chi_1^{L\alpha} = \sum_i g_i^L c_{il}^\alpha \qquad \chi_1^{L\beta} = \sum_i g_i^L c_{il}^\beta$$

$$\{\chi_s\} = \left\{ \begin{pmatrix} 0 \\ 0 \\ \chi_s^{S\alpha} \\ \chi_s^{S\beta} \end{pmatrix} \right\}$$

$$\chi_s^{S\alpha} = \sum_i g_i^S c_{is}^{\alpha} \qquad \chi_s^{S\beta} = \sum_i g_i^S c_{is}^{\beta}$$

Each non-zero component of these basis spinors consist of a linear combination of the basis functions from set $\{g_i^L\}$ or set $\{g_i^S\}$.

Apart from the structure as described and the constraint to gaussian primitives, there are no restrictions on the choice of the basis spinors. In practice the basis set is chosen to be kinetically balanced (Visscher et al., 1990) with respect to the large component: the small component basis set includes all functions $\alpha.p\,\chi_l$. The coefficients of the l-1 and l+1 basis functions resulting from $\sigma \cdot p$ operating on a large component function will in practice have separate variational coëfficients.

DACRE AND ELDER METHOD

The program package uses the Dacre & Elder method (Dacre, 1970; Elder, 1963). This method consists of the following steps:

- calculate the two-electron repulsion integrals over the (contracted) gaussian functions;
- reduce the integral list by using the spatial symmetry operations of the point group (skip integrals which are zero by symmetry and calculate only the non zero integrals)
- build the skeleton Fock matrix using the reduced integral list
- complete the Fock-matrix by applying the symmetry operations to the skeleton Fock matrix
- transform the Fock matrix to the symmetry adapted basis.

The contributions of the non-unique integrals are generated by applying the symmetry operations to the skeleton Fock matrix. This will work only if the density matrix is symmetric with respect to all applied symmetry operations. In practice, this means that also the Fock matrix must be symmetric with respect to the same symmetry operations, and the energy expression upon which the SCF procedure is based must treat all sub-representations from a given representation in the same manner.

The obvious advantages of the Dacre and Elder scheme are that only the reduced list of integrals is needed, and that these integrals will be real.

A disadvantage is that only closed shell and symmetrically averaged open shell systems can be handled easily.

AVERAGE OF CONFIGURATION OPEN SHELL METHOD

Let k and l refer to closed shell orbitals, let m and n refer to open shell orbitals, and define $Q_{ij} \equiv J_{ij} - K_{ij}$ in which J and K are the usual coulomb and exchange integrals. Define the one electron Dirac operator as $h^D = c\alpha \cdot p + \beta mc^2 + V$, and its matrix representation in the orbital basis as H^D_{ij}.

Consider the following energy expression:

$$E = \sum_k H^D_k + \frac{1}{2}\sum_{kl} Q_{kl} + f\left[\sum_m H^D_m + \frac{1}{2}af\sum_{mn} Q_{mn} + \sum_{km} Q_{km}\right]$$

This energy expression is almost the same as the one Roothaan (1960) proposed, the differences being that in this energy expression the summations run over *spin*-orbitals, and that this energy expression contains only one (instead of two) coupling coefficient.

We restrict a and f in this expression in such a way that the energy expression corresponds to the average energy (McWeeny, 1989) of all single determinants formed from a specified set of open shell orbitals and a specified number of open shell electrons; all subrepresentations belonging to a representation occur in the set of open shell orbitals so that the density and Fock matrices will automatically be totally symmetric. Presently it is only possible to specify one a and one f (so all open shell orbitals will have the same fractional occupation number).

As usual, the SCF equations are derived by locating the stationary point of the expectation value of the total energy <E> with the orbital orthogonality constraint $\langle i | j \rangle = \delta_{ij}$.

Define

$$\alpha = \frac{1-a}{1-f}$$

$$D^C_{pq} = \sum_k c^*_{kq} c_{kp} \qquad D^O_{pq} = f\sum_m c^*_{mq} c_{mp}$$

$$Q^C_{pq} = \sum_{rs}(pq\|rs)D^C_{sr} \qquad Q^O_{pq} = \sum_{rs}(pq\|rs)D^O_{sr}$$

$$L^C_{pq} = \sum_{rs}\left[S_{pr}D^C_{rs}Q^O_{sq} + Q^O_{pr}D^C_{rs}S_{sq}\right] \quad L^O_{pq} = \sum_{rs}\left[S_{pr}D^O_{rs}Q^O_{sq} + Q^O_{pr}D^O_{rs}S_{sq}\right]$$

$$F_c = H^D + Q_c + Q_o + \alpha L_o \qquad F_o = H^D + Q_c + aQ_o + \alpha L_c$$

For the closed shells we get, analogous to Roothaan,

$$F_c|k\rangle = \eta_k|k\rangle$$

and for the open shells

$$F_o|m\rangle = \eta_m|m\rangle$$

We get the following expression for the total energy:

$$E = \text{Tr}[H^D D^C] + \frac{1}{2}\text{Tr}[Q^C D^C] +$$
$$\text{Tr}[H^D D^O] + \frac{1}{2}a\text{Tr}[Q^O D^O] +$$
$$\text{Tr}[Q^O D^C]$$

In the SCF process, in each iteration both the closed-shell Fock matrix and the open-shell Fock matrix are diagonalized. From the two sets of eigenvectors we select the open shell orbitals either by eigenvalue or by overlap with a set of user specified vectors. The overlap selection is necessary in order to be able to create open shells with orbital energies lower than that of the highest closed shell (for example, states containing core holes or low lying f-shells in Lanthanides). After the selection, the set of occupied orbitals is re-orthogonalized.

EXAMPLE OF AVERAGING

As an example of the averaging method, consider a carbon atom (with configuration $1s^2\ 2s^2\ 2p^2$). Relativistically, the p orbitals are split into the $p_{3/2}$ and the $p_{1/2}$ orbitals (or into the E_{1u} and F_u orbitals if we use the Dirac double group corresponding to the O_h point group). Consider the configuration $1s^2 2s^2 2p_{3/2}^2$ where we have four open shell orbitals available for 2 electrons. This gives rise to six determinants each of which yields a different energy expression. The average of these six expressions yields the desired totally symmetric average energy expression.

By this construction, the Fock matrices and the resulting density matrices will be total symmetric and we will be able to use the Dacre & Elder method without further precautions.

TEST CALCULATIONS ON THE C AND SN ATOM

The average of configuration open shell method has been implemented in the MOLFDIR package, and test calculations on the C and Sn atom have been performed.

For the tests on the C atom, we used a large component basis (10s, 5p) contracted to (8s, 4p); the small component basis consists of (5s, 10p, 5d) and is left uncontracted. For the tests on the Sn atom, we used a large component basis (15s, 11p, 6d) contracted to (11s, 9p, 5d); the small component basis consists of (11s, 17p, 11d, 6f) contracted to (10s, 13p, 10d, 5f). Both total basis sets are kinetically balanced with respect to the their large component sets. The contraction coefficients are taken from an uncontracted relativistic Fock Dirac calculation.

In table 1 we list the total energies for the p^2 - average from our results and from Desclaux (1973) for comparison. In this table NR stands for non-relativistic and FD stands for Fock Dirac. The large discrepancy with respect to Desclaux' numerical results in the non-relativistic total energy of Sn is due to the fact that we have contracted the basis according to the relativistic solutions.

In fig.1, fig. 2 and table 2 the results of calculations on several averages and states which can be handled by the program are presented. Note that some averages (for example, the $1s^2 2s^2 2p_{1/2}^1 2p_{3/2}^1$) cannot be handled yet since we have the restriction that all open shell orbitals should have the same fractional occupation number.

After performing the calculation on the first average (the level in the first column) we have used the average orbitals from this calculation to start the SCF calculation in the next column. In the figures, the relaxation energy (the change in total SCF energy by optimizing the orbitals) is given in parentheses. Likewise, in the third column the number in parentheses is the relaxation energy with respect to the results from the second column.

Table 1. Atomic average of configuration open shell results for the p2 configuration.

	This work	Desclaux (1973)
Carbon NR:	-37.658 au	-37.660 au
Carbon FD:	-37.674 au	-37.676 au
Carbon diff:	-0.016 au	-0.016 au
Tin NR:	-6018.19 au	-6022.92 au
Tin FD:	-6174.92 au	-6176.14 au
Tin diff:	-156.73 au	-153.22 au

Table 2. Total SCF energies for several averages and states of the Sn and C atom in atomic units.

	Carbon	Tin
$(s, p)^4$	-37.329	-6174.645
$(p)^4$	-36.928	-6174.335
$(p_{3/2})^4$	-36.909	-6174.304
$(p_{1/2})^2 (p_{3/2})^2$	-36.925	-6174.343
$(s, p_{3/2})^4$	-37.444	-6174.723
$s^2 (p)^2$	-37.674	-6174.915
$s^2 (p_{1/2})^2$	-37.655	-6174.921
$s^2 (p_{3/2})^2$	-37.671	-6174.904

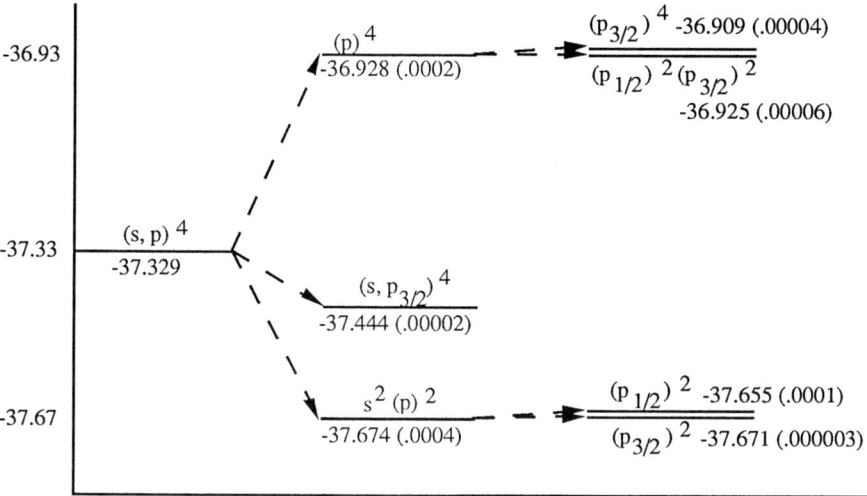

Figure 1. Total energies for the carbon atom in atomic units; between parentheses: relaxation energy with respect to average orbitals in previous column.

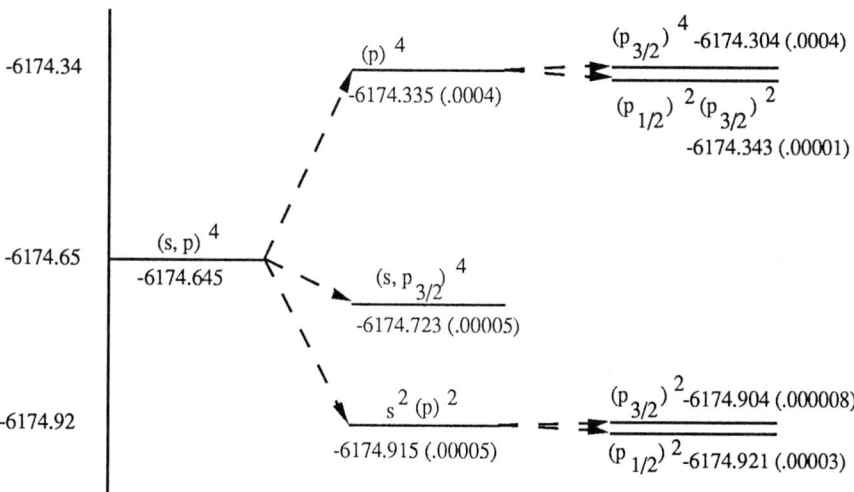

Figure 2. Total energies for the tin atom in atomic units; between parentheses: relaxation energy with respect to average orbitals in previous column.

One expects that in the case of the carbon atom, where the orbital energies of the $p_{1/2}$ and $p_{3/2}$ orbitals are almost identical, the ground state will not be the singlet $p_{1/2}^2$ state since the singlet coupling will lead to an increase in total energy. In the case of the tin atom, the orbital energy of the $p_{1/2}$ is much lower than the orbital energy of the $p_{3/2}$ orbital due to a large spin-orbit coupling, which over compensates the rise in energy due to the singlet coupling, so one expects the $p_{1/2}^2$ state to be the ground state. From the data in table 2 and figures 1 and 2 it is clear that our results confirm these expectations.

It is interesting to note that all relaxation energies are extremely small, which justifies our use of average orbitals in these calculations.

CONCLUSIONS

We have found that we can efficiently use the Dacre & Elder scheme for performing self consistent average of configuration open shell calculations. Such calculations generate average orbitals for further CI studies, and yield interesting and useful information by themselves. The average of configuration formalism is shown to work extremely well for the test systems chosen.

Currently, we are (the program is) working on some more interesting systems and more production runs are planned in the near future. The code necessary to perform the projections to specific states by CI calculations in the open shell manifold is being developed.

ACKNOWLEDGEMENT

This investigation was supported by The Netherlands Foundation for Chemical Research (SON) and the Netherlands Foundation for Fundamental Research on Matter (FOM) with financial aid from the Netherlands Organisation for Scientific Research (NWO).

LITERATURE

Aerts, P. J. C. , "Towards Relativistic Quantum Chemistry", thesis, Groningen (1986).

Aguilar-Ancono, A., Gazquez J. L., and Keller J., Approximate relativistic calculations within the one-center approximation for the series CH_4 to PbH_4, Chem. Phys. Lett. **96**, 200 (1983).

Dacre, P. D., On the use of symmetry in SCF calculations, Chem. Phys. Lett., 7, 47 (1970).

Desclaux, J. P., Relativistic Dirac-Fock expectation values for atoms with $Z = 1$ to $Z = 120$, Atomic data and Nuclear data tables, 12, 311 (1973).

Desclaux J. P., and Pyykkö, P., Relativistic and non-relativistic Hartree-Fock one-centre expansion calculations for the series CH_4 to PbH_4 within the spherical approximation, Chem. Phys. Lett. **29**, 534 (1974).

Elder, M., Use of Molecular Symmetry in SCF Calculations, Int. J. of Quantum Chem., **7**, 75 (1973).

Fernandez, J., Arriau, J., and Dargelos, A., Jahn-Teller distortion in the SnH_4^+ ion, Chem. Phys., **94**, 397 (1985).

Piggelen, H. U. van, "AB INITIO Calculations on the electronic states of $4f^n$ ions with applications to $[EuO_6]^{9-}$", thesis, Groningen (1978).

Pyykkö, P., "Relativistic Theory of Atoms and Molecules; A Bibliography 1916 - 1985.", Springer-Verlag, Berlin (1986).

Roothaan, C. C. J., Self-Consistent Field Theory for Open Shells of Electronic Systems, Rev. Mod. Phys., **32**, 179 (1960).

Visscher, L., Aerts, P. J. C., and Visser, O., General contraction in four-component relativistic Hartree-Fock calculations, IN: this publication, (1990)

McWeeny, R., "Methods of Molecular Quantum Mechanics", Academic Press, London (1989).

General contraction in four-component relativistic Hartree-Fock calculations

L. Visscher, P.J.C. Aerts and O. Visser

Laboratory for Chemical Physics
State University of Groningen
Nijenborgh 16
9747 AG Groningen
The Netherlands

ABSTRACT

General contraction is a well known means to reduce the computational effort for calculating electronic wave functions using basis set expansion techniques. In contrast to fixed contraction, optimal flexibility is available in constructing the "best" basis functions from basis set primitives. For relativistic four-component Dirac-Hartree-Fock calculations (and beyond), the variational space may consist of separate basis functions for the "Large" and "Small" component bases. The choice, and therefore also the contraction, of the Small component basis relative to the Large component basis is non-trivial. In particular in the case that "Kinetic Balance" is used to define the Small component basis relative to the Large component basis contraction imposes a severe problem, because the better the contraction of the Large component, the worse the kinetically balanced Small component counterpart becomes. Solutions to this dilemma are provided and have been implemented. Figures on reduction in the computational effort are given.

1. INTRODUCTION

In the eighties a growing interest has been developed into relativistic effects in chemical compounds (atoms, molecules and the solid state). In order to be able to calculate the relativistic electronic wave functions of atoms and molecules computer programs [1] have been devised that could tackle a specific category of systems, mainly atomic programs, both based on numerical integration of the Relativistic Hartree-Fock equations or on basis set expansion, and programs for linear molecules. Programs for general molecules have been very few, the only known production program to calculate complete all-electron four component Hartree-Fock-Dirac SCF wave functions with kinetic balance has been reported in 1985 [2]. Main reason is of course the tremendous computational effort to produce accurate solutions for such systems. Yet such calculations are becoming more and more feasible, both due to improved computer hardware and to theoretical developments. Here it is shown, that general contraction can be employed elegantly to reduce both the Large and Small component basis set sizes, yet keeping both sufficient variational freedom and kinetic balance.

2. KINETIC BALANCE

Although the Dirac operator and the Schrödinger operator are operators in different mathematical spaces, a correspondence exists between the solutions of the Dirac equation for large c, and the solutions of the Schrödinger equation. Without taking too much mathematical scrutiny, this is easily demonstrated by using the "method of the small components".

$$\{c \, \alpha.\mathbf{p} - \beta m_0 c^2 + q\phi\} \Psi = W \Psi \tag{1}$$

is the Dirac equation. Shifting the energy by the rest mass energy m_0c^2 of the electron and defining Ψ_L and Ψ_S as the Large component and Small component of the positive energy solutions respectively, this equation, written in bi-spinor form becomes:

$$\begin{bmatrix} q\phi & c\sigma.\mathbf{p} \\ c\sigma.\mathbf{p} & q\phi - 2m_0c^2 \end{bmatrix} \begin{bmatrix} \Psi_L \\ \Psi_S \end{bmatrix} = E \begin{bmatrix} \Psi_L \\ \Psi_S \end{bmatrix} \tag{2}$$

which in the limit $c \to \infty$, becomes

$$\{q\phi + (2m_0)^{-1}(\sigma.\mathbf{p})^2\} \Psi_L = E \Psi_L \tag{3}$$

This is equal to the Schrödinger equation for the two-component Ψ_L.

$$\{q\phi + (2m_0)^{-1}\mathbf{p}^2\} \Psi_L = E \Psi_L \tag{4}$$

since

$$(\sigma.p)(\sigma.p) = p^2 \qquad (5)$$

Thus Ψ_L, the upper part of the Dirac solution, equals the Schrödinger solution Ψ_{NR} when c is infinite.

If a basis set expansion is used to approximate the solutions of this equation and basis sets $\{\phi_L\}$ and $\{\phi_S\}$ are chosen to expand Ψ_L and Ψ_S then the matrix representation of the product $(\sigma.p)(\sigma.p)$ in equation (5) no longer equals the matrix representation of p^2, unless $\{\phi_S\}$ is carefully related to the basis $\{\phi_L\}$. In particular, for a basis $\{\phi_S\}$, defined by

$$\{\phi_S\} = \{\sigma.p\ \phi_L\} \cup \{\phi_i, i=1,k\} \qquad (6)$$

equation (5) can be mimicked exactly:

$$[\sigma.p]_{LS} [\sigma.p]_{SL} = [p^2]_{LL} \qquad (7)$$

provided the so constructed basis functions remain regular.

In practise, when using gaussian basis functions $r^n e^{-\alpha r^2}$ in the Large component, we use the kinetically balanced Small component $r^{n-1}e^{-\alpha r^2}$ (if n >0) and $r^{n+1}e^{-\alpha r^2}$ gaussian basis functions as separate basis functions. This extends the basis $\{\phi_S\}$ beyond strict kinetic balance, which is established by chosing $\{\phi_S\} = \{\sigma.p\ \phi_L\}$. In the strict kinetic balance approach one would have to take only one combination of the $r^{n-1}e^{-\alpha r^2}$ and $r^{n+1}e^{-\alpha r^2}$ gaussian basis functions.

Using kinetic balance guarantees that for $c \to \infty$ the Schrödinger limit is obtained even in a basis set approximation, but it does not guarantee a sufficiently good approximate solution for a finite value of c. In fact strict kinetic balance cramps the basis set more as relativistic effects grow. On the other hand, letting kinetic balance slip for, let's say, the most relativistic orbitals while keeping it for the intermediate and valence orbitals is tempting, but one may lose control over the process as a whole, risking spurious solutions to intrude.

3. GENERAL CONTRACTION

Basis set contraction is the technique by which primitive basis functions are combined into compound basis functions. The gain is a smaller number of improved basis functions at the expense of loss of variational freedom. The technique is applied frequently, in particular for molecular systems, using Gaussian primitives. In the usual sequential contraction the basis is partitioned in groups of functions, which are then combined into compound functions, using each primitive only once in any compound function. If, occasionally, a primitive is used more then once no use is made of this fact for reducing the subsequent calculations of integrals.

In a general contraction scheme [3], the primitives may (or may not) be partitioned into groups, but within each group each primitive may be used in more than one combination with the other primitives up to the number of primitives in that group. In subsequent integral calculations use is made explicitly of the fact that the primitives are the same for all combinations in a group. In this way, the primitives in the basis set can be exploited better. The necessary additional transformations during the calculation of the integrals over the contracted basis take hardly more computational effort, especially when the transformations can be carried out in core memory, which nowadays is usually large enough for this purpose.

4. CONTRACTION OF THE LARGE COMPONENT BASIS

Whether or not general contraction is applied, contraction of the large component basis for a Dirac-based four-component electronic wave function calculation is straightforward and analogous to the corresponding non-relativistic Schrödinger basis. In principle the same considerations apply.

Our approach is to base the contraction coefficients for a molecular calculation on the results of previous atomic calculations in large basis sets. By using the occupied atomic orbitals as basis functions one keeps a good description of the inner orbitals in the molecule. For describing the binding region one can either add diffuse functions or use atomic virtuals.

In this approach some differences occur with respect to a non-relativistic calculation. Implementation independent is the fact, that the innermost atomic solutions of the Dirac equation (using a point charge for the nucleus) are divergent at the origin. Representing such functions approximately by gaussian basis functions takes a lot of them. This situation becomes worse the larger the nuclear charge. This calls for heavier contraction involving more primitives (in total) than in a corresponding non-relativistic calculation.

Implementation dependent is the fact, that radially equal primitives will be used in different spin-angular symmetries (i.e. $j = l \pm 1/2$, such as $p_{1/2}$ and $p_{3/2}$, $d_{3/2}$ and $d_{5/2}$ etc.), so that some "best" contraction for a $j = l - 1/2$ ("$p_{1/2}$") is not at all the optimum contraction for a $j = l + 1/2$ ("$p_{3/2}$") basis function. General contraction can, in principle, be used to construct "best" basis functions for both $j = l - 1/2$ and $j = l + 1/2$ symmetries, using different contractions on the same primitives.

In our program an overlap criterion is used to decide whether the radial parts of the j = l - 1/2 and j = l + 1/2 are significantly different. If they are the two functions are used as separate basis functions.

5. CONTRACTION OF THE SMALL COMPONENT BASIS

Kinetic balance for basis sets as defined in equation (6) is in principle independent of the fact that contraction is used. But applying the prescription for constructing a kinetically balanced basis set from a given Large component basis may cause unexpected problems if contracted gaussians are used instead of gaussian primitives. This has to do with the singular character of the innermost atomic orbitals. Suppose one systematically contracts one large component basis set function to represent the large component part of the $1s_{1/2}$ orbital for successively improved (i.e. enlarged) basis sets. Continuing this process one should end up with successive improvements of the large component of the exact $1s_{1/2}$ orbital. The kinetically balanced small component function defined by $\sigma.p 1s_{1/2}$, however, diverges as $r^{(\gamma-2)}$ (with $\gamma = (\kappa^2 - Z^2/c^2)^{1/2}$) and is therefore too singular to be used for calculating matrix elements (including the norm of the orbital). So the large components of nearby exact solutions cannot be used to construct the small component by kinetic balance.

In practice the situation is not at all so severe, but the better the individual contracted large component functions, the "worse" the individual kinetically balanced small component functions become, as is shown in table 1. Those results confirm our earlier results [2] but are in contradiction with those recently reported by Ishikawa et al. [4].

Table 1. Matrix elements and energy of a Hydrogen-like system with Z=50 as a function of the number of basis functions in the contraction and of contraction method.

# prim.	$<\Psi_L\|-Z/r\|\Psi_L>$	$i<\Psi_S\|c\sigma.p\|\Psi_S>$	$<\Psi_S\|-Z/r\|\Psi_S>$	$E_{1s_{1/2}}$
12 a	-2685.10	7379.05	-3400.28	-1311.70
25 a	-2685.11	7379.23	-3477.87	-1314.24
12 b	-2685.10	7359.12	-2685.02	-1294.62
25 b	-2685.11	7359.14	-2685.10	-1294.63
exact	-2685.11	7359.15	-2685.11	-1294.63

a: $\Psi_S \sim c\sigma.p\Psi_L$.
b: Ψ_S is the Small component part of the Dirac solution in the uncontracted basis

General contraction can help here too. Because usually more than two primitives are involved in a contraction, one even reduces the basis set size (relative to not contracting) if for each large component function two small component functions are included in the small component basis. One of these can be chosen kinetically balanced (securing kinetic balance for the basis as a whole), the other arbitrary, but orthogonal (or any linear combination of such solutions).

This way one keeps all the benefits of overall kinetic balance, while reducing the small component basis according to the same guide-lines as used for reducing the large component basis by the contraction.

In our approach the overlap is calculated between the small component parts (b) of the atomic SCF orbitals, calculated using an uncontracted basis, and the small component functions (a) which kinetically balance the present large component basis. If the overlap is large, only the kinetically balanced functions are used, if the overlap is too small both the atomic solutions and the kinetically balanced functions are used as separate variational small component functions. In this way one retains the complete atomic solution in the basis.

6. AN EXAMPLE

As an example to illustrate the method described above we have chosen two calculations on group IV tetrahydrides. More details of the bondlength optimization calculations on this type of molecules will be presented elsewhere.

The symmetry of this type of molecules is T_d and calculations on the molecules were carried out in this symmetry.

Atomic calculations were carried out in O_h symmetry, which is the highest symmetry point group implemented in the *molecular* program.

SiH4

Basis I : Conventional contracted bases :
Si :   Large   12(9) s, 8(6) p   Small   8 s, 12 p, 8 d
H :   Large   6(3) s   Small   6 p

Basis II : General contracted bases constructed from basis I :
Si :   Large   12(6) s, 8(4) p   Small   8(4) s, 12(7) p, 8(5) d
H :   Large   6(3) s   Small   6(3) p

Table 2. Total number of two electron integrals (SiH_4)

	(LL|LL)	(SS|LL)	(SS|SS)
Basis I	19932	401166	2513002
Basis II	7284	89809	359447
Reduction	63 %	78 %	86 %

Table 3. Total SCF energies in atomic units, Si-H distances in Å

	Si atom	SiH4 (1.46)	SiH4 (1.48)	SiH4 (1.50)
Basis I	-289.439098	- 291.824204	-291.825209	-291.825013
Basis II	-289.438389	- 291.822274	-291.823446	-291.823411

SnH4

Basis I : Uncontracted basis :
Sn :   Large   15 s, 11 p, 6 d   Small   11 s, 15 p, 11 d, 6 f
H :   Large   6 s   Small   6 p

Basis II : Conventionally contracted basis constructed from basis I :
Sn :   Large   11 s, 9 p, 5 d   Small   10 s, 13 p, 10 d, 5 f
H :   Large   3 s   Small   3 p

Basis III : Minimal general contracted basis constructed from basis I :
Sn :   Large   5 s, 8 p, 4 d   Small   11 s, 14 p, 11 d, 6 f
H :   Large   2 s   Small   2 p

Basis III$_v$ is basis c plus the first s and p virtual orbitals.
Basis III$_d$ is basis c with one diffuse s and one diffuse p uncontracted.

Table 4. Total number of two electron integrals (SnH$_4$)

	(LL|LL)	(SS|LL)	(SS|SS)
Basis I	$5.4 \cdot 10^5$	$5.0 \cdot 10^6$	$1.3 \cdot 10^7$
Basis II	$2.1 \cdot 10^5$	$2.1 \cdot 10^6$	$5.6 \cdot 10^6$
Basis III$_v$	$7.3 \cdot 10^4$	$1.4 \cdot 10^6$	$6.7 \cdot 10^6$
Basis III$_d$	$7.3 \cdot 10^4$	$1.1 \cdot 10^6$	$4.8 \cdot 10^6$

Table 5. Total SCF energies in atomic units, Sn-H distance in Å

	Sn atom	SnH4 (1.73)
Basis I	-6175.20410	-6177.4868
Basis II	-6174.92100	-6177.1939
Basis III	-6175.20456	
Basis III$_v$		-6177.4800
Basis III$_d$		-6177.4815

7. ACKNOWLEDGEMENT

This investigation was supported by the Netherlands Foundation for Fundamental Research on Matter (FOM) and The Netherlands Foundation for Chemical Research (SON) with financial aid from the Netherlands Organisation for Scientific Research (NWO).

8. REFERENCES

1. P. Pyykkö, "Relativistic Theory of Atoms and Molecules, a Bibliography 1916-1985", Springer-Verlag, Berlin (1986).
2. P. J. C. Aerts, W. C. Nieuwpoort, On the use of gaussian basis sets to solve the Hartree Fock Dirac equation. I. Application to one electron systems, Chem. Phys. Lett., **113**, 165 (1985).
3. R. C. Raffenetti, General contraction of Gaussian atomic orbitals : Core valence, polarization, and diffuse basis sets; Molecular integral evaluation, J. of Chemical Physics, **58**, 4452 (1973).
4. Y. Ishikawa, H. Sekino, R.C. Binning Jr., Effects of basis set contraction in relativistic calculations on Neon, Argon and Germanium, Chem. Phys. Lett., **165**, 237 (1990).

ACCURATE RELATIVISTIC DIRAC-FOCK AND MBPT CALCULATIONS ON ARGON WITH BASIS SETS OF CONTRACTED GAUSSIAN FUNCTIONS

Yasuyuki Ishikawa and R. C. Binning, Jr.

Department of Chemistry and
Chemical Physics Program
The University of Puerto Rico
Rio Piedras, PR   00931   USA

INTRODUCTION

Most electronic structure calculations on molecules are done with basis sets of Gaussian type functions (GTF). Multi-center integrals over GTF are easily evaluated. However, Gaussians behave correctly at neither short nor long range; longer expansions in GTF than in exponential functions are required to attain similar accuracy. Much of the additional length is due to functions added simply to represent the region very near the nucleus. Primitive functions which represent this region may be grouped for the purpose of self-consistent field (SCF) and correlation correction procedures; the grouping is known as contraction. Contraction reduces integral storage requirements and time needed for SCF and post-SCF steps. Extensively characterized sets of contracted GTF for molecular calculations (Dunning and Hay, 1977; Wilson, 1987) exist.

The approach to contraction for molecular calculations has been articulated by Pople et al. (1980), Huzinaga (1984), and Dunning (1971). Functions which represent the core region are extensively contracted while valence functions are left uncontracted to provide flexibility in describing electronic changes which attend processes of bond formation or ionization. Even extensively contracted basis sets can, when corrected for truncation of the polarization series and for valence shell correlation effects, yield accurate bond dissociation and ionization energies (Pople and Curtiss, 1988).

In nonrelativistic atomic calculations which employ the basis set expansion method, exponential or Slater type functions (STF) are employed. STF can reproduce the cusp condition on the wavefunction at the origin, and they decay properly at long range. In relativistic atomic calculations, on the other hand, GTF basis sets have advantages. In heavy atoms the point nucleus approximation is often discarded in favor of a finite nuclear model. With the uniformly charged sphere model the wavefunction near the origin is Gaussian (Ishikawa and Quiney, 1987). Basis sets of GTF used with that model can be more compact than those used with a particle point model (Ishikawa et al., 1985). In addition basis sets of GTF do not exhibit the tendency to linear dependency shown by basis sets of STF (Wilson, 1988). Atomic calculations usually aim at accounting in some way for correlation effects. When those effects are treated by diagrammatic many-body perturbation theory (MBPT), the large virtual orbital coefficients characteristic of near linear dependence can introduce significant round-off effects into sums for third- and higher order diagrams.

The technique of basis set contraction has not been widely employed in atomic calculations. Atomic calculations which aim at high accuracy employ the largest practical basis sets. Yet, provided the error introduced by basis set contraction is kept to a fraction of the absolute error in an atomic calculation, contracted basis sets should provide a useful means of curtailing computation time. Basis sets of GTF contain functions of high exponent which are needed to describe inner core dynamics in DF calculations but which are not needed to describe correlation effects. This hypothesis has motivated us to study the use of contracted functions in accurate calculations.

Ishikawa et al. (1990) have recently reported calculations on neon, argon and germanium with basis sets derived for molecular calculations. Some results from that study appear in Table I. The basis sets employed are McLean and Chandler's (1980). The contracted basis sets reproduce the relativistic correction well. However, they give relatively poor values for second and third order MBPT energy corrections, some in error by more than 30%. These basis sets have been derived to reproduce valence, but not core or core-valence, interactions. Caution must be taken in using them in calculations which must recover core correlation energy. Basis sets contracted for accurate calculations on atoms must rely on slightly different criteria than those contracted for molecular calculations. A basis set must have flexibility in the core as well as the valence region and must contain functions which have significant amplitude in all regions in which electrons are to be correlated.

To test the feasibility of doing accurate DF and MBPT calculations on atoms with contracted GTF basis sets, we have performed HF SCF and second and third order MBPT calculations on ground state argon with moderately large well-tempered (Matsuoka and Huzinaga, 1987) GTF basis sets of s- and p-type functions. The calculations have been repeated with several contracted versions of the basis sets. Finally, DF and second order MBPT calculations have been done with each basis set augmented by a partial-wave expansion in functions of up to h-type.

In a number of atomic calculations the major source of error may be laid to truncation of the partial-wave expansion, rather than, for instance, truncation of the perturbation series. The point at which the series in higher angular momentum functions is ended is ordinarily determined by the time available for computation. Basis sets contracted such that energy loss due to contraction is a minor source of error, may permit a longer partial wave expansion, hence a more accurate energy determination than would have been possible with an uncontracted basis set.

Table I. Dirac-Fock (DF) and nonrelativistic limit (NR) SCF, and second and third order MBPT energies (a.u.), for argon atom in basis sets derived for molecular calculations.

Basis sets	(12s9p1d)	[6s4p1d]	[6s5p1d]
E_{DF}	−528.67101	−528.65732	−528.66164
E_2	− 0.31058	− 0.18595	− 0.20094
E_3	− 0.01031	− 0.01117	− 0.01136
E_{NR}	−526.80815	−526.80226	−526.80648
E_2	− 0.30927	− 0.18578	− 0.20085
E_3	− 0.01039	− 0.01118	− 0.01137

COMPUTATIONAL

The N-electron Hamiltonian for the DF and MBPT calculations is the Dirac-Coulomb Hamiltonian. In its use the assumption is made that negative energy states are filled. In the MBPT calculations the Hamiltonian is partitioned (Møller and Plesset, 1934) as

$$H = H_0 + H'$$

with

$$H_0 = \sum_{i=1}^{N} F_D(i) \quad \text{and} \quad H' = H - \sum_{i=1}^{N} F_D(i)$$

F_D is a DF one-electron Hamiltonian. Basis sets were balanced as outlined by Ishikawa et al. (1985) and Quiney et al. (1987). Application to contracted GTF is a direct extension of the application to uncontracted basis sets. The large component radial orbitals, $P_k(r)$, are expanded in a basis of contracted GTF, $\{X^L_{ki}(r)\}$:

$$P_k(r) = \sum_i c^L_{ki} X^L_{ki}(r).$$

The $\{c^L_{ki}\}$ are the expansion coefficients.

$$X^L_{ki}(r) = \sum_p d_{kip} f^L_{kp}(r)$$

The $\{d_{kip}\}$ are the contraction coefficients, and the $\{f^L_{kp}\}$ are the primitive GTF. The small component orbitals, $Q_k(r)$, are similarly expanded in a basis $\{X^S_{ki}\}$;

$$Q_k(r) = \sum_i c^S_{ki} X^S_{ki}(r),$$

Table II. Exponents of GTF basis functions employed in the reported calculations. Exponents of higher angular momentum functions used to generate the virtual orbitals are also noted.

Exponent	s	p	d	f	g	h
1192038.6	X					
176715.15	X					
40309.619	X					
11089.577	X					
3461.2322	X	(X)[a]				
1163.6413	X	X				
409.93944	X	X	X			
153.08437	X	X	X	X		
60.637511	X	X	X	X	X	
25.040039	X	X	X	X	X	X
10.550746	X	X	X	X	X	X
4.571726	X	X	X	X	X	X
2.015611	X	X	X	X	X	X
0.836350	X	X	X	X	X	X
0.332530	X	X	X			
0.125515	X	X				

[a] Included in the (16s12p) basis set and in contracted sets derived from it.

and X^S_{ki} is related to X^L_{ki} as

$$X^S_{ki} = (d/dr + k/r) X^L_{ki}$$

as the condition of balance.

The finite nucleus model described by Ishikawa et al. (1985) was employed in the relativistic calculations; a point nucleus model was assumed in the nonrelativistic calculations. A V^N potential was used in the determination of the virtual orbitals in the relativistic calculations, and Goldstone diagrams (Goldstone, 1957) were summed to compute perturbation corrections. Summation was done within the subspace of the positive energy branch to eliminate the "continuum dissolution" problem, and the no-pair approximation has been used (Sucher, 1980). The value of the speed of light used in the relativistic calculations was 137.0370 a.u.

Contraction coefficients were optimized using the HF SCF energy as criterion. Optimization was done iteratively on individual coefficients. Contractions were performed on s- and p-functions of the (16s11p) basis set indicated in Table II, and on p-functions of the (16s12p) basis set. In each case, the contracted basis set was derived by grouping the functions of highest exponents.

DISCUSSION OF RESULTS

Table III shows the results of nonrelativistic calculations

on ground state argon. The (16s11p) and (16s12p) basis sets yield energies 1 - 2 millihartrees shy of the Hartree-Fock limit, which has been estimated as -526.8175 a.u. (Froese-Fischer, 1977) and -526.818 a.u. (Desclaux, 1973). The calculated second order perturbation correction, E_2, is about a millihartree above its s,p limit, for which there is no exact estimate. Contraction of the s-functions to [14s] and to [12s] results in degradations of SCF energy and in MBPT energies which are negligible compared to the error arising from truncation of the basis. Contraction of the p-set from (11p) or (12p) to [10p] results in larger, still minor, errors in E_2 and E_3. The effects of s- and p- contractions is approximately additive.

Results of relativistic DF and second and third order MBPT calculations are presented in Table IV. Desclaux (1973) has reported an energy of -528.683 hartree as the numerical Dirac-Fock limit for ground state argon. As expected, contraction of the s-functions of the basis set has a slightly greater effect on the DF energy than on the HF. Relativistic E_2 and E_3, however, are no more affected than nonrelativistic. Contraction of the (12p) set to [10p] has a greater effect on correlation corrections than does contraction of the (16s) to [12s]. The absolute error in relativistic E_2 and E_3 arising from contraction of the p-functions is about the same as the absolute error in nonrelativistic E_2. The relative error is, therefore, smaller in the relativistic case. A calculation in which the partial-wave expansion is truncated at g- instead of h-type functions is reported to indicate the magnitude of error in truncating the basis in that respect. The contraction error found in even the most severe contraction we report here is small compared to the error arising from truncation of the partial-wave series.

The error introduced by contracting the p-functions is not reduced by expanding the primitive basis set from (11p) to (12p) before contraction. Contraction of the larger primitive basis set does, of course, result in lower SCF energies. Dyall (1990) has suggested that the relatively large error induced by contraction of the p-set may derive from the fact that these functions must not only describe the 2p orbital properly, but must also serve to correlate the 1s electrons. That being the situation, correlation might be improved by increasing the number of contracted primitives by one, and adding to the basis a function solely for 1s correlation. An alternative would be to contract the p-functions in two ways, one to describe the 2p orbital, the other to optimally correlate the 1s.

We have explored the first of these alternatives. The (16s12p) basis was contracted to [14s9p]. Then an additional p-function of exponent 583.39666, chosen to optimize E_2, was added to produce the basis set labelled [14s10p*] in Table III. The nonrelativistic second and third order MBPT energies displayed in Table III do not improve upon those obtained with the [14s10p] basis set of conventional construction. The idea presented in the comment merits further exploration, but the indication from this single calculation is that contraction of the s-set does not reduce the correlation energy significantly because the functions contracted are not in the active space; the p-functions are, and contraction is restricting the space.

Table III. Calculated nonrelativistic SCF (HF), second and third order MBPT energies (in a.u.) of Ar in several basis sets.

Uncontracted basis set	Contracted basis set	HF	E_2	E_3
16s11p		-526.8158985	-0.2042687	-0.0032086
	14s11p	-526.8158985	-0.2042686	-0.0032086
	12s11p	-526.8158977	-0.2042615	-0.0032088
	16s10p	-526.8158985	-0.2040601	-0.0032128
	14s10p	-526.8158985	-0.2040600	-0.0032128
	12s10p	-526.8158977	-0.2040532	-0.0032130
16s12p		-526.8164240	-0.2042993	-0.0032075
	16s10p	-526.8164240	-0.2040564	-0.0032128
	14s10p	-526.8164240	-0.2040562	-0.0032128
	12s10p	-526.8164232	-0.2040495	-0.0032130
	14s10p*	-526.8164239	-0.2040311	-0.0032129

Table IV. Relativistic SCF (DF) and MBPT energies (in a.u.) for Ar calculated with several basis sets.

Uncontracted basis set	Contracted basis set	DF	E_2	E_3
16s11p9d7f6g5h		-528.681482	-0.682257	-0.012869
	14s11p9d7f6g5h	-528.681468	-0.682256	-0.012869
	12s11p9d7f6g5h	-528.681181	-0.682250	-0.012870
	16s10p9d7f6g5h	-528.681467	-0.682039	-0.012877
16s11p9d7f6g			-0.669685	
16s12p9d7f6g5h		-528.682406		
	16s10p9d7f6g5h	-528.682376	-0.682042	
	14s10p9d7f6g5h	-528.682364	-0.682042	
	12s10p9d7f6g5h	-528.682074	-0.682036	

The contractions reported here are modest, but they save enough time to permit a longer partial-wave series for second and third order MBPT calculations. The initial expectation that the s-set can be extensively contracted with negligible loss in accuracy is supported by this study. The p-set contains fewer high exponent functions, but a contraction can be made which induces an error of less than a percent in the correlation calculations. It will clearly be feasible to augment the higher angular momentum functions used to generate the virtual space with a contracted set of high-exponent GTF. The result will be correlation energy estimates which are more accurate than those to be obtained with uncontracted basis sets.

References

Desclaux, J.P., 1973, Relativistic Dirac-Fock expectation values for atoms with Z = 1 to Z = 120, At. Data Nucl. Data Tables, 12:311.

Dunning, T.H. and Hay, P.J., 1977, Gaussian basis sets for molecular calculations, in: " Methods of Electronic Structure Theory," H.F. Schaefer, ed., Plenum, New York.

Dunning, T.H., 1971, Gaussian basis functions for use in molecular calculations. III. Contraction of (10s6p) atomic basis sets for first-row atoms, J. Chem. Phys., 55:716.

Dyall, K.G., 1990, Comment accompanying this article.

Froese-Fischer, C., 1977, "The Hartree-Fock Method for Atoms," Wiley, New York.

Goldstone, J., 1957, Derivation of the Brueckner many-body theory, Proc. Roy. Soc. London Ser. A, 239:267.

Huzinaga, S., ed., 1984, "Gaussian Basis Sets for Molecular Calculations," Elsevier, Amsterdam.

Ishikawa, Y., Baretty, R., and Binning, R.C., 1985, Relativistic Gaussian basis set calculations on one-electron ions with a nucleus of finite extent, Chem. Phys. Lett., 121:130.

Ishikawa, Y. and Quiney, H.M., 1987, On the use of an extended nucleus in Dirac-Fock Gaussian basis set calculations, Int. J. Quantum Chem., Quantum Chem. Symp., 21:523.

Ishikawa, Y., Sekino, H., and Binning, R.C., 1990, Effects of basis set contraction in relativistic calculations on neon, argon and germanium, Chem. Phys. Lett., 165:237.

Krishnan, R., Binkley, J.S., Seeger, R., and Pople, J.A., 1980, Self-consistent molecular orbital methods. XX. A basis set for correlated wavefunctions, J. Chem. Phys., 72:650.

Matsuoka, O. and Huzinaga, S., 1987, Relativistic well-tempered Gaussian basis sets, Chem. Phys. Lett., 140:567.

McLean, A.D. and Chandler, G.S., 1980, Contracted Gaussian basis sets for molecular calculations. I. Second row atoms, Z = 11 - 18, J. Chem. Phys., 72:5639.

Møller, C. and Plesset, M.S., 1934, Note on an approximate treatment for many-electron systems, Phys. Rev., 46:618.

Pople, J.A. and Curtiss, L.A., 1988, Theoretical thermochemistry. 4. Ionization energies and proton affinities of AH_n species (A - Li to B and Na to Al); geometries and enthalpies of formation of their cations, J. Phys. Chem., 92:894.

Quiney, H.M., Grant, I.P., and Wilson, S., 1987, The Dirac equation in the algebraic approximation, Phys. Scr., 36:460.

Sucher, J., 1980, Foundations of the relativistic theory of many-electron atoms, Phys. Rev. A, 22:348.

Wilson, S., 1987, Basis sets, Adv. Chem. Phys., 67:439.

Wilson, S., 1988, Relativistic molecular structure calculations, in: "Methods in Computational Chemistry, Volume 2: Relativistic Effects in Atoms and Molecules," S. Wilson, ed., Plenum, New York.

COMMENT ON "ACCURATE RELATIVISTIC DIRAC–FOCK AND MBPT CALCULATIONS ON ARGON WITH BASIS SETS OF CONTRACTED GAUSSIAN FUNCTIONS" BY Y. ISHIKAWA AND R. C. BINNING Jr

K G Dyall

Eloret Institute, 3788 Fabian Way, Palo Alto
California 94303, USA[1]

Much of the 1s correlation comes from terms involving p functions, so that the inner p basis functions are playing a dual rôle. If they are contracted to describe the 2p orbital, the 1s correlation will be adversely affected. Have you tried contracting the p functions for the 2p, and then adding an extra tight p function for 1s correlation, or, contracting the innermost 3 or 4p functions to 2 functions, one for the 2p orbital, and one for 1s correlation (by maximizing the radial overlap, for example).

---

[1] Mailing address: NASA Ames Research Center, MS RTC 230–3, Moffett Field, CA 94035–1000, U.S.A.

RELATIVISTIC MANY—BODY PERTURBATION THEORY OF ATOMIC AND MOLECULAR ELECTRONIC STRUCTURE

Stephen Wilson

Rutherford Appleton Laboratory
Chilton, Oxfordshire, OX11 0QX, England

1. Introduction

Over the past thirty years, the many—body perturbation theory has emerged as one of the principal techniques for attacking the electron correlation problem in atoms and molecules. When formulated within the algebraic approximation (that is, the approximation in which the single particle state functions are parametrized by expansion in some finite set of basis functions transforming the integro—differential Hartree—Fock equations into a set of algebraic equations for the expansion coefficients) the many—body perturbation expansion (Wilson and Silver 1976, Wilson 1984 and references therein) is a particularly accurate, flexible and efficient procedure.

For systems which contain heavy atoms, the non—relativistic quantum mechanics on which almost all contemporary treatments of electron correlation are based, is inadequate because the mean speed of core electrons is a substantial fraction of the speed of light, so that a fully relativistic electronic structure theory is required. The proper treatment of relativity and quantum electrodynamic effects will demand increasing attention in the years ahead, since these may be more important than electron correlation in heavy elements and are thus an essential ingredient of an *ab initio* electronic structure studies for the lower rows of the Periodic Table. A great deal of research over the past decade has been directed towards the development of a relativistic many—body perturbation theory formulated within the algebraic approximation. As illucidated in sections 2 and 3, problems of principle have been largely resolved and the remaining problems are primarily of a computational nature.

In this paper, we address some of the considerable computational demands of the relativistic electron correlation problem, giving particular attention to the molecular case. Relativistic molecular structure calculations are sometimes regarded as two orders of magnitude more demanding than the corresponding atomic structure calculations (Kim 1989). Even non—relativistic calculations of the electronic structure of molecular systems containing heavy atoms demand considerable computational resources.

This work compliments a series of papers devoted to the computational implementation of the non—relativistic many—body perturbation theory of electron correlation effects in atoms and molecules within the algebraic approximation. The papers in the series (Silver 1978ab, Wilson 1978, Wilson and Silver 1979, Wilson and Saunders 1980, Moncrieff et al 1989, Baker et al. 1990a, see also Wilson 1981, 1983, 1985, Baker *et al* 1990b) have been concerned with the design of efficient algorithms for the non–relativistic electron correlation problem.

The outline of this paper is as follows:– In section 2, we briefly describe the background to the relativistic electron correlation problem. An overview of the relativistic many–body perturbation theory is given in section 3 and some recent results described. In section 4 and 5, we discuss two of the computational aspects of relativistic many–body perturbation theory calculations. In section 4, we describe the a technique for the exploitation of the numerical linear dependence which is always present in the matrix of two–electron integrals. This technique can be used to reduce both the processing times and the storage requirements for relativistic electronic structure calculations. The use of parallel processing techniques is considered in section 5. By employing both vector and parallel processing execution rates in excess of 2000 million floating point operations per second can be achieved in non–relativistic many–body perturbation theory calculations on the most powerful contemporary computers. Finally, a summary is given in section 6.

2. The relativistic electron correlation problem

Following the development, in the mid 1950's, of the many–body perturbation theory, by Brueckner (1955), Goldstone (1957), Hugenholtz (1957) and others (see, for example, March, Young and Samphanthar (1967)), from the quantum field theory of Feynman (1949), Dyson (1949) and Wick (1950), the first applications to atoms were reported, in the early 1960's by Kelly (1963). He generated numerical solutions to the radial Hartree–Fock equations using finite differences, which he then used as a reference for a perturbative treatment of correlation effects. The integration over the continuum states formed the most demanding part of these calculations.

In the late 1960's, the first applications of the many–body perturbation theory to the molecular electronic structure problem were also reported by Kelly (1969). He performed molecular calculations on systems containing only one non–hydrogenic atom, which was used as an expansion centre, and treated the hydrogen nuclei as additional perturbations.

Applications of the method to arbitrary molecular systems awaited the introduction of basis set expansion techniques, that is the algebraic approximation, of the type which are traditional in molecular electronic structure theory. This was achieved in the mid–1970's by Kaldor (1975) and by Silver and his coworkers (Wilson and Silver (1976, 1977) and references therein).

Until the late 1970's, the "standard" approach to the correlation problem in molecules was the method of configuration mixing or configuration interaction. However, the rapid increase of the dimension of the hamiltonian matrix with increasing number of electrons in the system and number of basis functions, limited the applicability of the method to small systems and small basis sets unless the configuration expansion was severely truncated. A perturbative analysis indicates that only double excitations contribute to the energy in second and third order and thus limited "double excitation" configurations was a widely used approach. Today many–body perturbation theory has emerged as the standard approach. "Nowadays, many thousands of quantum chemistry calculations are performed on small and medium–sized molecules within the framework of perturbation theory. Such programs have become standard tools for many research workers ..." (Knowles, Somasundram, Handy and Hirao, 1985)

However, it soon became apparent that the major source of error in these calculations was attributable to basis set truncation effects. Research was first directed towards even–tempered and universal basis sets (Silver and Wilson 1978) and then systematic sequences of even–tempered basis sets (Wilson, 1980a, 1980b) and their completeness properties. For atoms and small molecules the errors associated with truncation of the basis set were reduced to the level where relativistic corrections became important for all but the lightest atoms and systems containing them (Wilson and Silver 1980, 1982).

Relativistic effects are almost always introduced into molecular electronic structure studies by means of perturbation theory. This approach is frequently employed in atomic studies also. This was the approach which we began to explore in the early 1980s (Cooper and Wilson, 1982ab), but then became convinced that it is essential to use methods which build in relativity at the orbital level from the outset if reliable results are required. At that time, some Dirac–Hartree–Fock calculations were performed for atoms but they employed finite difference methods which preclude their application to arbitrary molecular systems (Grant et al, 1980). Furthermore, because finite difference calculations do not afford a representation of the complete spectrum, a description of electron correlation effects was obtained by using a multiconfiguration wave function.

By invoking the algebraic approximation, that is by using a finite basis set, the relativistic electronic structure methods can not only be applied to molecules but also a representation of the complete Dirac spectrum is obtained allowing many–body perturbation theory treatment of correlation effects with all its theoretical and computational advantages. However, it has been known since the pioneering work of Kim (1968) that the use of basis set expansion techniques in relativistic calculations requires particular care. In a series of papers with the general title "The Dirac equation in the algebraic approximation" (Dyall, Grant and Wilson 1984abc, Quiney, Grant and Wilson 1985, 1987ab, Wood, Grant and Wilson 1984, Laaksonen, Grant and Wilson 1988), we have investigated this problem in detail.

The most frequently employed realization of the algebraic approximation in non–relativistic molecular calculations is the L.C.A.O. approximation in which molecular orbitals are approximated by some Linear Combination of Atomic Orbitals. In the relativistic case considerable care is required in the implementation of the algebraic approximation if the so–called "finite basis set disease" is to be avoided and a clean separation of the two branches of the spectrum is to be maintained. It emerges that this separation can be achieved if a spectral method is employed. According to the recent monograph by Mercier (1989), the term "Spectral methods is the name given to a numerical approach to the solution of partial differential equations. In this approach the solution to the equation is approximated by a truncated series of special functions which are eigenfunctions of some differential operator.". The key to the successful implementation of the algebraic approximation in relativistic calculations is to use basis functions which are themselves solutions of a one–electron Dirac equation. Now the separation into the two branches of the Dirac spectrum is determined both by the finite value of the speed of light and the strength of the external field. (Problems occur for superheavy atomic nuclei with very high Z (Greiner, Muller and Rafelski 1985) but are not considered here.) The admissible forms for the radial basis functions are constrained by these requirements. The basis sets which seems to have the most general applicability to relativistic atomic structure theory consist of the exponential–type functions

$$f_i^T(r) = r^\gamma \exp(-\zeta_i r)\left\{ -(1-\delta_{0n})L_{n-1}^{2\gamma}(2\zeta_i r) \pm \frac{(N_n - \kappa)}{(n + 2\gamma)} L_n^{2\gamma}(2\zeta_i r) \right\}$$

$n = 0$ for $\kappa < 0$ $i = 1, 2, ..., M_\kappa$
$n = 1$ for $\kappa > 0$

where the upper sign is taken for the large component (T=L) and the lower sign for the small component (T=S).

$$\gamma^2 = (\kappa^2 - (Z/c)^2)$$

and

$$N_n^2 = n^2 + 2n\gamma + \kappa^2$$

$L_n^\alpha(z)$ is the generalized Laguerre polynomial. The exponential parameters $\{\zeta_i\}$ are chosen so as to provide a sequence of basis sets which becomes complete in some well defined limit (Klahn and Bingel 1977ab, Schmidt and Ruedenberg 1979, Wilson 1980ab)). The non-relativistic limit of the relativistic exponential-type function basis set is a "kinetically balanced" set of exponential-type functions (Stanton and Havriliak 1984, Quiney 1988, Wilson 1988). The dimension of the radial basis for symmetry-type κ is M_κ.

3. Relativistic many-body perturbation theory

In our first relativistic many-body perturbation theory calculation employing the algebraic approximation (Quiney, Grant and Wilson, 1985), it was demonstrated in a study of a simple hydrogenic model problem that the discrete representation of the complete spectrum, comprising the bound states and the positive- and negative-energy continua, provided by the analytic basis set method greatly facilitates the evaluation of the algebraic sum-over-states expressions corresponding to each of the diagrams (Quiney, Grant and Wilson 1989abc).

In further work, we presented calculations within the independent electron model using both the Dirac-Coulomb and the Dirac-Breit hamiltonians (Quiney, Grant and Wilson 1987ab, 1989abc). It has been known for many years that the frequency-independent Breit interaction may be incorporated into the self-consistent field equations for atoms, but this approach has until recently been regarded as too demanding computationally in finite-difference calculations; first-order perturbation theory has been the method of choice. In the algebraic approximation, we have demonstrated that the self-consistent treatment of the Breit interaction is preferred to the usual method which treats the Breit interaction as a first-order perturbation to the Dirac-Coulomb operator.

Relativistic many-body calculations based on the Dirac-Coulomb hamiltonian and including energy shifts which arise through the creation of virtual electron-positron pairs have been reported by Quiney (1988) and Quiney, Grant and Wilson (1987b, 1989ab). Very recent calculations (Quiney, Grant and Wilson 1990), which are discussed further below, reveal that the terms in the perturbation expansion corresponding to the self-consistent modification of the one-electron orbitals due to the Breit interaction result in the most significant relativistic many-body effect.

The principal advantage of the algebraic approximation technique described here is that the two-electron integrals over the basis functions which are required in many-body theory may be evaluated both efficiently and accurately. The transformation of the integrals over basis functions into relativistic matrix elements may be performed using linear algebra techniques, so that the scheme may be programmed for high performance on vector and parallel processing computers (Quiney 1990, Wilson 1990). Many of the developments embodied in non-relativistic electronic structure computer codes (Wilson 1990) are transferred to the relativistic method without substantial modification. We suggest that this method has advantages over the alternatives (e.g. Johnson and Sapirstein 1986, Salomonsson and Öster 1989) on the grounds of simplicity and computational efficiency.

The negative-energy states are taken to be filled in the true vacuum state. The events which occur in a relativistic many-body system are conveniently described within the particle-hole formalism in which the negative energy continuum and

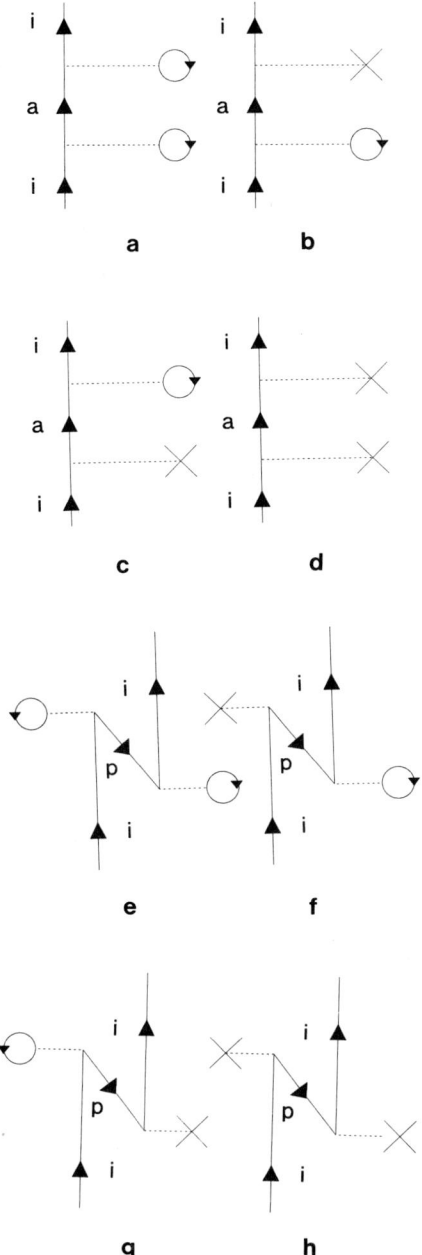

Figure 1. *Diagrammatic representation of some second−order energy contributions involving single excitations. Upwards directed lines represent single particle states in the positive energy branch of the spectrum whilst downward directed lines represent single particle states in the negative energy branch. The horizontal dashed line terminated by a cross represent a one−electron time−independent potential, such as the Dirac−Hartree−Fock (−Coulomb/−Breit) potential. The horizontal dashed line is taken to represent either the instantaneous Coulomb interaction or the sum of the Coulomb interaction and frequency−independent Breit interaction. The indices $i,j,k,...$ label occupied positive energy states and the indices $a,b,c,...$ label unoccupied states whilst the indices $p,q,r,...$ label negative energy states.*

occupied positive energy states are taken to be below the Fermi level. Unlike the non–relativistic particle–hole formalism, the relativistic extension involves no restriction on particle number because virtual excitations from the negative energy sea must be considered. If we take the electron–electron interaction to be instantaneous, that is either just the Coulomb interaction or the sum of the Coulomb interaction and the frequency–independent Breit interaction, then the diagrammatic representation of the second–order energy for a closed–shell system is given in Figures 1 and 2. In Figure 1 the second order diagrams involving singly excited intermediate states are shown whilst in Figure 2 the second order diagrams which involve doubly excited intermediate states are displayed. In order to keep the time ordering of events obvious when working beyond the no virtual pair approximation, it is most convenient to use "open" Goldstone diagrams. In these diagrams, upwards directed lines are used to represent single particle states in the positive–energy branch of the spectrum and downward directed lines represent states in the negative–energy branch. All of the diagrams given in Figures 1 and 2 arise in second order when an arbitrary single determinantal function is employed as a reference for the perturbation expansion.

Diagrams (a)–(d), which arise in the no virtual pair approximation, describe single excitations creating a hole in the positive–energy branch below the Fermi level and a particle above the Fermi level. Diagrams (e)–(h), which represent effects that can be said to go beyond the no virtual pair approximation, describe the formation of a virtual electron–positron pair by creation of a hole in the negative–energy branch of the spectrum. Diagrams (i)–(l) involve double excitations. Diagram (i) arises in the no virtual pair approximation, diagrams (j) and (k) describes the creation of one virtual pair, and diagram (l) the creation of two virtual pairs.

For calculations using a bare nucleus reference function, the diagrams containing the single–particle potential, represented by a cross, do not arise. For calculations based on a Dirac–Hartree–Fock–Coulomb or Dirac–Hartree–Fock– Breit reference function there is not an exact cancellation between diagrams containing "bubbles" and "crosses" because the polarization of the negative–energy sea is not accommodated within the self–consistent field potential. Labzovskii (1971) pointed out that although the familiar cancellation of single–particle insertions occurs within the positive–energy branch of the Dirac spectrum when the Dirac–Hartree–Fock (–Coulomb/–Breit) method is employed, this cancellation does not occur in the negative–energy branch.

The algebraic expressions corresponding to each of the diagrams shown in Figures 1 and 2 can be written in terms of the one–electron integrals

$$<\mu|V|\nu>, \quad \mu,\nu = i,j,k,\ldots,a,b,c,\ldots,p,q,r,\ldots$$

in which V is some potential (for example, the Dirac–Hartree–Fock(–Breit) potential), the two–electron integrals

$$<\mu\nu|\hat{O}|\sigma\tau>, \quad \mu,\nu,\sigma,\tau = i,j,k,\ldots a,b,c,\ldots,p,q,r,\ldots$$

where $\hat{O}$ is the electron–electron interaction operator including exchange, that is,

$$\hat{O} = (I-P_{12})\, g_{12}$$

where I and the identity and P_{12} permutes the coordinates of electrons 1 and 2. g_{12} is the electron–electron interaction, which in the Dirac–Coulomb hamiltonian is simply

$$g_{12} = 1/r_{12},$$

and in the Dirac–Breit hamiltonian has the form

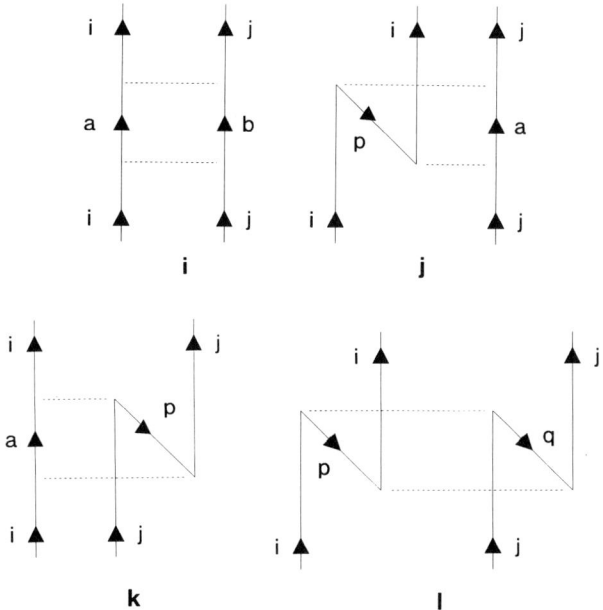

Figure 2. *Diagrammatic representation of some second−order energy contributions involving double excitations. Upwards directed lines represent single particle states in the positive energy branch of the spectrum whilst downward directed lines represent single particle states in the negative energy branch. The horizontal dashed line is taken to represent either the instantaneous Coulomb interaction or the sum of the Coulomb interaction and frequency−independent Breit interaction. The indices $i,j,k,\ldots$ label occupied positive energy states and the indices $a,b,c,\ldots$ label unoccupied states whilst the indices $p,q,r,\ldots$ label negative energy states.*

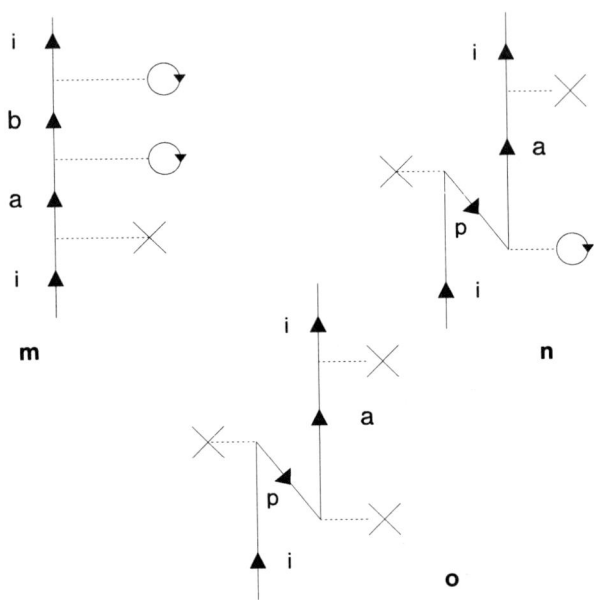

Figure 3. *Diagrammatic representation of some third−order energy contributions involving single excitations. Upwards directed lines represent single particle states in the positive energy branch of the spectrum whilst downward directed lines represent single particle states in the negative energy branch. The horizontal dashed line terminated by a cross represent a one−electron time−independent potential, such as the Dirac−Hartree−Fock (−Coulomb/−Breit) potential. The horizontal dashed line is taken to represent either the instantaneous Coulomb interaction or the sum of the Coulomb interaction and frequency−independent Breit interaction. The indices $i,j,k,...$ label occupied positive energy states and the indices $a,b,c,...$ label unoccupied states whilst the indices $p,q,r,...$ label negative energy states.*

$$g_{12} = 1/r_{12} + B_{12}$$

where B_{12} is the Breit interaction

$$B_{12} = -(2R)^{-1}[\alpha_1 \cdot \alpha_2 + R^{-2}(\alpha_1 \cdot R)(\alpha_2 \cdot R)]$$

$\epsilon_i, \ldots, \epsilon_a, \ldots, \epsilon_p, \ldots,$ are the orbitals energies. We then have for diagrams (a)–(d), which arise in the "no pair" approximation, the following expressions

$$E(a) = \sum_{ijka} \frac{\langle ij|\hat{O}|aj\rangle \langle ak|\hat{O}|ik\rangle}{\epsilon_i - \epsilon_a}$$

$$E(b) = \sum_{ija} \frac{\langle ij|\hat{O}|aj\rangle \langle a|\hat{V}|i\rangle}{\epsilon_i - \epsilon_a}$$

$$E(c) = \sum_{ija} \frac{\langle i|\hat{V}|a\rangle \langle aj|\hat{O}|ij\rangle}{\epsilon_i - \epsilon_a}$$

$$E(d) = \sum_{ia} \frac{\langle i|\hat{V}|a\rangle \langle a|\hat{V}|i\rangle}{\epsilon_i - \epsilon_a}$$

For diagrams (e)–(h), which are single excitation diagrams involving the creation of one virtual electron–positron pair, we have

$$E(e) = -\sum_{ijkp} \frac{\langle pj|\hat{O}|ij\rangle \langle ik|\hat{O}|pk\rangle}{\epsilon_i - \epsilon_p}$$

$$E(f) = -\sum_{ijp} \frac{\langle pj|\hat{O}|ij\rangle \langle i|\hat{V}|p\rangle}{\epsilon_i - \epsilon_p}$$

$$E(g) = -\sum_{ijp} \frac{\langle p|\hat{V}|i\rangle \langle ij|\hat{O}|pj\rangle}{\epsilon_i - \epsilon_p}$$

$$E(h) = -\sum_{ip} \frac{\langle p|\hat{V}|i\rangle \langle i|\hat{V}|p\rangle}{\epsilon_i - \epsilon_p}$$

For the double excitation diagrams (i)–(l), we have

$$E(i) = \frac{1}{4}\sum_{ijab} \frac{\langle ij|\hat{O}|ab\rangle \langle ab|\hat{O}|ij\rangle}{\epsilon_i + \epsilon_j - \epsilon_a - \epsilon_b}$$

$$E(j) = -\sum_{ijap} \frac{\langle pj|\hat{O}|ia\rangle \langle ia|\hat{O}|pj\rangle}{\epsilon_i + \epsilon_j - \epsilon_a - \epsilon_p}$$

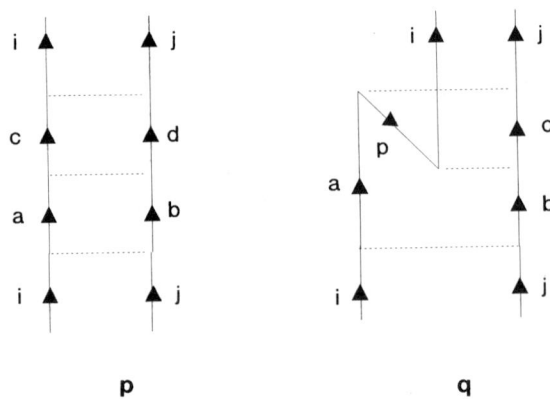

Figure 4. *Diagrammatic representation of some third–order energy contributions involving double excitations. Upwards directed lines represent single particle states in the positive energy branch of the spectrum whilst downward directed lines represent single particle states in the negative energy branch. The horizontal dashed line is taken to represent either the instantaneous Coulomb interaction or the sum of the Coulomb interaction and frequency–independent Breit interaction. The indices i,j,k,... label occupied positive energy states and the indices a,b,c,... label unoccupied states whilst the indices p,q,r,... label negative energy states.*

$$E(k) = -\sum_{ijap} \frac{\langle ip|\hat{0}|aj\rangle \langle aj|\hat{0}|ip\rangle}{\epsilon_i + \epsilon_j - \epsilon_a - \epsilon_p}$$

$$E(l) = \frac{1}{4}\sum_{ijpq} \frac{\langle pq|\hat{0}|ij\rangle \langle ij|\hat{0}|pq\rangle}{\epsilon_i + \epsilon_j - \epsilon_p - \epsilon_q}$$

Figures 3 and 4 show some typical third order diagrams. The diagrams in Figure 3 involve single excitations whilst those in Figure 4 are associated with doubly excited intermediate states. Diagram (m) arises in the no virtual pair approximation. Diagrams (n)–(o) represent effects beyond the no virtual pair approximation. Diagrams (p) arises in the no virtual pair approximation and diagram (q) represents effects beyond the no virtual pair approximation.

Very recently, we have reported relativistic and non–relativistic many–body perturbation theory calculations using systematically constructed basis sets of exponential–type functions for the ground state of the argon atom (Quiney, Grant and Wilson 1990). The relativistic calculations employed both the Dirac–Coulomb and the Dirac–Breit hamiltonian.

At the independent electron model level, the difference between the Dirac–Hartree–Fock–Coulomb energy and the Dirac–Hartree–Fock–Breit energy was found to be to 0.132365 hartree. The difference between the sum of Dirac–Hartree–Fock–Coulomb energy and the first–order Breit energy and the Dirac–Hartree–Fock–Breit energy is only 40 μhartree. The effect of the Breit interaction is most pronounced in the spinor energies and in the shapes of the spinors as reflected in the expectation values of r^k. For the inner shell spinors, changes of about 0.1% are observed in the expectation values of r^k when the Breit interaction is included. A similar conclusion was reached by Lindroth et al (1989) in studies of hyperfine interactions.

In Figure 5, a partial wave analysis of the second–order pair energies obtained from non–relativistic many–body perturbation theory calculations for the argon atom ground state using basis sets of s p and d symmetry are presented. Note that the ordinate has a logarithmic scale in microhartree. In Figures 6 and 7, a partial wave

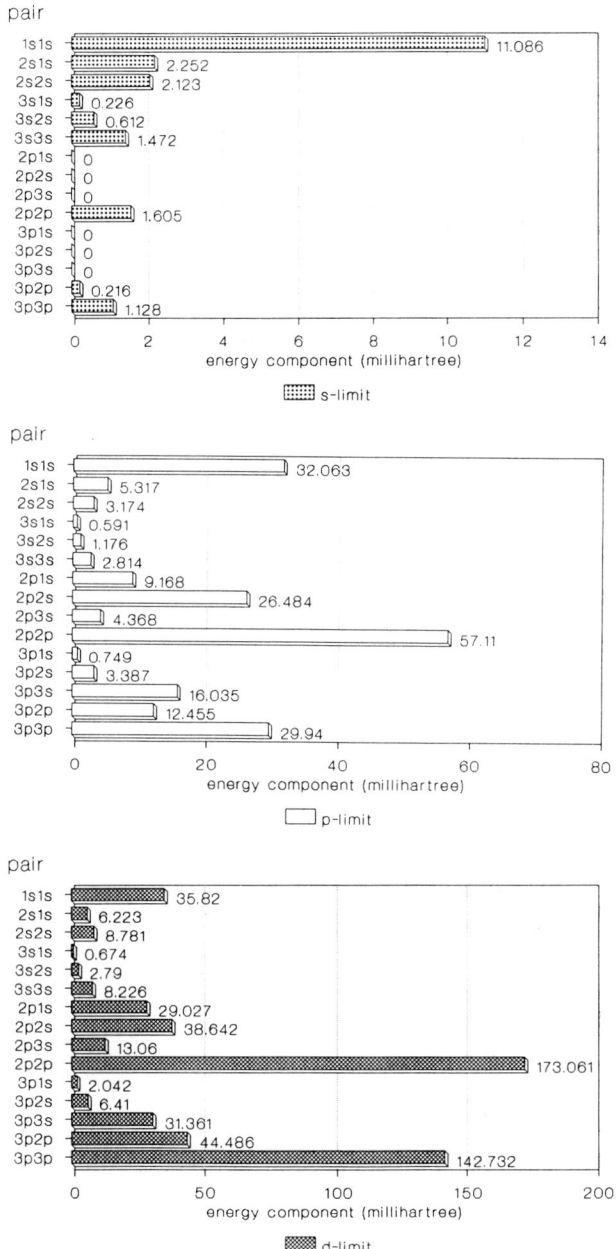

Figure 5. *Partial wave analysis of the second−order pair correlation energies for the ground state of the argon atom from calculations employing the Schrödinger hamiltonian and a basis set containing functions of s, p and d symmetry. (Based on the work of Quiney, Grant and Wilson (1990))*

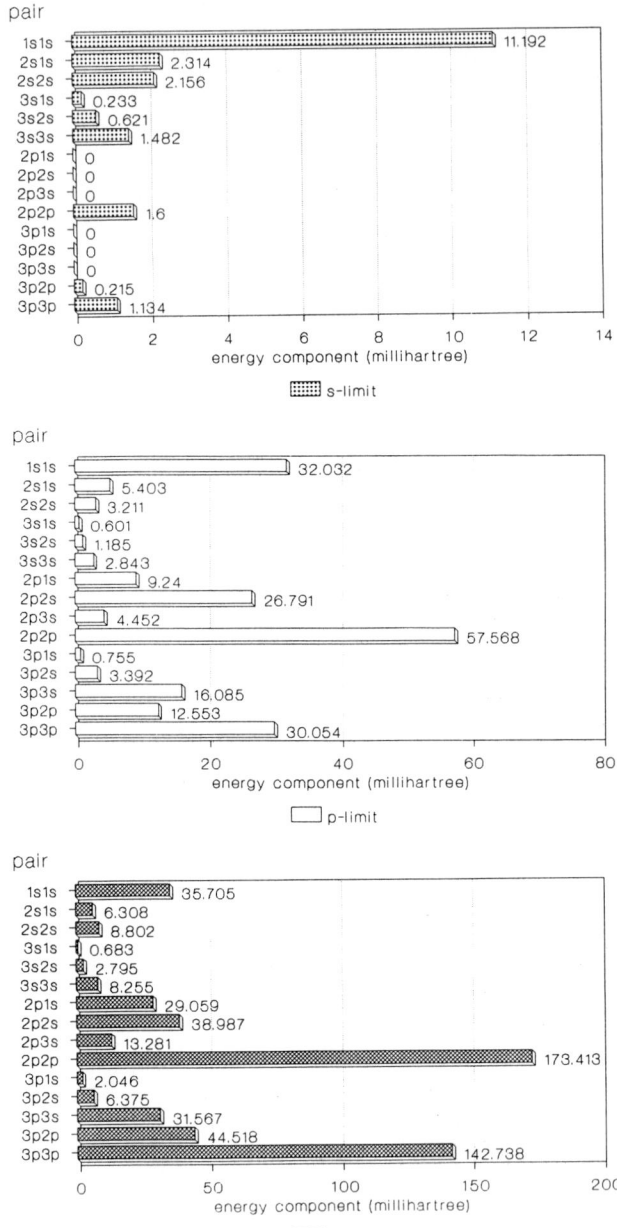

Figure 6. *Partial wave analysis of the second—order pair correlation energies for the ground state of the argon atom from calculations employing the Dirac—Coulomb hamiltonian and a basis set containing functions of s, p and d symmetry. (Based on the work of Quiney, Grant and Wilson (1990))*

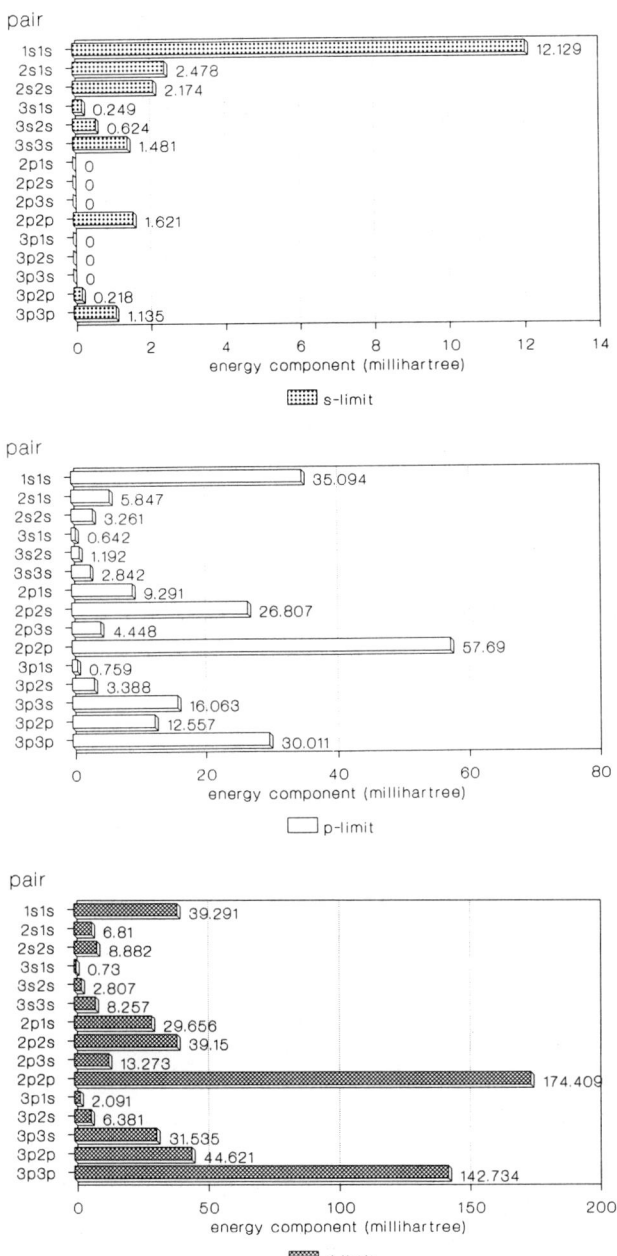

Figure 7. *Partial wave analysis of the second−order pair correlation energies for the ground state of the argon atom from calculations employing the Dirac−Breit hamiltonian and a basis set containing functions of s, p and d symmetry. (Based on the work of Quiney, Grant and Wilson (1990))*

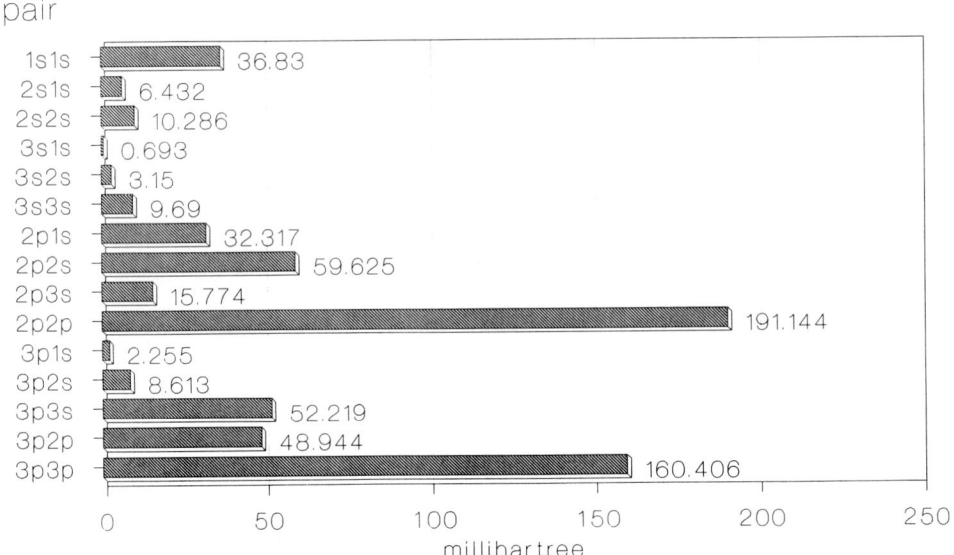

Figure 8. *Second–order pair correlation energies for the ground state of the argon atom from calculations employing the Schrödinger hamiltonian and a basis set containing functions of s, p, d and f symmetry. (Based on the work of Quiney, Grant and Wilson (1990))*

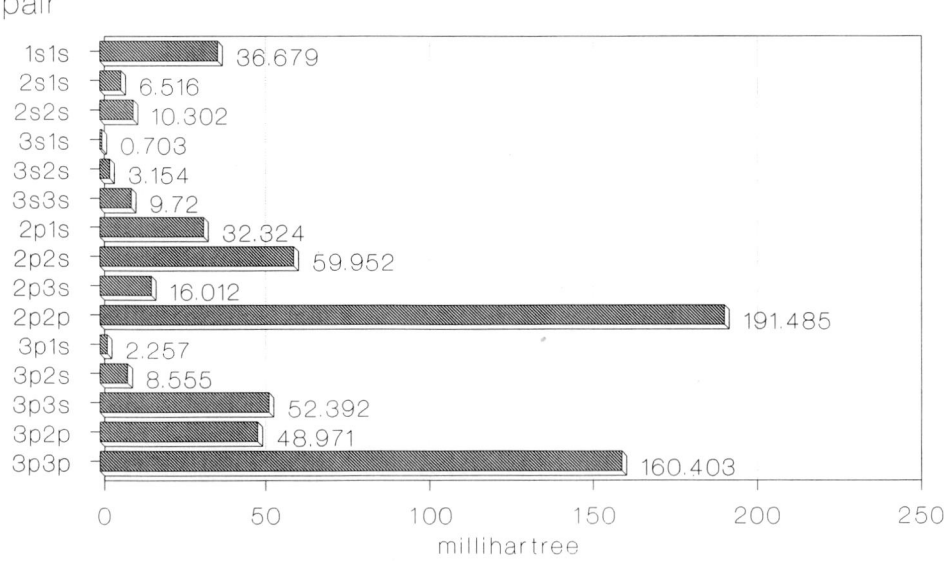

Figure 9. *Second–order pair correlation energies for the ground state of the argon atom from calculations employing the Dirac–Coulomb hamiltonian and a basis set containing functions of s, p, d and f symmetry. (Based on the work of Quiney, Grant and Wilson (1990))*

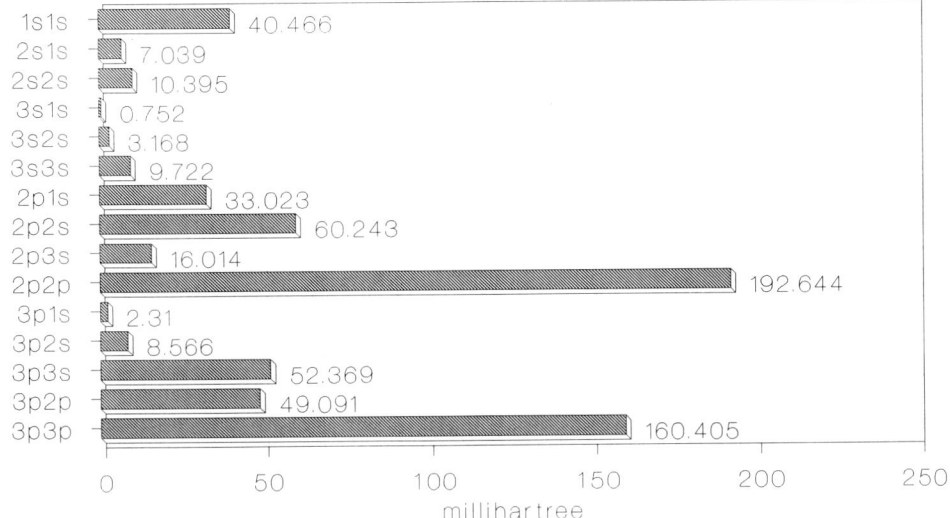

Figure 10. *Second–order pair correlation energies for the ground state of the argon atom from calculations employing the Dirac–Breit hamiltonian and a basis set containing functions of s, p, d and f symmetry. (Based on the work of Quiney, Grant and Wilson (1990))*

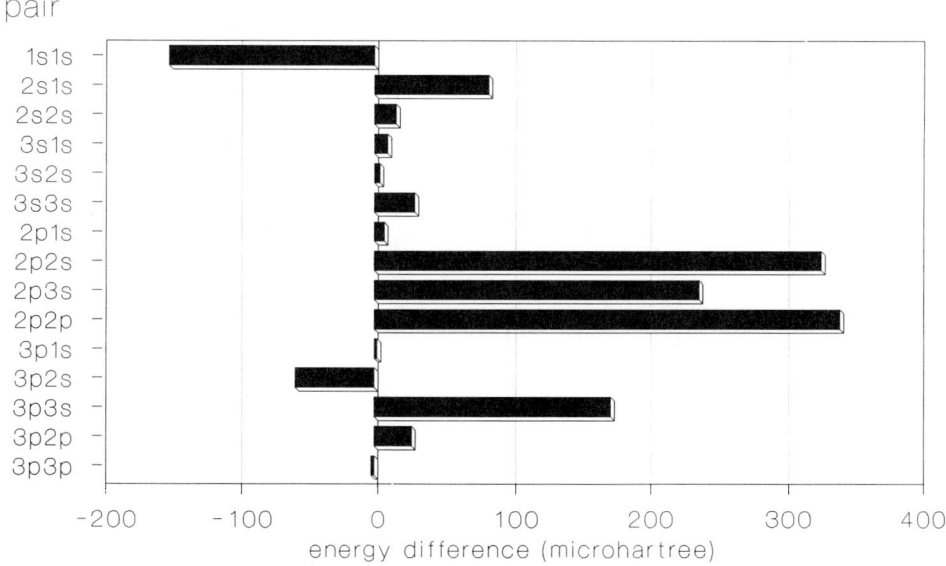

Figure 11. *Differences between second–order pair correlation energies for the ground state of the argon atom from calculations using the Schrödinger hamiltonian and those using the Dirac–Coulomb hamiltonian. The basis sets employed in these calculations contained exponential–type functions of s, p, d and f symmetry type. (Based on the work of Quiney, Grant and Wilson (1990))*

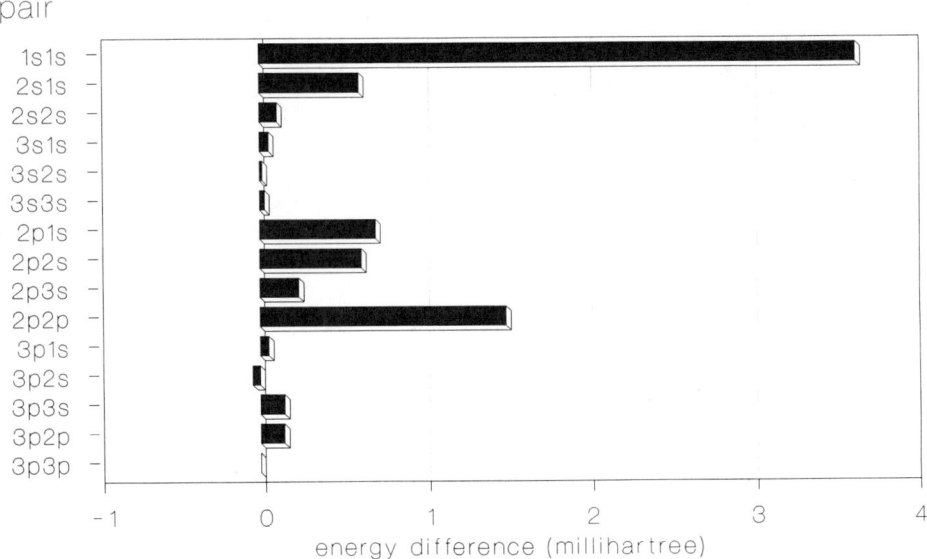

Figure 12. *Differences between second–order pair correlation energies for the ground state of the argon atom from calculations using the Schrödinger hamiltonian and those using the Dirac–Breit hamiltonian. The basis sets employed in these calculations contained exponential-type functions of s, p, d and f symmetry type. (Based on the work of Quiney, Grant and Wilson (1990))*

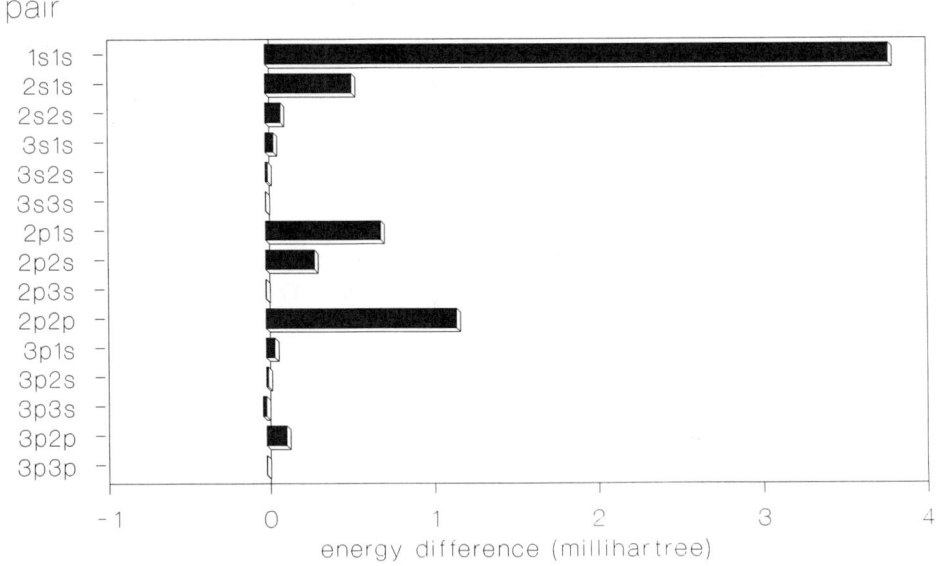

Figure 13. *Differences between second–order pair correlation energies for the ground state of the argon atom from calculations using the Dirac–Coulomb hamiltonian and those using the Dirac–Breit hamiltonian. The basis sets employed in these calculations contained exponential-type functions of s, p, d and f symmetry type. (Based on the work of Quiney, Grant and Wilson (1990))*

analysis of the second–order pair energies obtained from relativistic many–body perturbation theory calculations using the Dirac–Coulomb hamiltonian and the Dirac–Breit hamiltonian, respectively. Both relativistic calculations used the no virtual pair approximation.

Second–order pair correlation energies obtained from a non–relativistic calculation and from relativistic calculations employing basis sets of functions with s, p, d and f symmetry (the largest basis sets considered by Quiney, Grant and Wilson (1990)) are displayed in Figures 8, 9 and 10, respectively. In these Figures, the ordinate has a linear scale in millihartree.

The differences between the second–order pair correlation energies for the ground state of the argon atom from calculations using the Schrödinger hamiltonian and those using the Dirac–Coulomb hamiltonian are shown in Figure 11. The basis sets employed in these calculations contained exponential–type functions of s, p, d and f symmetry type. In Figure 12, the differences between the second–order pair correlation energies from calculations using the Schrödinger hamiltonian and those using the Dirac–Breit hamiltonian are shown. The differences between the second–order pair correlation energies from calculations using the Dirac–Coulomb hamiltonian and those using the Dirac–Breit hamiltonian are shown in Figure 13.

Comparing the results of the relativistic calculation using the Dirac–Coulomb hamiltonian with the non–relativistic results, the effects of relativity on the correlation energy can be explained in terms of competing effects due to shielding and the relative phases of the large and small components of the relativistic wave functions (Lindroth 1987, Quiney, 1987). However, the Dirac–Coulomb hamiltonian accounts for only a small fraction of the total relativistic correction to the correlation energy, though it is, of course, this hamiltonian that has been used as a reference in all previous studies of relativistic correlation effects. The bulk of the relativistic correction to the correlation energy comes from single–particle insertions (one Coulomb and one Breit photon diagrams) which are not in the Dirac–Hartree–Fock–Coulomb basis. These diagrammatic terms are summed through all orders in the Dirac–Hartree–Fock–Breit approximation, which has the additional advantage that evaluation of terms in the relativistic many–body perturbation series is simplified, particularly in higher orders.

The terms in the partial wave expansion for for the relativistic shift associated with s^2 pairs vary approximately as $1/l^2$, where l is the orbital angular momentum of the intermediate states. This should be compared with the $1/l^4$ behaviour of the partial wave expansion for s^2 pairs in the non–relativistic case. The effects of relativity on the valence spinors are almost entirely accounted for by the Dirac–Coulomb hamiltonian. For the core spinors and for the interaction between the core and the valence spinors, it is necessary to use the Dirac–Breit hamiltonian.

The total non–relativistic second–order many–body perturbation theory energy at the f–limit is -638680 μhartree. Within the no virtual pair approximation, the Dirac–Coulomb second–order correlation energy is -639424 μhartree, so that the relativistic correction is -744 μhartree. This difference arises from the use of relativistic single–particle states, but the same two–body interaction. The corresponding relativistic calculation using the Dirac–Breit operator gives a total second–order energy of -646206 μhartree, which represents a relativistic shift of -7526 μhartree, which is an order of magnitude larger than the Dirac–Coulomb many–body correction. Although roughly half of the Dirac–Breit many–body correction is associated with the $1s^2$ pair, the remainder arises through subtle interactions between the core and the valence orbitals.

Quiney, Grant and Wilson (1990) conclude that the terms in the perturbation expansion corresponding to the self–consistent modification of the one–electron orbitals due to the Breit interaction result in the most significant relativistic many–body effect. Furthermore, this studies underlines that fact that the relativistic many–body perturbation theory using the algebraic approximation is a tractable

computational step	power of n
integral evaluation	
one–electron integrals	2
two–electron integrals	4
self–consistent–field iterations	
Fock matrix construction	4
Fock matrix diagonalization	3
integral transformation	
one–electron	3
two–electron	5
many–body perturbation theory	
second order	4
third order	6
fourth order	7

Figure 15. *Deep red complex formed by the interaction of rhodium (Rh^{III}) chloride solution with stannous chloride. The computational demands which would arise in a calculation on this system are considered in detail in the text.*

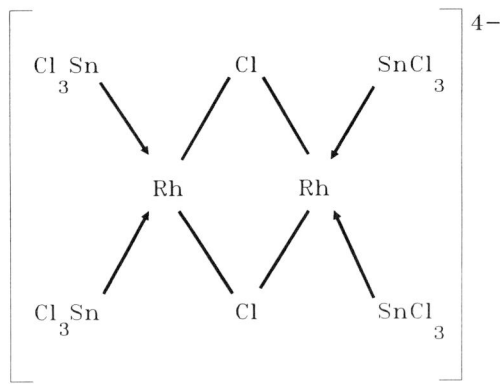

Figure 14. *Computational demands of the various stages of a typical electronic structure calculation performed within the algebriac approximation. For each stage the dependence on the number of basis functions, n, is given.*

scheme giving a systematic approach to the calculation of relativistic correlation effects in atoms and opens up the possibility of molecular applications.

4. Linear dependence in the two–electron integral matrix

The computational demands of the various stages of a molecular electronic structure calculation are quite considerable both in terms of the processing time used and the storage requirements. The number of two–electron integrals increases as the fourth power of the number of basis functions which in turn influences all subsequent phases of the computation. The dependence of each stage on the number of basis functions, n, is summarized in Figure 14.

Consider, by way of example, the computational demands that would arise in a calculation for the deep red complex formed when rhodium chloride and stannous chloride solutions are mixed; the complex is shown in Figure 15. It contains 20 atoms and 532 electrons. Each of the atoms can be accurately described by a basis set containing say 18 functions of s symmetry plus functions of higher angular symmetry which, for simplicity, we ignore in the present discussion. The number of two–electron integrals arising in a non–relativistic calculation or in the [LL|LL] block of a relativistic calculation is

$$M = m(m+1)/2, \text{ where } m = n(n+1)/2.$$

Putting n = 18 gives the number of two–electron integrals arising in the atomic case: M = 14,706. Putting n = 20 x 18 = 360 gives the number of integrals arising in the molecular case:– M = 2,111,232,690. When the number of basis functions increases by a factor of 20 the number of two–electron integral increases by a factor in excess of 143,562. In a relativistic calculation, the problem is compounded by the fact that integral blocks of the type [LL|SS], [LS|LS], [LS|SL] and [SS|SS] have to be considered in addition to the [LL|LL] block (Wilson 1988).

The traditional approach to the construction of basis sets for molecular electronic structure calculations has been *(i)* to optimize the exponents, ζ_k, for atomic self–consistent field calculations; that is, using the object function

$$E_{scf} = E_{scf}(\zeta_1, \zeta_2, \ldots, \zeta_n)$$

find those exponents which minimize the energy; *(ii)* add polarization functions to describe the deformation of the atom in the molecule; *(iii)* add correlation functions. The procedure is somewhat empirical and it is not surprising that much of the art of molecular electronic structure calculations lies in the selection of appropriate basis functions.

The first step towards a more rational approach to basis set construction was taken by Ruedenberg and his coworkers (Ruedenberg, Raffenetti and Bardo 1973), who, building on earlier work by Reeves (1963), introduced the even–tempered basis set. In these basis sets the orbital exponents are taken to form a geometric series

$$\zeta_{lk} = (\alpha_l)(\beta_l)^k, \; \alpha_l > 0, \; \beta_l > 1, \; k = 1, 2, \ldots, n$$

The parameters α_l and β_l are optimized for each atom and each l–shell. In later work it was shown that there is little loss in accuracy if α and β are taken to be independent of l. As Raffenetti pointed out in 1973

The introduction of even–tempered orbitals has opened the possibility of determining analytic SCF functions for heavy atoms with relatively modest computing effort

In 1978, Silver and Wilson employed even–tempered basis sets in electron correlation studies. Such calculations demand larger basis sets than self–consistent field studies. These large basis sets have a flexibility which enables them to used for different atoms and differing molecular environments. This leads to the concept of a universal basis set. Many–body perturbation theory calculations using universal basis sets have yield some of the most accurate correlation energies for small molecules (Wilson and Silver, 1980,1982).

Most molecular electronic structure calculations reported in the 1970's employed a single basis set selected in some *ad hoc* fashion. With the advent of even–tempered and universal basis sets the resulting improvement in the accuracy of electronic structure calculations gave impetus to the study of the convergence of calculated energy with increasing size of basis set. Ruedenberg and his coworkers introduce the idea of a systematic sequence of even–tempered basis sets for use in atomic self–consistent–field studies. They generated even–tempered exponents by means of the empirical prescription

$$\ln \ln \beta_1 = b_1 \ln n_1 + b_1'$$

$$\ln \alpha_1 = a_1 \ln(\beta_1-1) + a_1'$$

which ensure that

$$\zeta_{min} = \alpha\beta \longrightarrow 0, \qquad \zeta_{max} = \alpha\beta^n \longrightarrow \infty, \qquad \beta \longrightarrow 1$$

since

$$\alpha \longrightarrow 0, \qquad \beta \longrightarrow 1, \qquad \beta^n \longrightarrow \infty$$

The parameters a_1, a_1', b_1 and b_1' are determined from atomic self–consistent–field calculations. The use of systematic sequences of even–tempered basis sets of Gaussian–type functions was found to afford atomic self–consistent field energies which were often more accurate than those given by finite difference calculations.

In correlation energy calculations, not only is an accurate representation of the occupied orbitals required but also of the virtual spectrum. The use of universal systematic sequences of even–tempered basis sets can be shown to lead to

$$\alpha_1(n_1) = \{\beta_1(n_1) - 1)/(\beta_1(n_1-1) - 1)\}^{a_1} \alpha_1(n_1-1)$$

$$\ln \beta_1(n_1) = \{n_1/(n_1-1)\}^{b_1} \ln \beta_1(n_1-1)$$

The set of exponential–type, $\exp(-\zeta_k r)$, or Gaussian–type, $\exp(-\zeta_k r^2)$, basis functions generated according to the prescriptions given above can be shown to be asymptotically complete in the limit $n \longrightarrow \infty$ since the generalized Muntz–Szasz theorem given by Klahn (1985) holds; that is

$$\lim_{n \to \infty} \sum \zeta_k(n)/(1 + \{\zeta_k(n)\}^2) = \infty$$

It is easily demonstrated that for fixed α and β this limit is finite. In Figure 16, the logarithms of the orbital exponents obtained with fixed α and β are compared with those generated by means of the recursions given above. Each of the rows in this Figure corresponds to a basis set in the sequence and is labelled by n, the number of functions it contains. It should be noted that for the exponents generated by the above recursions that the range of the exponents increases with n, the lower limit decreasing to zero and the upper limit becoming infinite. Furthermore, the distribution of the exponents between these two limits becomes increasingly dense as n is increased.

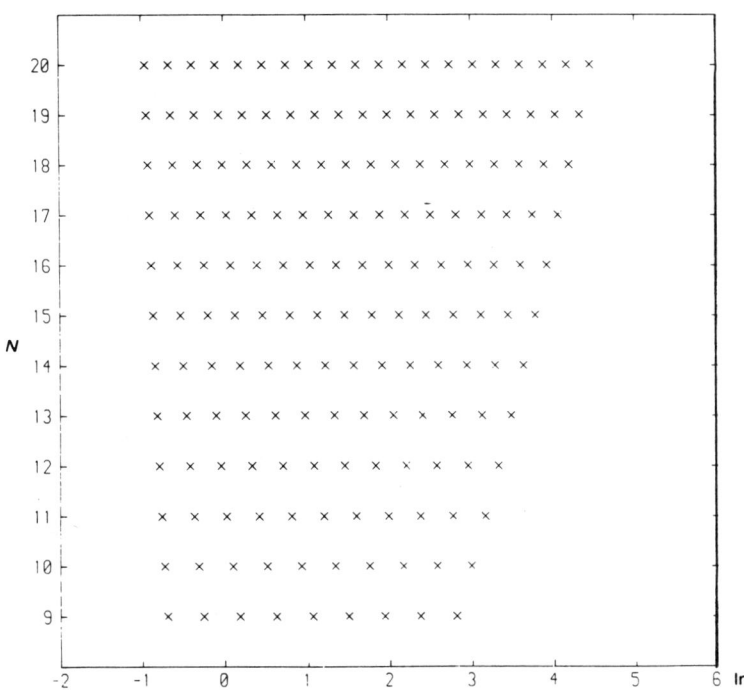

Figure 16. *Logarithm of the orbital exponents generated by using a universal systematic sequence of even−tempered basis functions. Each row of crosses in this figure corresponds to a given basis set in the sequence and is labelled by N, the number of functions it contains. Each of the crosses in a given row corresponds to an orbital exponent. (Based on the work of Wilson (1990).*

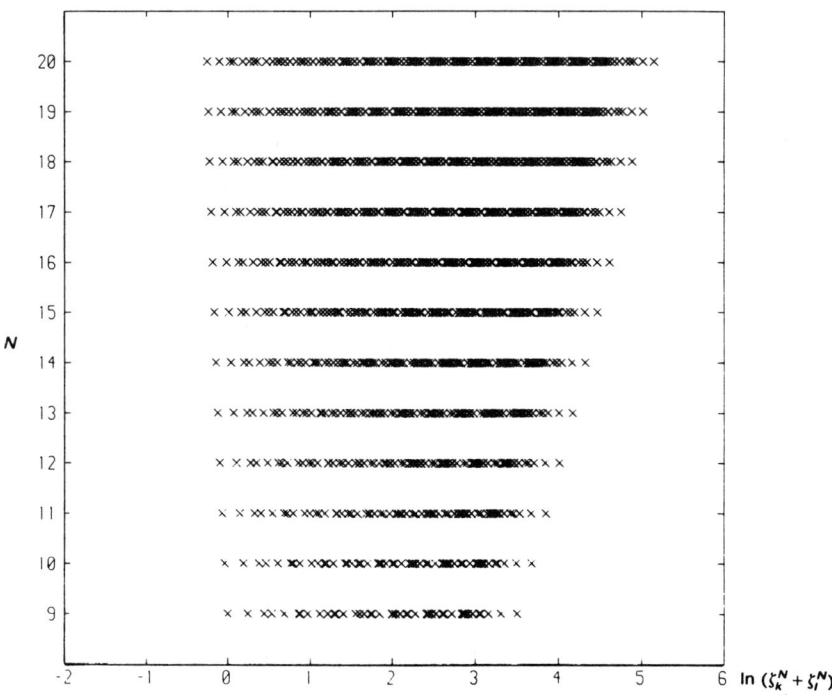

Figure 17. *Logarithm of the charge distribution exponents generated by using a universal systematic sequence of even−tempered basis functions. Each of the rows of crosses in this figure corresponds to a given basis set in the systematic sequence and is labelled by N, the number of functions is contains. Each of the crosses in a given row corresponds to a charge distribution exponent generated from the orbital exponents used in the construction of Figure 16. (Based on the work of Wilson (1990).*

Now the two-electron integrals take the form

$$\int\int dr_1\, dr_2\, \varphi_i(r_1)\, \varphi_k(r_2)\, (1/r_{12})\, \varphi_j(r_1)\, \varphi_l(r_2)$$

where r_{12} is the interelectron distance and φ_k is a basis function. These integrals involve two charge distributions of the form

$$\rho_{jk}(r) = \varphi_j(r)\, \varphi_k(r)$$

In the case of an even–tempered basis set of exponential–type functions centred on the same point, these charge distributions assume the form

$$\exp(\zeta_j\, r)\, \exp(\zeta_k\, r) = \exp(-\{\zeta_j + \zeta_k\}\, r)$$

In Figure 17, the logarithm of the exponent arising in each of the charge distributions, $\ln\{\zeta_j + \zeta_k\}$, is plotted for the same universal systematic sequence of even–tempered basis sets that was employed in the generation of Figure 16. Each row in Figure 17 corresponds to a particular basis set in the sequence and is labelled by the number of functions, N, it contains. Figure 17 suggests a great deal of linear dependence in the two–electron integral matrix. It should noted that whilst the exponents for the basis functions range from $\alpha\beta$ to $\alpha\beta^n$, the exponents for the charge distributions range from $2\alpha\beta$ to $2\alpha\beta^n$. However, there are $n(n+1)/2$ charge distribution exponents compared with only n orbital exponents, which implies that $n(n-1)/2$ of the former are redundant.

If the charge distribution exponents are defined as follows

$$\zeta_{jk}(n) = \zeta_j(n) + \zeta_k(n)$$

then the set of charge distributions can be shown to be asymptotically complete in the limit $n \longrightarrow \infty$ since the generalized Muntz–Szasz theorem again holds

$$\lim_{n \to \infty} \sum \zeta_{jk}(n)/(1 + \{\zeta_{jk}(n)\}^2) = \infty$$

n of the charge distribution exponents can be deleted and asymptotic completeness is maintained. One choice is to retain those distributions for which j=k giving

$$\lim_{n \to \infty} \sum \zeta_{jj}(n)/(1 + \{\zeta_{jj}(n)\}^2) =$$

$$\lim_{n \to \infty} \sum 2\zeta_j(n)/(1 + 4\{\zeta_j(n)\}^2) = \infty$$

Other possible choices are considered elsewhere (Wilson 1990).

The two–electron integrals may be arranged in a symmetric, positive definite matrix of dimension $(n(n+1))/2$

$$V_{ij,kl} \qquad i \geq j \quad k \geq l \quad (ij) \geq (kl)$$

The computational linear dependence amongst the charge distributions involved in the two–electron integral matrix can be exploited by a Cholesky decomposition in which the matrix $V_{ij,kl}$ is written in the form

$$V_{ij,kl} = \sum_{mn} L_{ij,mn}\, L^{\dagger}_{mn,kl}$$

where **L** is a lower triangular matrix and $\mathbf{L}^\dagger$ is its transpose. If linear dependence is present in **V** then the summation index mn will run for considerably less than its full range 1 to $(n(n+1))/2$. An approximation to **V** may be written

$$V'_{ij,kl} = \sum_{mn}^{\nu} L_{ij,mn} L^\dagger_{mn,kl}$$

where ν is the effective numerical rank of the two–electron integral matrix.

In Figure 18, the effective numerical rank of the two–electron integral matrix is shown for a universal systematic sequence of one–centre even–tempered basis sets of Gaussian–type functions of s symmetry. The upper curve shows the full rank of the two–electron integral matrix, that is

$$m = (n(n+1))/2$$

as a function of the size of the basis set on a logarithmic scale. The lower curves show the effective numerical rank, ν, when a tolerance of 10^{-t}, $t = 4,5,6,7,8,9,10$ is demanded.

The total number of two–electron integrals arising in a calculation employing a basis set of n functions is

$$M = (m(m+1))/2$$

The number of elements of **L** which have to be stored for a two–electron integral matrix of effective numerical rank ν is

$$N = M - (m-\nu)(m-\nu+1)/2$$

and the percentage storage requirements are thus

$$P = [N/M] \times 100\%.$$

For a tolerance of 10^{-8}, P is shown as a function of the number of basis functions, n, in Table 1. It can be seen that as the basis set is systematically extended the percentage of the integral matrix that is actually required decreases monotonically. For a basis set of twenty functions the integral storage requirements are reduced to less than 25% of the length of the original integral list. The larger the basis set the more accurate is the calculation and the greater then percentage reduction in the computer storage demands.

Linear dependence in the two–electron integral matrix can be exploited in both non–relativistic and relativistic electronic structure calculations. It can reduce the number of integrals that have to be evaluated and stored in the integral evaluation phase of the calculation. It can reduce the number of integrals that have to be processed in subsequent self–consistent field and many–body perturbation theory calculations.

The reductions both in computer time in generating and processing the two–electron integral list are even more significant in relativistic calculations than they are in non–relativistic counterparts. Firstly, the number integrals arising in the relativistic case is significantly greater than the non–relativistic case and thus a larger number of integral evaluations can actually be avoided. Secondly, in relativistic electronic structure calculations, the integrals involving the small component basis functions are not required to such a high accuracy as those involving the large component basis function for a given accuracy in the final calculation energy or expectation value.

The most demanding phase of a low order many–body perturbation theory calculation is the four–index transformation of the two–electron integrals

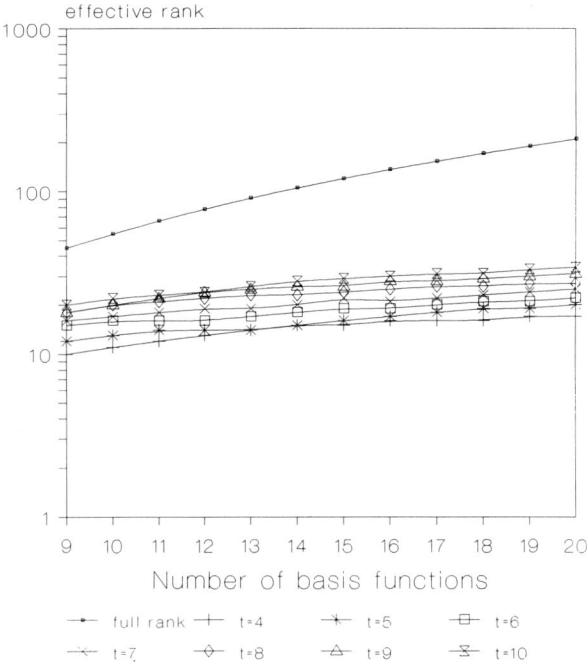

Figure 18. *Effective numerical rank of the two−electron integral matrix for a one−centre even−tempered basis set of Gaussian−type functions of s symmetry.*

Table 1

Comparison of the total number of two–electron integrals, M, which arise in a calculation with n basis functions, with the number of elements, N, of the lower triangular matrix **L** required for a tolerance of 10^{-8}. P is the percentage storage requirements given by [N/M] x 100%

n	N	M	P
9	657	1035	63.5%
10	910	1540	59.1%
11	1176	2211	53.2%
12	1485	3081	48.2%
13	1840	4186	44.0%
14	2162	5565	38.8%
15	2604	7260	35.9%
16	3100	9316	33.3%
17	3653	11781	31.0%
18	4121	14706	28.0%
19	4779	18145	26.3%
20	5319	22155	24.0%

$$V_{pq,rs} = \sum_{ijkl} c_{pi}\, c_{qj}\, c_{rk}\, c_{sl}\, V_{ij,kl}$$

which is most efficiently performed as a series of partial transformations

$$V^1_{ij,ks} = \sum_{l} c_{sl}\, V_{ij,kl}$$

$$V^2_{ij,rs} = \sum_{k} c_{rk}\, V^1_{ij,ks}$$

$$V^3_{iq,rs} = \sum_{j} c_{qj}\, V^2_{ij,rs}$$

$$V_{pq,rs} = \sum_{i} c_{pi}\, V^3_{iq,rs}$$

Introducing a Cholesky decomposition of the two–electron integral matrix, we can write the four–index transformation in the form

$$V_{pq,rs} = \sum_{mn} \sum_{ijkl} c_{pi}\, c_{qj}\, c_{rk}\, c_{sl}\, L_{ij,mn}\, L_{kl,mn}$$

Exploiting the computational linear dependence of the two–electron integral matrix restricts the summation over mn

$$V'_{pq,rs} = \sum_{mn}^{\nu} \sum_{ijkl} c_{pi}\, c_{qj}\, c_{rk}\, c_{sl}\, L_{ij,mn}\, L_{kl,mn}$$

which may be written in the form

$$V'_{pq,rs} = \sum_{mn}^{\nu} L_{pq,mn}\, L_{rs,mn}$$

$L_{pq,mn}$ is given by the two–index transformation

$$L_{pq,mn} = \sum_{ij} c_{pi}\, c_{qj}\, L_{ij,mn}$$

the computational demands of which scale as n^3.

5. Parallel processing in many–body perturbation theory

The performance of high quality electronic structure calculations, particularly for molecules, has always depended on the effective exploitation of *state-of-the-art* computing machines. The importance of vector processing in many–body perturbation theory calculation was recognized at an early stage (Wilson and Saunders 1980, Guest and Wilson 1981). Each of the summations which arise in the sum–over–states perturbations expression for the energy coefficients can be written as a matrix multiplication which can be very efficiently performed on machines such

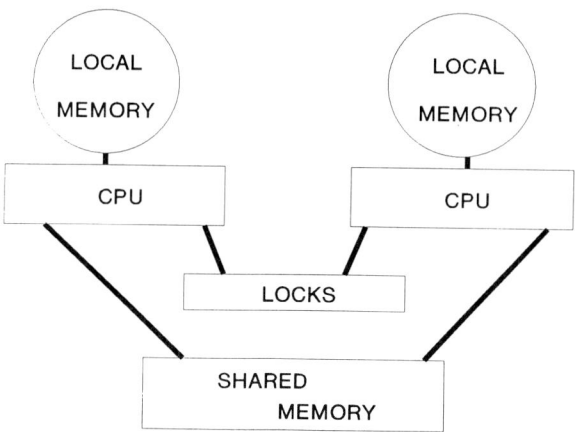

Figure 19. *Schematic diagram of a shared memory parallel system.*

as the CRAY X–MP and Y–MP using the machine coded routine MXMA which accumulates results in a vector register rather than constantly reading then writing intermediate results to memory which is the case when FORTRAN code is employed. For example, in non–relativistic calculations of the fourth–order triple excitation energy components, which are particularly demanding since they scale as the seventh power of the number of basis functions, the computational kernel takes the form

```
CALL MXMA(D2,1,NVIRT,D1,1,NVIRT,T1,1,NVIRT, NVIRT,NVIRT,NOCC)
CALL MXMA(E2,1,NVIRT,D1,1,NVIRT,T2,1,NVIRT, NVIRT,NVIRT,NOCC)
CALL MXMA(D2,1,NVIRT,E1,1,NVIRT,T3,1,NVIRT, NVIRT,NVIRT,NOCC)
CALL MXMA(E2,1,NVIRT,E1,1,NVIRT,T4,1,NVIRT, NVIRT,NVIRT,NOCC)
```

where D1, D2, E1 and E2 are direct and exchange two–electron integrals, including denominator factors in the case of D1 and E1, and T1, T2, T3 and T4 are intermediates.

In addition to a vector processing capability, modern supercomputers increasingly employ parallel processing. The CRAY X–MP has a maximum of four processors whilst the CRAY Y–MP can have up to eight vector processors working in parallel. The linked diagram theorem of many–body perturbation theory ensures that the many–electron problem is effectively decoupled into a series of smaller problems each of which can be treated on a separate processor. Many–body perturbation theory calculations are therefore well suited to a parallel processing environment. The effective exploitation of the parallel processing capabilities of modern supercomputers is clearly of central importance for the effective implementation of a relativistic many–body perturbation theory of molecular electronic structure.

Now the benefits of parallel processing on the CRAY X–MP and Y–MP systems have until recently largely been seen in terms of what might be termed the "$1/N_{cpu}$ rule", where N_{cpu} denotes the number of processors in the system (Taylor and Bauschlicher 1987, Taylor and Bauschlicher and Schwenke 1989). In this model, the use of more than $1/N_{cpu}$ th of a scarce resource (for example, memory) requires that one uses more one processor, otherwise it is considered that other jobs will be unlikely to saturate the remaining processors because of shortage of the resource. Taken to an extreme, if a particular computer program uses all of the available memory, then it should try to use all of the available processors. An argument in favour of parallel processing even when memory requirements are small has recently been presented (Baker et al 1990a). It has been estimated that it is desirable to have about two executing processes per processor to ensure that no idle time accumulates, largely because of wait times for input/output. This is very difficult to ensure given the type of workload (many large memory heavily input/output bound jobs) often encountered. A single processor–bound small memory four–way parallel processing job, with a dynamically balanced load and running at low scheduling priority is capable of consuming all potentially idle cycles on all the processors on CRAY X–MP and CRAY Y–MP system, with little disturbance to the flow of other work, and a greatly improved overall throughput. A similar argument has recently been presented for the use of CRAY "micro–tasking" facilities (Bieterman 1989).

On shared memory architecture parallel computers, static load balancing of the work assigned to each process is usually far from optimal. This is not only because it is often impossible to divide the computation in such a way that the load is exactly balanced but also because each processor is not normally dedicated to a single user; instead its computing power is shared out amongst a number of users. The effective computing power delivered to a given user (or a parallel process of that user) by a given processor is largely unpredictable, because it depends upon what other users are doing at any given time. Thus instead of requiring that all processors carry out an identical workload, it is better to require that each processor be given a workload commensurate with its effective power; the latter is far from constant with time, so that it is necessary to feed work to the processors at a rate which is similarly time dependent. This may be accomplished by the technique of "dynamic load balancing", using a "global index" under control of a "lock".

To illustrate these ideas consider the following simple example. Suppose the sequential version of the code has the form:

```
      DO 1 I = 1,T
      ... Execute task I
   1  CONTINUE
```

Using CRAY X–MP and CRAY Y–MP multi–tasking software the dynamically balanced parallel version using N processes would be of the form shown in Figure 20.

In this code, NCPUMAX should be initialized to a value greater than or equal to N. ITASK is a 3 by NCPUMAX array, each column of which is used by the CRAY system to identify a given process and is initialized in DO 9. LOCKI is a "lock" used to control unique access to the global index, I, the latter being initialized to 1. The call to LOCKASGN is required to declare that LOCKI will be used as a lock. A call to TSKSTART initiates a parallel process ("forking"), the corresponding call to TSKWAIT creates a "barrier" causing the parent process to wait until the indicated parallel process is complete. TSKSTART/TSKWAIT thus correspond to the system process creation/termination mechanism. During the parallel phase of the job N copies of subroutine PROCESS are in execution. On the CRAY system PROCESS is made re–entrant, circumventing the need to actually make the copies. PROCESS then has the form shown in Figure 21.

In the code displayed in Figure 21, LOCKON is called to acquire unique access to the global index I which is being protected by LOCKI; if another parallel copy of PROCESS calls LOCKON whilst such unique access is granted, the second copy will be halted until the first copy releases access with a call to LOCKOFF. If I is greater than T then all tasks have been completed, and after releasing LOCKI, PROCESS returns to the parent process, the latter checking for this by its call to TSKWAIT. Otherwise a local copy of the global index is taken in ITEMP, and the global index updated, after which LOCKI is released. The task indicated by ITEMP (note that

```
      PARAMETER (NCPUMAX=8)
      COMMON /TASKID/ ITASK(3,NCPUMAX)
      COMMON /GLOBAL/ I, LOCKI, T
      EXTERNAL PROCESS
*
*     Initialize multitasking variables
*
      T=? : number of tasks
      N=? : number of processes
      CALL LOCKASGN(LOCKI)
      I=1
      DO 1 K=1,N
      ITASK(1,K)=3
      ITASK(3,K)=K
    1 CONTINUE
*
*     Create N-1 out-of-line processes and
*       one in-line process - FORK
*
      DO 2 K=2,N
      CALL TSKSTART(ITASK(1,K),PROCESS)
    2 CONTINUE
      CALL PROCESS
*
*     Wait until all processes have finished -
*       BARRIER
*
      DO 3 K=2,N
      CALL TSKWAIT(TASK(1,K))
    3 CONTINUE
```

Figure 20. *Calling sequence for dynamic balance. (Based on the work of Baker, Moncrieff, Saunders and Wilson (1990))*

```
      SUBROUTINE PROCESS
      COMMON /GLOBAL/ I, LOCKI, T
    1 CONTINUE
      CALL LOCKON(LOCKI)
      IF(I.GT.T) THEN
      CALL LOCKOFF(LOCKI)
      RETURN
      ELSE
      ITEMPI=I
      I=ITEMPI+1
      CALL LOCKOFF(LOCKI)
      ENDIF
      ...
      Execute Task ITEMP
      ...
      GOTO 1
      END
```

Figure 21. *Dynamically balanced parallel routine. (Based on the work of Baker, Moncrieff, Saunders and Wilson (1990))*

the global index itself is not used in this phase since it may be updated by another parallel copy of PROCESS) is then carried out, after which the routine branches to the first executable statement to see if work remains. Obviously, if task execution is controlled by a nest of loops (in the present work there is a four–fold nest of loops over the tasks), then all loop indices associated with the nest should be treated as a shared locked data structure (Saunders 1990) and accessed under control of a lock. The overheads associated with the above implementation are T O_r, where O_r is the time to acquire/release LOCKI. This is of the order of a few microseconds on the CRAY X–MP system. Of course, if the individual tasks are small, then contention for access to the lock will be correspondingly large, leading to a reduction in the average width of parallelism; however all the suggested methods have problems in this small task size limit. When the task size is large (as in the present work), so that the overheads can be largely ignored, the decisive advantage of the procedure described above over alternative strategies is that each process generates only as much work as its corresponding processor is able to deal with (it is dynamically balanced), and is for this reason the chosen strategy in the present work. When the processors are dedicated, all will perform an identical amount of work, as in the statically balanced case. In fact, given a reasonably large number of tasks, dynamic balance produces an approximately even loading over dedicated processors even if the tasks are not of equal size. The major disadvantage of dynamic balance is that it requires the use of locks, hardware support for which is unavailable (and software emulations expensive probably requiring that all processing of a lock be carried out by one processor which communicates its results to the other processors by message passing) on all current independent memory machines. Its use is thereby restricted to shared memory parallel architectures, which all provide hardware supported locks. We here observe that current shared memory systems such as are available from CRAY, IBM, Convex, Alliant, Sequent, Encore and FPS are capable of running each processor in multi–user mode (and thus require dynamic balance and the associated locks for optimal performance). However, the operating systems of the independent memory machines (exemplified by those available from Intel, Meiko or Suprenum) are presently capable of running a given processor in only single–user mode; it seems to us that the ease of use of such machines in a (as yet unavailable) multi–user per processor environment would be greatly facilitated by the provision of hardware locks, and program portability between shared and distributed memory systems thereby greatly enhanced.

The programming strategy described above has been employed in the evaluation of the most computationally demanding of the fourth order terms in the non–relativistic many–body perturbation expansion, namely, those involving triply excited intermediate states. In a dedicated environment, execution rates in excess of 820 MFLOPS have been observed on the four processor CRAY X–MP/416 computer, whilst on the CRAY Y–MP/832, which has eight processors, execution rates of 2,230 MFLOPS have been obtained. Executions rates of this order allow the efficient use of large basis sets which in turn is reflected in the improved accuracy of calculation. The dynamic load balancing technique will clearly enable the efficient implementation of relativistic many–body perturbation theory.

6. Conclusions

The relativistic many–body perturbation theory using the algebraic approximation affords a tractable scheme yielding a systematic approach to the calculation of relativistic correlation effects in atoms and molecules.

In this paper, we have considered, in detail, some of the computational problems which arise in the relativistic many–body perturbation theory when the algebraic approximation is invoked.

In section 4, we have shown how the numerical linear dependence in the two–electron integral matrix can be exploited in molecular electronic structure calculations. Although useful in both non–relativistic and relativistic studies, the techniques described in section 4 are particularly useful in the relativistic case because of the greater number of two–electron integrals which arise.

In section 5, we have shown how the parallel processing capabilities of modern supercomputers can be effectively harnessed in many–body perturbation theory calculations. We have shown that each diagrammatic term in the many–body perturbation expansion can be decomposed into a set of partial contributions which can be computed by an algorithm structured around the matrix multiply operation, so that conventional vector processors can be driven nearly optimally. We have also shown that the evaluation of the partial contributions may proceed entirely independently, so that the theory is well suited for implementation on a parallel processing system with either a distributed or shared memory. "Static" and "dynamic" procedures for balancing the load amongst the parallel processors have been considered, and we have concluded that static balance is probably best for independent memory architectures in their present state of development. However, in the case of typical shared memory systems, which normally operate in multi-user mode, dynamic balance is to be preferred, since it allows one to partition the work amongst the processors according to their ability to deal with it. In a dedicated environment close to linear speed–up with number of processors was observed, there being little degradation due to memory bank conflicts or other overheads of the parallel implementation, and rate of computations of greater then 2,230 Mflops were observed on a CRAY Y–MP/832 machine.

It is clear that the techniques described in this paper can be exploited in relativistic electronic structure calculations based on the many–body perturbation expansion. They may be used in atomic calculations which go beyond the application to the argon atom ground state described in section 3 and take account of finite nuclear size effects, the Lamb shift, and higher order terms in the relativistic many–body perturbation theory. They will be essential in applications to molecular systems if adequate precision is to be obtained. It has recently been demonstrated (Wells and Wilson 1989, Thompson and Wilson 1990) that sub μhartree accuracy can be attained in non–relativistic molecular structure calculations.

References

Baker, D.J., Moncrieff, D., Saunders, V.R., and Wilson, S., 1990a, Comput. Phys. Commun. *(in press)*
  Diagrammatic many–body perturbation expansion for atoms and molecules. VII. Experiments in vector and parallel processing for fourth–order energy terms involving triply excited intermediate states.
Baker, D.J., Moncrieff, D., and Wilson, S., 1990b, in *Supercomputational Science*, edited by R.G. Evans and S. Wilson, (Plenum, New York)
  Vector processing and parallel processing in many–body perturbation theory calculations of electron correlation effects in atoms and molecules.
Bieterman, M., 1989, CRAY CHANNELS, Summer issue, page 10.
  The impact of microtasked applications in a multiprogramming environment
Brueckner, K.A., 1955, Phys. Rev. **100**, 36
  Many–body problem for strongly interacting particles. II. Linked cluster expansion.
Cooper, D.L., and Wilson, S., 1982a, J. Phys. B: At. Mol. Phys. **15**, 493
  Ab initio calculation of atomic spin–orbit coupling constants using universal systematic sequences of even–tempered basis sets
———, 1982b, J. Chem. Phys. **76**, 6088
  Ab initio calculation of molecular spin–orbit coupling constants using universal systematic sequences of even–tempered basis sets

Dyall, K.G., Grant, I.P., and Wilson, S., 1984a, J. Phys. B: At. Molec. Phys. **17**, 493
  Matrix representation of operator products
——, 1984b, J. Phys. B: At. Molec. Phys. **17**, L45
  The Dirac equation in the algebraic approximation. I. Criteria for the choice of basis functions and minimum basis set calculations for the hydrogen atom.
——, 1984c, J. Phys. B: At. Molec. Phys. **17**, L45
  The Dirac equation in the algebraic approximation. II. Extended basis set calculations for hydrogenic atoms.
Dyson, F.J., 1949, Phys. Rev. **75**, 1736
  The S–matrix in quantum electrodynamics.
Feynman, R., 1949, Phys. Rev. **76**, 769
  Space–time approach to quantum electrodynamics.
Furry, W.H., 1951 Phys. Rev. **81** 115
  On bound states and scattering in positron theory
Goldstone, J., 1957, Proc. Roy. Soc. A **239**, 267 (1957)
  Derivation of the Brueckner many–body theory
Grant, I.P., and Quiney, H.M., 1988 Adv. At. Mol. Phys. **23** 37.
  Foundations of atomic structure theory
Grant, I.P., Mackenzie, B., Mayers, D.F., and Pyper, N.C., 1980, Comput. Phys. Commun. **21**, 207
  An atomic multiconfigurational Dirac–Fock package
Greiner W, Muller B and Rafelski J, 1985, Quantum Electrodynamics of Strong Fields (Berlin:Springer)
Guest, M.F., and Wilson, S., in: *Supercomputers in Chemistry*, ed. P. Lykos and I. Shavitt (American Chemical Society, Washington D.C., 1981) p.1.
Hugenholtz, N.M., 1957, Physica **23**, 481
  Perturbation theory of large quantum systems
Johnson W.R. and Sapirstein J. 1986, Phys. Rev. Lett. **57** 1126
  Relativistic second–order correlation energies of helium–like systems
Kaldor, U., 1975, J. Chem. Phys. **62**, 4634.
Kelly, H.P., 1963, Phys. Rev. **131**, 684
  Correlation effects in atoms
Kelly, H.P., 1969, Phys. Rev. Lett. **23**, 455
Kim, Y.–K., 1968, Phys. Rev. Phys. Rev. **154**, 17
  Relativistic self–consistent field theory for closed shell atoms
——, 1989, remark made at Relativistic Atomic Structure Meeting, Grenoble, October, 1989.
Klahn, B, 1985, J. Chem. Phys. **83**, 5749
  A generalization of the Muntz–Szasz theorem to floating exponents with applications to Gauss– and Slater–type functions.
Klahn, B., and Bingel, W.A., 1977a Theoret. chim. Acta. **44** 9
  The convergence of the Rayleigh–Ritz method in quantum chemistry. I. The criteria of convergence
—— 1977b Theoret. chim. Acta. **44** 27
  The convergence of the Rayleigh–Ritz method in quantum chemistry. II.Investigation of the convergence for special systems of Slater, Gaussian and two–electron functions
Knowles, P.J., Somasundram, K., Handy, N.C., and Hirao, K., 1985, Chem. Phys. Lett. **113** 8
  The calculation of higher–order energies in the many–body perturbation series
Laaksonen, L., Grant, I.P., and Wilson, S., 1988, J. Phys. B: At. Mol. Opt. Phys. **21**, 1969.
  The Dirac equation in the algebraic approximation. VI. Molecular self–consistent field studies using basis sets of Gaussian–type functions
Labzovskii, L.N., 1971 Sov. Phys. JETP **32** 94.
  Electron correlation in the relativistic theory of atoms
Lindroth E 1987, thesis, Chalmers University of Technology and University of Göteborg.

Lindroth E, Mårtensson–Pendrill A–M, Ynnerman A and Öster P 1989 J Phys B: At. Mol. Opt. Phys. **22** 2447
  Self–consistent treatment of the Breit interaction, with application to the electric dipole moment in thallium
March, N.H., Young, W.H., and Sampanthar, S., 1967, The many–body problem in quantum mechanics, (Cambridge University Press)
Mercier, B., 1989, Lecture Notes in Physics **318** (Berlin: Springer)
  An introduction to the numerical analysis of spectral methods
Moncrieff, D., Baker, D.J., and Wilson, S., 1989, Comput. Phys. Commun. **55** 31
  Diagrammatic many–body perturbation expansion for atoms and molecules. VI. Experiments in vector processing and parallel processing for second–order energy calculations
Quiney H.M., 1987, D. Phil. thesis, University of Oxford.
  Finite basis set studies of the Dirac equation
Quiney H.M., 1988 in *Relativistic effects in atoms and molecules*, ed. S.Wilson, Meth. Comput. Chem. **2** 223 (New York:Plenum)
  Relativistic Many–Body Perturbation Theory
Quiney H.M., 1990 in *Supercomputational Science*, edited by R.G. Evans and S.Wilson, (New York:Plenum)
Quiney H.M., Grant I.P. and Wilson S., 1985 J. Phys. B: At. Mol. Phys. **18** 2805
  The Dirac equation in the algebraic approximation. III. Diagrammatic perturbation theory applied to a model problem.
Quiney H.M., Grant I.P. and Wilson S., 1987a J. Phys. B: At. Mol. Phys. **20** 1413
  The Dirac equation in the algebraic approximation. V. Self–consistent field studies including the Breit interaction.
Quiney H.M., Grant I.P. and Wilson S., 1987b Physica Scripta **36** 460
  The Dirac equation in the algebraic approximation.
Quiney H.M., Grant I.P. and Wilson S., 1989a in Numerical Determination of the Electronic Structure of Atoms, Diatomic and Polyatomic Molecules, ed. M.Defranchesi and J.Delhalle,NATO Advanced Research Institute, Versailles, April 1988, (Dordrecht:Reidel)
  On the accuracy of the algebraic approximation in relativistic electronic structure calculations
Quiney H.M., Grant I.P. and Wilson S., 1989b in Many–Body Methods in Quantum Chemistry, Lecture Notes in Chemistry **52** 331, ed. U. Kaldor, (Berlin:Springer)
  On the relativistic many–body perturbation theory of atomic and molecular electronic structure
Quiney H.M., Grant I.P. and Wilson S., 1989c J. Phys. B: At. Mol. Opt. Phys. (Letters) **22** L15.
  On the accuracy of Dirac–Hartree–Fock calculations using analytic basis sets
Quiney H.M., Grant I.P. and Wilson S., 1990, J. Phys. B: At. Mol. Opt. Phys. (Letters) **23** L271
  Relativistic many–body perturbation theory using analytic basis functions
Raffenetti, R.C., 1973, J. Chem. Phys. **59**, 5936
  Even–tempered atomic orbitals. II. Atomic SCF wavefunctions in terms of even–tempered exponential bases.
Reeves, C.M., 1963, J. Chem. Phys. **39**, 1
Salomonson S and Öster P, 1989, Phys. Rev. **A40** 5559
Saunders, V.R., 1990, in *Supercomputational Science*, edited by R.G. Evans and S. Wilson, (Plenum, New York)
  Parallel processing on shared memory multi–user systems
Schmidt MW and Ruedenberg K 1979 J. Chem. Phys. **71** 3951
  Effective convergence to complete orbital bases and to the atomic Hartree–Fock limit through systematic sequences of Gaussian primitives.
Silver, D.M., 1978a, Comput. Phys. Commun. **14** 71
  Diagrammatic many–body perturbation expansion for atoms and molecules. I. General organization.

Silver, D.M., 1978*b*, Comput. Phys. Commun. **14** 81
>Diagrammatic many–body perturbation expansion for atoms and molecules. II. Second–order and third–order ladder energies.

Stanton, R.E., and Haviliak, S., 1984, J. Chem. Phys. **81** 1910
>Kinetic balance : A partial solution to the problem of variational safety in Dirac calculations.

Taylor, P.R., and Bauschlicher, C.W., 1987, Theoret chim Acta **71** 105
>Strategies for obtaining the maximum performance from current supercomputers

Taylor, P.R., Bauschlicher, C.W., and Schwenke, D.W., 1989, in "Concurrent computation in chemical calculations", Meth. Comput. Chem. **4** 63
>Chemical Calculations on Cray Computers

Thompson, J.W., and Wilson, S., 1990a, J. Phys. B: At. Mol. Opt. Phys. **23** 2205
>On the accuracy of the algebraic approximation in molecular electronic structure calculations. II. Comparison of diatomic molecule self–consistent field calculations using basis sets of elliptical functions with fully numerical Hartree–Fock studies

Wells, B.H., and Wilson, S., 1986, J. Phys. B: At. Mol. Opt. Phys. **19** 2411
>Second–order correlation energy of the argon atom using basis sets of Gaussian–type functions.

Wells, B.H., and Wilson, S., 1989, J. Phys. B: At. Mol. Opt. Phys. **22** 1285
>On the accuracy of the algebraic approximation in molecular electronic structure calculations. I. Calculations for H_2^+, HeH^{2+}, H_2 and HeH^+ uisng basis sets of atom centred Gaussian–type functions

Wick, G.C., 1950, Phys. Rev. **80**, 268.
>The evaluation of the collision matrix

Wilson, S., 1978, Comput. Phys. Commun. **14** 91
>Diagrammatic many–body perturbation expansion for atoms and molecules. III. Third–order ring energies

Wilson, S., 1978, in *Correlated Wavefunctions*, Proceeding of a Study Weekend, December 1977, edited by V.R. Saunders, Daresbury Laboratory.
>Application of many–body perturbation theory to molecules

Wilson S., 1980a, Theoret. chim. Acta. **57** 53
>Systematic sequences of even–tempered Gaussian primitives in electron correlation calculations using many–body perturbation theory.

Wilson S., 1980b, Theoret. chim. Acta. **58** 31
>Universal systematic sequence of even–tempered Gaussian primitive functions in electron correlation studies

Wilson, S., 1981, in *Proceedings of Fifth Seminar on Computational Problems in Quantum Chemistry*, Groningen, eds. P.Th. van Duijknen and W.C. Nieuwpoort.

Wilson, S., 1983, in: *Methods in Computational Molecular Physics*, eds. G.H.F. Diercksen and S. Wilson, (Reidel, Dordrecht,)
>Diagrammatic many–body perturbation theory for atoms and molecules.

Wilson, S., 1984, *Electron correlation in molecules*, Clarendon Press, Oxford.

Wilson, S., 1985, Comput. Phys. Repts. **2**, 389
>Diagrammatic many–body perturbation expansion for atomic and molecular electronic structure

Wilson, S., 1987, Adv. Chem. Phys. **67** 439.
>Basis sets

Wilson S., 1988, in *Relativistic effects in atoms and molecules*, ed. S.Wilson, Meth. Comput. Chem. **2** 73 (New York:Plenum)
>Relativistic molecular structure calculations

Wilson S., 1989, in *Concurrent computation in chemical calculations*, ed. S.Wilson, Meth. Comput. Chem. **3** 1 (New York:Plenum)
>Parallel computers and concurrent computation in the chemical sciences

Wilson S., 1990 in *Supercomputational Science*, edited by R.G. Evans and S.Wilson, (New York:Plenum)
>Numerical Recipes for Supercomputers

Wilson, S., and Saunders, V.R., 1980, Comput. Phys. Commun. **19** 293
    Diagrammatic many—body perturbation expansion for atoms and molecules. V. Fourth—order diagrams involving triply excited states.
Wilson, S., and Silver, D.M., 1976, Phys. Rev. A**14**, 1949
    Algebraic approximation in many—body perturbation theory.
——, 1977, J. Chem. Phys. **66**, 5400
    Diagrammatic perturbation theory: Many—body effects in the $X^1\Sigma^+$ states of first—row and second—row diatomic hydrides.
——, 1979, Comput. Phys. Commun. **17** 47
    Diagrammatic many—body perturbation expansion for atoms and molecules. IV. Fourth—order lined diagrams involving quadruply excited states.
——, 1980, J. Chem. Phys. **72**, 2159
    Diagrammatic perturbation theory: An application to N_2, CO and BF using a universal basis sets.
——, 1982, J. Chem. Phys. **77**, 3674
    Diagrammatic perturbation theory: An application to the LiH and FH molecules using a universal basis set.
Wood, J., Grant, I.P., and Wilson, S., 1985, J. Phys. B: At. Mol. Phys. **18**, 3027
    The Dirac equation in the algebraic approximation. IV. Application of the partitioning technique.

SOLID STATE

RELATIVISTIC DENSITY-FUNCTIONAL THEORY FOR ELECTRONS IN SOLIDS

B.L. Gyorffy
H.H. Wills Physics Laboratory, University of Bristol
Tyndall Avenue, Bristol   BS8 1TL

J.B. Staunton
Department of Physics, University of Warwick
Coventry   CV4 7AL

H. Ebert
Siemens A.G. ZFE TPH 11, Paul-Gossen St Strasse 100
D-8520 Erlagen, Germany

P. Strange
Department of Physics, University of Keele
Staffordshire   ST5 5BG

B. Ginatempo
Instituto di Fisica Teorica, Universitá di Messina
I-98166 Messina, Italy

1. Introduction

Practising solid state physicists are accustomed to using two dramatically different approaches to describing the electron 'glue' that holds the atomic nuclei together. The first pictures the solid as made up of atoms and talks of outer, valence, electrons hopping from site to site on an infinite lattice. This mode of thought leads to tight-binding models defined in language of local atomic like orbitals (LCAO) and the terms of discussions are very close to the parlance of quantum chemistry [McWeeny 1989]. The other approach begins with a model in which the positive charges of the nuclei are smeared out into a uniform, positively charged background for the negatively charged, interacting many electron system and proceeds to correct this, 'jellium', model by reintroducing the point charges of the nuclei [Ashcroft and Mermin 1976]. Usually, this second step is accomplished using the Density Functional Theory of Kohn and Sham [Kohn, W. and Vashista, P. 1983] and the notion of an atomic like object never enters the discussion. In a book dedicated to the common problems of atomic, molecular and solid state physics it may seem perverse to follow the second line of thought. Nevertheless, it should be excused on the practical grounds that much of the most interesting modern applications of Relativistic Quantum Mechanics in condensed matter physics uses this second language and hence a useful introduction to the solid state oriented contributions will have to do likewise. Moreover, it is hoped that even the readers more at home within the conceptual framework of quantum chemistry may find the change in the habit of thought and the mode of presentation an interesting diversion.

To motivate an interest in effects of relativity on the electronic structure of solids, consider the non-interacting jellium model. Recall that the single particle states are plain waves, $\sim e^{i\vec{k}\cdot\vec{r}}$, and, according to the Pauli exclusion principle, in the ground state, they are filled up to the Fermi surface as shown in Fig. 1. The velocity of each electron is $\vec{v}_{\vec{k}} = \frac{1}{m}\hbar\vec{k}$ and hence there can be only two electrons in the system which are stationary. They are the spin up and spin down electrons in the $\vec{k} = 0$ state. The rest must be moving. In fact, at the Fermi surface, they move with the velocity $v_F = \frac{\hbar}{m}\sqrt[3]{3\pi^2 n}$, where n is the density of electrons. Thus, for a sufficiently dense system of degenerate Fermions the most interesting particles, namely those on the Fermi surface, move with the relativisitic velocities $v_F \sim c$. Evidently, a good dimensionless measure of how close we are to the relativistic limit is

$$\beta = v_F / c = \frac{\hbar}{mc}(3\pi^2 n)^{1/3} \qquad (1)$$

To hint at what may happen as β goes to 1 we note that according to Eq. 1, for $\beta \sim 1$, the particles are within their Compton wave length, $\lambdabar = \frac{\hbar}{mc}$, of each other. This is a dramatic circumstance because, as was stressed by Landau [Beresteskii et al. 1971], a relativistic particle cannot be localized on a length scale less than $\lambdabar$ and hence on such length scale the Dirac wave-function cannot be interpreted as a probability amplitude [Beresteskii et al. 1971].

Another example relativistic effects is the famous Chandrasekart limit of stellar stability [Cottingham & Greenwood 1987]. To see the point recall that in non relativistic quantum mechanics the energy per particle of a homogenous liquid in a uniform, neutralizing, positively charged background is given by [Fetter and Walecka 1975]

$$\frac{E^{NR}}{N} = \frac{e^2}{2a_o}\left[\frac{2.21}{r_s^2} - \frac{.91}{r_s} + \frac{2}{\pi}(1-\ln 2)\ln r_s + \text{constant} + O(r_s)\right] \qquad (2)$$

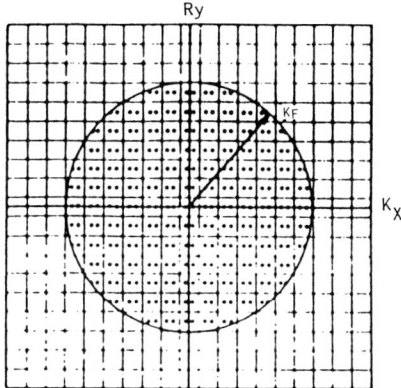

Fig. 1 Fermi sphere with radius K_F in k-space. Each volume element of size $(2\pi)^3/V_0$ contains two states which, inside the Fermi sphere at T=0, are occupied by electrons with opposite spin.

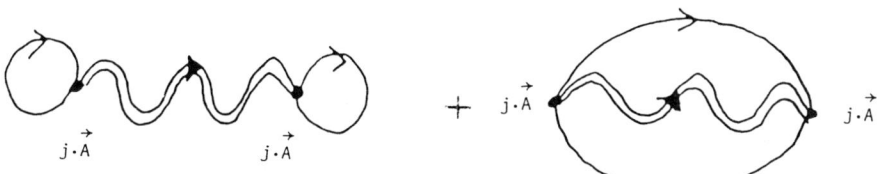

Fig. 2 Feynman diagrams of many-body perturbation theory second order in the electron-transverse photon vertex.

where $\frac{4\pi}{3}r_s^3$ is the average spherical free volume occupied by an electron and r_s is measured in units of the Bohr radius a_o (= .529 Å). Surprisingly, in the high density limit ($r_s \to 0$) the first, kinetic energy, term $\left(\frac{2.21}{r_s^2} \cong \frac{3}{5}\varepsilon_{k_F} \cong \frac{\hbar^2}{2m}k_F^2\right)$ dominate and it provides the ultimate, Pauli exclusion principle, resistance to the gravitational collapse of stars called white dwarfs. In the relativistic case the volume dependence of this term changes on account of the relativistic dispersion relation

$$\varepsilon_{k_F}^R = \sqrt{\hbar^2 k_F^2 + m^2 c^4} \cong c\hbar k_F \sim \frac{1}{r_s} \qquad (3)$$

This softening against collapse leads to the famous result that stars with M > 1.44 M_o are unstable [Cottingham & Greenwood 1987].

Although the average density in a typical metal is such that $r_s\sim 5$, and this corresponds to a Fermi velocity of $v_F \cong 10^8 cm/sec$ only, near the atomic nuclei the local density can be considerably higher than that. In fact, at the nucleus, in $In(Z=50)$, according to standard band theory calculations, n(o) = 6.2 10^{29} cm^{-3}. This implies that, locally, $\beta_{In}\sim 1$ [Ramana and Rajogopal 1983]. Thus, it is not surprising that relativistic effects abound in atomic, molecular and solid state physics of elements with Z>50.

Many of these are small 1-10% quantitative effects like the shrinking of the inner core states and the subsequent relaxation of the outer orbitals in atoms, or the shift of the *s-p* bands relative to the *d-* and *f*-bands in transition and rear earth metals respectively [Koeling and McDonald 1983]. Others are qualitatively new features such as the coupling of the spin and orbital degrees of freedoms. Even when very small, as in the case for spin-orbit coupling in Fe, Co and Ni, these can give rise to surprising new phenomena such as magneto crystalline anisotropy, magneto optical effects and various manifestations of the Fano effect.

The first of these will be treated, in this volume by Staunton et al. [1991]. The second and third will be the subjects of the articles by Ebert et al. [1991a] and Ebert et al [1991b] respectively. The solid state analogues of the Fano effect will be discussed by Ginatempo and Gyorffy [1990]. Thus, the present article can be red as a pedagogical introduction to these four contributions. In turn these articles provide ample motivation for studying the 'electron glue', which binds the atomic nuclei of condensed matter together, within the framework of Relativistic Quantum Mechanics.

The complete, fully relativistic theory of an interacting many electron system such as the one in a solid, is Quantum Electrodynamics (QED). While this theory is very well understood in the limit of a vacuum with a few electrons and positrons [Beresteskii et al. 1971] for the dense system of electrons at hand it is far from being a closed book. Fortunately, for solid state problems of current interest we may bypass such interesting questions as pair creation and radiative corrections [Vshivtsev et al. 1989] and proceed along the well known lines of arguments familiar from the non-relativistic treatment of the problem. In particular, we adopt the methods of the Density Functional Theory and confine our interest to the well trodden path of the Local Density Approximation (LDA) [Kohn & Vashishta 1983].

For clarity, we shall present our brief review of this approach to the problem in three instalments. In Section 2 we summarise the basic ideas of LDA within the Relativistic Density Functional Theory. In Section III we describe the multiple-scattering method for solving the basic, Kohn-Sham-Dirac, equation of this theory. In a final, fourth, section we list a number of general issues of current interest.

2. Density functional theory

Density Functional Theory is the formal foundation of all modern 'band theory' calculations [Kohn & Vashishta 1983]. In short it replaces the full complexity of the Fock-space formulation of the many electron problem by a self-consistent field one-electron-problem. Surprisingly, it does so without introducing any approximations [Kohn & Hohenberg 1964]. Of course, as in any other approach to the problem, to render the procedures of theory tractable approximations have to be made. Within the framework of the Density Functional Theory the necessary simplifications take the form of mathematical recipies for calculating the potential function in the effective one electron problem from the knowledge of the charge densities $n_\sigma(\vec{r})$ for electrons with spin projections σ. In this, as well as in the campaignon papers, on applications, we shall be working in the well tried Local Density Approximation (LDA). For general orientation we shall now outline the theory and give a brief description of the LDA.

2.1. The non-relativistic theory

For simplicity we shall deal with the paramagnetic case at zero temperature (T=0) only. Then the theory consists of the following two parts.

The first is a theorem that the ground state energy is a unique functional $E[n(\vec{r})]$ of the charge density $n(\vec{r})$ and that the minimum of this, E_o, at $n_o(\vec{r})$, is the ground state energy of the system. Furthermore, $n_o(\vec{r})$ is the ground state charge density.

The second part is the construction and minimization of the functional $E[n(\vec{r})]$. This leads to the Kohn-Sham's equations which are the Euler-Lagrange equation of the above variational problem. They are given by [Kohn & Sham 1983]

$$(-\nabla^2 + \sum_i v(r - \vec{R}_i;[n(\vec{r})])\psi_n(\vec{r}) = \varepsilon_n \psi_n(\vec{r}) \tag{2.1}$$

where the local potential wells are described by the effective potential function

$$v(\vec{r} - \vec{R}_i;[n(\vec{r})]) = \frac{Z_e^2}{r - \vec{R}_i} + e^2 \int dr' \frac{n(\vec{r}')}{|\vec{r} - \vec{r}'|} + v_{xc}(\vec{r} - \vec{R}_i;[n(\vec{r})]) \tag{2.2}$$

and the exchange correlation potential is the functional derivative of the exchange correlation energy $E^{xc}[n(\vec{r})]$. Namely,

$$v_{xc}(\vec{r} - \vec{R}_i;[n(\vec{r})]) = \left(\frac{\delta E^{xc}[n(\vec{r})]}{\delta n(\vec{r})}\right) \tag{2.3}$$

Eqns. II.1, 2 and 3 has to be solved self-consistently: one assumes a charge density $n(\vec{r})$, calculates $v_{xc}(\vec{r} - \vec{R}_i;[n(\vec{r})])$ and then solves Eqn. II.1 for the wave functions $\psi_n(\vec{r})$, from these one is to calculate the charge density by evaluating the sum over the occupied states up to $\varepsilon_n \leq \varepsilon_F$ as follows:

$$n(\vec{r}) = \sum_n^{\varepsilon_n \leq \varepsilon_F} |\psi_n(r)|^2 \tag{2.4}$$

where ε_F is the Fermi energy, finally the calculated $n(\vec{r})$ is compared with the initial charge density and if they do not agree within a prescribed limit the calculation is repeated with an improved initial $n(\vec{r})$. This procedure is continued until convergence.

Note the remarkable achievement of the Kohn-Hohenberg theorem: it has replaced the complex many-body problem in Fock-space by a self consistent field one electron problem without making an approximation. Of course, the above algorithm is not complete until an explicit form is specified for the exchange correlation functional. In this approach to the problem of interacting electrons it is at this point where principle approximations are made. In this talk I will always have in mind the local density of approximation

$$E_{xc}^{LD}[n] = \int dr \varepsilon_{xc}^{LDA}(n(\vec{r}))n(\vec{r}) \tag{2.5}$$

where the energy density $E_{xc}^{LD}(n)$ is the same function of $n(\vec{r})$ as in the homogeneous electron liquid with a uniform positively charged background. This quantity is now well known from quantum Monte Carlo calculations [Caperley 1980].

Thus the Density Functional Theory approach does hark back to the jellium model of a solid and can be seen as a successful attempt to restore realism to the theory by reintroducing the point ions and the attendant inhomogeneities. From the point of view of our present concerns it is a striking feature of this approach that the density appears to play an explicitly central role. Clearly, where $\beta(\vec{r}) = \frac{\hbar}{mc}\sqrt{3\pi^2 n(\vec{r})} \sim 1$ relativisitic effects will be important. The necessary modifications of the theory will be discussed in the next section.

2.2. The relativistic density functional theory

The first fully-fledged relativistic density functional theories were formulated by Rajagopal (1978) and MacDonald and Vosko (1979). The subject has been reviewed by Ramana and Rajagopal (1983), Rajagopal (1985) and, more recently, with a quantum chemistry bias by Parr and Yand (1989). Done with care, this theory is capable of encompassing the complete quantum electrodynamics of the dense electron system in solids (Ramana and Rajagopol 1983) but in fact only a small part of the whole formal edifice has been put to use so far. This introduction, as well as the following applications, will only concern this part. In fact we will always have the relativistic version of the Local Density Approximation (LDA) in mind.

In short the bases of the formalism is the theorem that the ground state energy is a unique functional of the four current $J_\mu(\vec{r})$ and spin-density $\vec{m}(\vec{r})$, and that the minimum of this functional, $E[J_\mu(\vec{r}), \vec{m}(\vec{r})]$, is the ground state energy. As in the non-relativistic case the Euler-Lagrange equations of this variational problem is a self-consistent one electron equation. The difference is that this Kohn-Sham equation is now of the form of the Dirac equation. Thus the relativistic analogue of Eqn. II.1 is

$$[c\frac{\hbar}{i}\vec{\alpha}\cdot\vec{\nabla} + mc^2(\beta-1) + \underset{\approx eff}{V}(\vec{r};[J_\mu])$$

$$+\vec{\Sigma}\cdot\vec{W}_{eff}(\vec{r};[J_\mu,\vec{m}])]\varphi_n(\vec{r}) = \varepsilon_n \varphi_n(\vec{r}) \tag{2.6}$$

where $\vec{\underline{\alpha}}$, $\underline{\beta}$, $\underline{1}$ are the 4x4 Dirac matrices in the standard representation (Landau et al. 1971), all quantities with double underline, like in $\underline{\sigma}$, are 4x4 matrices, $\varphi_n(\vec{r})$ is a 4-component Dirac wave function, $\underline{V}_{eff}(\vec{r})$ and $\underline{W}_{eff}(\vec{r})$ are the local electrostatic and exchange potentials respectively and $\vec{\underline{\Sigma}}$ is the usual 4x4 Dirac spin matrix. Furthermore,

$$\underline{V}_{eff}(\vec{r};[J_\mu]) = -\underline{1}(v_{ext}(\vec{r}) + e^2 \int dr'^3 \frac{n(\vec{r}')}{|r-r'|} + \frac{\delta E_{xc}}{\delta n(\vec{r})})$$

$$-\vec{\underline{\alpha}} \cdot (\vec{A}_{ext}(\vec{r}) + \frac{e}{c} \int dr'^3 \frac{\vec{J}(\vec{r}')}{|\vec{r}-\vec{r}'|} + \frac{\delta E_{xc}}{\delta \vec{J}(\vec{r})}) \quad (2.7)$$

and

$$\vec{\underline{W}}_{eff}(\vec{r}) = -\frac{e}{c}(\vec{\nabla} x \vec{A}_{ext}(\vec{r}) + \frac{\delta E_{xc}}{\delta \vec{m}(\vec{r})}) \quad (2.8)$$

where

$$n(\vec{r}) = \sum_n^{occ} \varphi_n^+(\vec{r}) \cdot \varphi_n(\vec{r})$$

$$\vec{J}(\vec{r}) = \sum_n^{occ} \varphi_n^+(\vec{r}) \cdot \vec{\underline{\alpha}} \cdot \varphi_n(\vec{r})$$

$$\vec{m}(\vec{r}) = \sum_n^{occ} \varphi_n^+(\vec{r}) \cdot \vec{\underline{\Sigma}} \cdot \varphi_n(\vec{r}) \quad (2.9)$$

Evidently, the LDA consists of using $E_{xc}(J_\mu, \vec{m})$ for the relativistic, spin polarized jellium. Unfortunately, and surprizingly, this is not as well known [Cortona et al. 1983, Eschrig et al. 1985] as the non-relativistic exchange amd correlation energy, E_{xc}^{NR}, for jellium [Ceperley 1980] and hence, frequently, E_{xc}^{NL} is used even in relativistic calculations.

The most striking new features of the above relativistic LDA potential is the contribution arising from the exchange of (transverse) photons between electrons. These are the consequences of the grand state energy diagrams with $\vec{j} \cdot \vec{A}$ vertecies and photon ($\approx$) lines shown in Fig. 2 and they are often referred to as Breit interactions even though in Breit's original paper [Breit 1925, Landau 1935] only a very approximate description of the full effect was considered. Although very small (~1 K°) the Briet interaction has important practical consequences due to its long range. For example its contribution to the total energy in the Hartree approximation:

$$E^H_{Breit} = \frac{1}{2} \int dr^3 \int dr'^3 \frac{1}{|\vec{r}-\vec{r}'|^3} [\vec{M}(\vec{r}) \cdot \vec{M}(\vec{r}')$$

$$-3 \frac{[(\vec{r}-\vec{r}') \cdot \vec{M}(\vec{r})][(\vec{r}-\vec{r}') \cdot \vec{M}(\vec{r}')]}{|\vec{r}-\vec{r}'|^2}] - \frac{4\pi}{3} \int dr^3 |\vec{M}(\vec{r})|^2 \qquad (2.10)$$

gives rise to shape anisotropy effects which are the root causes of domain structure and dominate the magnetism of thin films [Jansen 1988].

3. The multiple scattering method (KKR) for solving the Kohn-Sham-Dirac equations

Most well-established band theory methods (LMTO, LAPW, KKR etc.) have a relativistic version which can solve the self-consistant one particle problem specified by Eqns. 2.6,7,8,9. Here we shall recall, briefly, only the relativistic Korringa-Kohn-Rostoker method which is based on the multiple scattering approach to the problem [Korringa 1947, Lloyd and Smith 1972, Gyorffy and Stott, 1973].

The strategy of multiple scattering theory for solving a Dirac or Schrödinger equation for an ordered array of potential wells, such as one encounters in a crystalline solid, is to solve the one scatterer problem first, by calculating the corresponding scattering amplitude, and then use this result to find the energy eigen values and eigen functions for the infinite lattice.

For a spherically symmetric finite range scatterer, the partial wave scattering amplitude is given by

$$f_\kappa(\varepsilon) = \frac{1}{2i}(e^{i2\delta_\kappa(\varepsilon)} - 1) \qquad (3.1)$$

where κ is the relativistic rotational quantum number, e.g.
$\kappa = -2,-1,1,2$ \qquad (3.2)

which are equivalent to ℓ_{m_j} = p$_{3/2}$, s$_{1/2}$, p$_{1/2}$, d$_{3/2}$ and $\delta_\kappa(\varepsilon)$ is the phase shift of the outgoing spherical wave in the κ-th channel. It is to be calculated by matching the solutions of the Dirac equation inside the potential well to an incoming spherical wave, labelled by the spin angular quantum number κ, of unit amplitude and an outgoing spherical wave with the same κ and coefficient $f_\kappa(\varepsilon)$.

As an illustration of the general description above we show the d$_{3/2}$ and d$_{5/2}$ phase shifts corresponding to potential wells in an f.c.c. Pt metal in Fig.3. Such sharp resonances are characteristic of crystals with narrow bands such as those of transition or rear-earth metals. Note that the two d-resounces are split by the spin-orbit coupling. The extent of this splitting, 1.5 eV, determines the spin-orbit spliting of the d-bands which arise from them.

If the target is spin polarized the various spin-angular channels will be coupled and the scattering amplitude becomes a matrix $f_{\kappa\mu,\kappa'\mu'}(\varepsilon)$ [Strange et al. 1984]. This is due to the fact that the spin-orbit coupling term, $\vec{\sigma}\cdot\vec{L}$, is no longer a constant of the motion because it does not commute with the spin-external field coupling, $\vec{\sigma}\cdot\vec{B}$, contribution to the Hamiltonian. Under this circumstance it is useful to define the S matrix

$$S_{QQ'} = \delta_{Q,Q'} - 2if_{Q,Q'}(\varepsilon) \tag{3.3}$$

where Q stands for both κ and the magnetic quantum number μ, and diagonalize $S_{Q,Q'}$ with the help of the unitary rotation matrix $U_{Q,Q'}(\varepsilon)$ by writing

$$S_{QQ'} = \sum_{Q''} U_{QQ''} e^{i2\delta_{Q''}(\varepsilon)} U^{-1}_{Q''Q'} \tag{3.4}$$

The main point of this form is that the eigen values of the S-matrix are complex numbers of unit amplitudes and hence they can be written as $e^{i2\delta_Q(\varepsilon)}$. Namely, the unitarity of $S_{Q,Q'}$ allows us to define phase shifts even when it is not diagnonal.

Strange et al. (1984) studied the Pt potential whose phase shifts are shown in Fig.3 in a magnetic field B$_{eff}$. They found that in the $\ell=2$ channel all 10 phase shifts, defined as in Eqn. III.4, show a resonance whose position in energy, ε_r^Q, moved with increasing B$_{eff}$. Their results are shown in Fig.4. Evidently, these curves give an account of the Zeeman effect on the scattering states at positive energies.

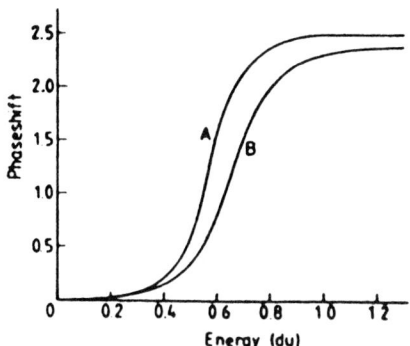

Fig. 3 The platinum $J=\frac{3}{2}$ (A) and $J=\frac{5}{2}$ (B) phaseshifts for l=2. (1 du=1.4 Ryd.)

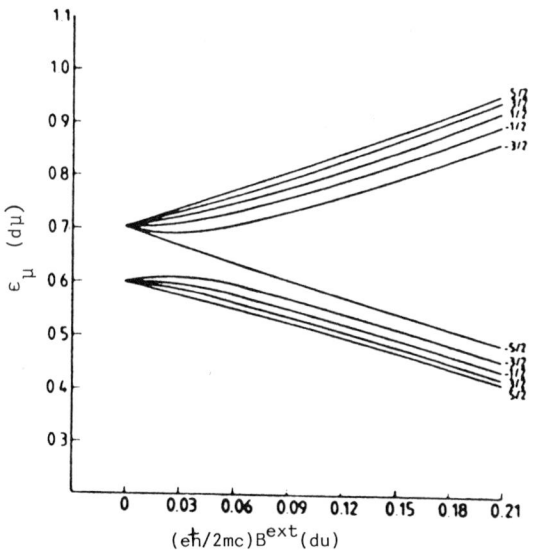

Fig. 4 Splitting of the d phase shifts in platinum as a function of magnetic field in the diagonal representation. (1 du= 1.4 Ryd.)

The above discussion dealt with the case of non-spherical symmetric scatterers where the asphericity was confined to spin-space. Many interesting effects are associated with potentials which lack spherical symmetry in both spin and real space. For instance, to describe the coupling between the quadrupole moment of atoms in solids and the spin polarization of the electrons would require the consideration of such scattering centres. Unfortunately, as yet, this sector of the theory has not been developed.

Once the individual scattering centres have been fully described the next step in a multiple scattering theory is to construct the conditions for the infinite array of scattering centres to have an energy eigen value. Following Korringa (1947) such conditions may be derived by demanding that the incident wave to each site should be equal to the sum of outgoing waves from all the other sites. For the case of one site per unit cell this leads to the well-known KKR condition:

$$\left\| -\sqrt{\varepsilon}\, f^{-1}_{Q,Q'}(\varepsilon) - G_{QQ'}(\vec{q};\varepsilon) \right\| = 0 \tag{3.5}$$

where $\left\| 0_{Q,Q'} \right\|$ stands for the instruction to take the determinant of the matrix $O_{Q,Q'}$ and $G_{Q,Q'}(\vec{q},\varepsilon)$ is the lattice Fourier transform of the relativistic KKR structure constants [Weiberger, 1990]. This latter quantity describes how free spherical waves propogate from site to site and is independent of the potential wells. Namely, it contains only geometrical information (the crystal structure) on the arrangement of the scatterers.

For a fixed wave vector $\vec{q}$ the KKR determinant in Eqn. III.5 vanishes at a number of energies $\varepsilon_{q,v}$. These are the Bloch wave energy eigen values. Finding these zeros is the principle computational task in applications of the KKR method.

To proceed to self-consistency one must also calculate the charge density $n(\vec{r})$ and the magnetization density $m(\vec{r})$. These quantities are readily obtained from the density-matrix

$$\underset{\approx}{n}(\vec{r},\vec{r}') = -\frac{1}{\pi}\mathrm{Im}\int d\varepsilon f(\varepsilon)\underset{\approx}{G}(\vec{r},\vec{r}';\varepsilon) \tag{3.6}$$

where the imaginary part of the Greens function $\underset{\approx}{G}(\vec{r},\vec{r}';\varepsilon)$, which is a 4x4 bispinor, is given by

$$\mathrm{Im}\, G(\vec{r},\vec{r}';\varepsilon) = \sum_{QQ'} Z_{\sim Q'}(\vec{r}_i;\varepsilon)\, \mathrm{Im}\, T_{QQ}^{i,i}(\varepsilon)\, Z_{\sim Q'}(\vec{r}_i;\varepsilon) \qquad (3.7)$$

This formula is valid within the i-th unit cell. The position vector $\vec{r}_i = \vec{r} - \vec{R}_i$ where $\vec{R}_i$ is the position of the i-th lattice site, $Z_Q(\vec{r};\varepsilon)$ is a solution, which is regular at the origin, of the Dirac equation in the i-th unit cell and

$$T_{Q,Q'}^{ij} = \sum_{\vec{q}}^{BZ} (-\sqrt{\varepsilon} f_{Q,Q'}^{-1} - G_{Q,Q'}(\vec{q},\varepsilon))^{-1} \qquad (3.8)$$

where the sum is over the Brillouin Zone.

Thus, given V_{eff}^{LDA} and W_{eff}^{LDA} in terms of $n(\vec{r},\vec{r}')$ with the help of Eqn. III.5,6,7 and 8 a self-consistent solution to the Kohn-Sham-Dirac equations given in Eqn. II.6 can be affected. Examples of results of such calculation will be given in the following chapters where application of the above theory are discussed. Nevertheless, for the sake of explicitness we show, in Fig. 5., a comparison between a relativistic and non-relativistic LDA calculation.

In conclusion we would like to stress that frequently relativistic effects cannot be calculated in low order perturbation theory using $\frac{1}{c^2}$ as an expansion parameter. For instance, let us consider the polarization of the photo-emitted electrons from the (111) face of an FCC Pt metal. Since Pt is not magnetic no such polarization would exist in a non relativistic calculation. Thus this polarization is completely a relativistic effect. We have calculated it for various values of the velocity of light. We show our results in Fig.6 as a function of $\frac{1}{c^2}$ which, experimentally, in our units, is 50. Evidently, the polarization is not a linear function of $\frac{1}{c^2}$ for all three energies. From the shape of these curves we deduce that first order perturbation theory can be expected to work only up to $\frac{1}{c^2} = 5.10^{-6}$ which is factor of 10 smaller than the experimental value.

4. Interesting relativistic effects in solids

In conclusion we should like to mention explicitly a few cases of relativistic effects in solids. The most studied examples are:

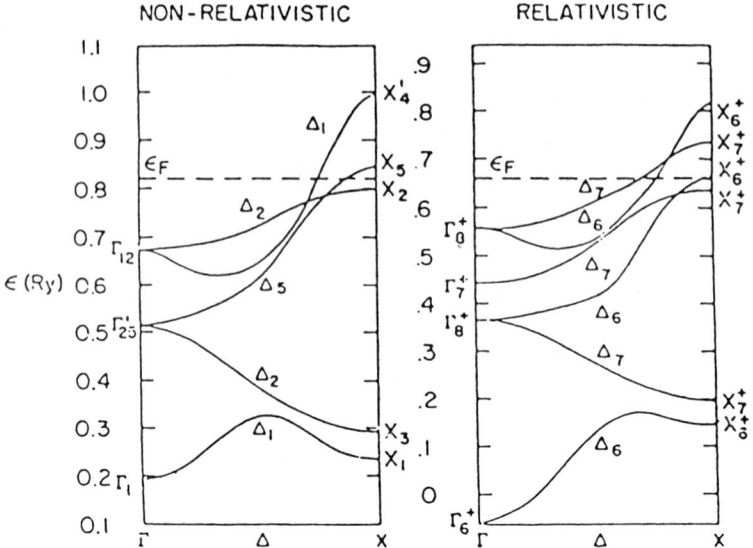

Fig. 5. Relativistic effects of energy bands of Pt. (Due to Mackintosh).

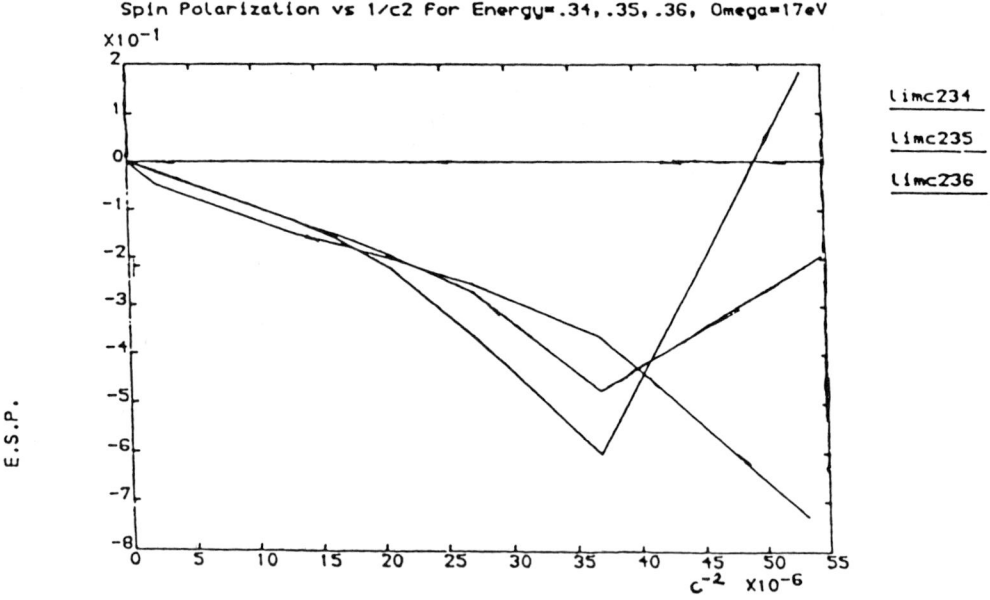

Fig. 6. Photo current anisotropy: $(I^+-I^-)(I^++I^-)$ are the spin averaged photo currents excited by right-handed and left-handed, circularly polarized incident radiation.

i) magneto crystalline anisotropy
ii) magneto optic (Faraday and Kerr) effects
iii) Fano effects
iv) magnetic scattering of X-rays

The first three of these will be treated at length in the following contributions. Due to the new very powerful X-beams, available from synchrotron sources, the fourth topic is of particular current interest and as such it has been well reviewed by Blume [1985].

There are two more, less well known, topics which we thought merit a note. They differ from those above in that they do not have analogues in chemistry. Moreover, they deal with phenomena whose significance has not yet been fully recognized. The first of these is

4.1. The relativistic correction to the London moment

The existence of gyromagnetic phenomena in solids, like the Barnett and the Einstein-deHaas effects, is stressed by Landau et al. [1984]. A particularly interesting example is the London moment of a rotating superconductor [London 1960]..

Consider a cylindrical superconductor rotating about is axis with angular velocity $\vec{\omega}$. Under this circumstance the statistical mechanics of the electrons have to be studied in the coordinate system rotating with the cylinder. In this frame of reference an electron experiences, in addition to the coulomb interaction with the ions and the other electrons, fictitious forces due to the rotation. These are the familiar centrifugal and Coriolis forces:

$$\vec{F} = -m\vec{\omega} x (\vec{\omega} x \vec{r}) - 2m(\vec{\omega} x \vec{v}) \tag{4.1}$$

where $\vec{v}$ is the velocity of the electron. As was argued by Rystephanic [1976] in a rotating superconductor both of these forces will be cancelled by a Lorentz force due to induced electric and magnetic forces:

$$\vec{F}_L = -e(\vec{E}_{ind} + \frac{1}{c}\vec{v}x\vec{B}_{ind}) \tag{4.2}$$

where

$$\vec{E}_{ind} = \frac{m}{e}\vec{\omega}x\vec{\omega}x\vec{r} \quad \text{and} \quad \vec{B}_{ind} = \frac{2mc}{e}\vec{\omega} \tag{4.3}$$

and -e is the charge of the electrons. The induced $\vec{B}$-field, $\vec{B}_{ind}$, is commonly referred to as the London moment. Evidently, it corresponds to an induced vector potential:

$$\vec{A}_{ind} = \frac{2mc}{e}(\vec{\omega} \times \vec{r}) \tag{4.4}$$

Solving the Ginzburg-Landau equations [Abrikosov, 1988] for this vector potential, leads to a surface current which is the physical origin of the London moment.

The principle general interest in the above phenomena is its consequence that the quantized flux within a rotating superconducting ring includes a contribution due to $\vec{A}_{ind}$. Namely, the flux quantization constraint becomes:

$$\oint (\vec{A} + \vec{A}_{ind}) d\vec{e} = n\frac{hc}{2c} \tag{4.5}$$

Consequently, by measuring the angular velocity difference $\Delta\omega (= \omega_{n+1} - \omega_n)$ between the ground states that has n+1 and n quanta respectively treading a rotating ring, one can measure the fundamental constant

$$\frac{\hbar}{m} = 2S\Delta\omega \tag{4.6}$$

where S is the area enclosed by the ring [Cabrera et al. 1982].

The relevance of this effect to relativistic considerations arises from the fact that there is a measurable relativistic correction to the above formula [Cabrera et al. 1982]:

$$\frac{\hbar}{m} \cong (1 + \frac{<T>_{av}}{mc^2} + ...)S\Delta\omega \tag{4.7}$$

where $<T>_{av}$ is the relativistic kinetic energy of the electrons. Thus an accurate (10 ppm) measurement of S and $\Delta\omega$ determines $<T>_{av}$. Evidently this is a remarkable circumstance since $<T>_{av}$ is not probed by any other experiments. The results of the first calculations of $<T>_{av}/mc^2$ by Cabrera et al. [1982], are shown in Table 1. Clearly, further more accurate calculations and experiments could make this technique a unique probe of the electronic structure near the atomic nuclei.

TABLE a) Estimates for the relativistic corrections $\gamma-1$ are calculated from the Herman-Skillman tables. For each valence electron the expectation value of the potential energy (V) is numerically integrated. Then using the tabulated nonrelativist valence energies E (NRL), $\langle T \rangle$ and $\gamma-1$ are computed. All energies are given in eV. The ratio of the Fermi energy μ to the calculated $\langle T \rangle$ is also shown as an estimate of the error in $\gamma-1$.

Element	Shell	No.	$-\langle V \rangle$	$-E$(NRL)	$\langle T \rangle$	$\gamma-1$ (ppm)	$\mu/\langle T \rangle$
Be	2s	2	23.6	8.1	15.5	30	0.92
Al	3s	2	43.3	10.0	33.3	65	0.35
	3p	1	23.1	4.8	18.2	35	0.64
Ti	3d	2	130.2	8.5	121.7	238	0.04
	4s	2	29.7	6.2	23.5	46	0.22
Nb	4d	4	114.5	6.1	108.3	212	0.04
	5s	1	30.6	5.4	25.1	49	0.16
In	5s	2	72.3	10.2	62.1	122	0.14
	5p	1	34.4	4.7	29.6	58	0.29
Sn	5s	2	111.7	12.5	99.2	194	0.10
	5p	2	62.6	5.9	56.7	111	0.18
Ta	5d	3	145.6	8.5	137.0	268	0.04
	6s	2	45.4	6.2	39.1	77	0.13
Pb	6s	2	101.1	12.1	89.0	174	0.11
	6p	2	55.6	5.7	49.9	98	0.19

TABLE b) Estimates of the total relativistic mass shifts for elemental superconductors based on the expectation values of the kinetic energy.

Element	T (eV)	$m\cdot/m-1$ (ppm[c])
Be	25[a]	20
Al	33[a]	40
Ti	73	120
Nb	92	150
In	51	80
Sn	78	130
Ta	111[b]	180
Pb	74[b]	120

[a]From OPW calculation.
[b]From Liberman calculation.
[c]To nearest 10 ppm.

4.2. The universality classes of metallic magnets

In general, a second order phase transition is characterized by a set of critical point exponents. For instance the mangetization M(T) of a ferromagnet goes to zero at the critical temperature T_c as $M(T) \sim (T-T_c)^\beta$. Similarly the susceptibility $\chi(T) \sim (T-T_c)^{-\gamma}$, coherence length $\xi(T) = (T-T_c)^{-\nu}$ and $C_H(T) \sim (T-T_c)^{-\alpha}$ respectively. Moreover, these exponents (β,γ,ν,α.....) form one of a relatively few, apparently, universal sets. These are then said to define the universality class of a phase transition. Or, in other words, the set of critical exponents which characterise it, depends only on the number of components the order parameter has (n), the spatial dimensionality of the system (d) and the range of the interaction [Wilson, K. 1975].

Experimentally, the critical exponents characterising the Currie-Point of Fe, Co and Ni are those associated with the Heisenberg Spin Hamiltonian

$$H = \sum_{ij} J_{i,j} \vec{S}_i \vec{S}_j \tag{4.2}$$

However, this cannot be the whole story since the spin carrying electrons are described by Relativistic Quantum Mechanics whose relevant symmetry properties differ greatly from the simple, spin-only, description afforded by H in Eq IV.2. In particular, it is already well known (theoretically) that the dipolar spin-spin interaction and the cubic and quartic spin space anisotropies induced by the spin-orbit coupling, change the universality class [Toulouse and Pfeuty, 1975]. However, the true universality class of the ferromagnetic metals is determined by Relativistic Quantum Mechanics and it remains to be fully investigated.

References

Abrikosov, A.A. 1988 "Fundamentals of the Theory of Metals" (North Holland 1988).
Ashcroft, N.W. and Mermin, N.D. (1976) "Solid State Physics" (Holt, Rinehart and Winston, New York).
Beresteskii, V.B., Lifshitz, E.H. and Pitaevskii (1971) "Relativistic Quantum Theory" (Oxford: Pergamon Press).
Blume, M.J. (1985) *J. Appl. Phys.* **57** 3615.
Breit, G. (1929) *Phys. Rev.* **34** 553.
Cabrera, B., Gutfreund, H. and Little, W.A. *Phys. Rev.* **B25**, 6644 (1982). See also Cabrera, B. and Peskin, M.E. (1989) *Phys. Rev.* **39** 6425.
Carperley, D.H. and Adler, B.J. 1980, *Phys. Rev. Lett.* **45** 566-569.
Cartona, P., Doniach, S. and Sommers, C. (1985) *Phys. Rev.* **A31** 2842.
Cottingham, W.N. and Greenwood, D.A. 1986, (Cambridge University Press).
Ebert, H. Drittler, B., Strange, P., Zeller, R. and Gyorffy, B.L. (1991a) (in this volume).
Ebert, H. Drittler, B., Zeller, R and Dederichs, P.H. (1991b) (in this volume).

Eschring, H., Seifert, G. and Ziesche, P. (1985) *Sol. Stat. Comm.* **56** 777.
Fetter, A.L. and Walecka, J.D. (1971) "Quantum Theory of Many-Particle Systems" (McGraw-Hill, New York).
Ginatempo, B. and Gyorffy, B.L. (1991) (in this volume).
Gyorffy, B.L. and Stott, M.J. (1973) "Bound Structure Spectroscopy of Metals and Alloys" eds. D.J. Fabian and L.M. Watson (Academic Press, London).
Jansen, H.J.F. (1988) *Phys. Rev.* **B38** 8022.
Koelling, D.D. and MacDonald, A.H. (1983) "Relativistic Effects in Atoms, Molecules and Solids" ed. E.E. Malli (Plenum Press 1983).
Kohn, W. and Vashishta, P. (1983) "Theory of the Inhomogeneous Electron Gas" eds. S. Lundqvist and N.H. March (Academic Press, New York 1983).
Korringa, J. (1947) Physica XIII 392.
Landau, L.D. (1935) *Phys. Z, Siwjet.* **8** 487.
Landau, L.D., Lifshitz, E.M. and Pitaevskiii (1984) "Electrodynamics of Continuous Media" 2nd Ed. (Pergamon Press, 1984).
Lloyd, P. and Smith, P.V. (1972) *Advances in Physics* **21** 69.
London, F. (1960) "Superfluids" Vol.1 p.78-83 (Dover, New York).
MacDonald, A.H. and Vosko, S.H. (1979) *J. Phys. C: Solid State Phys.* **12** 2977.
McWeeny, R. "Methods of Molecular Quantum Mechanics" Academic Press (London 1989).
Parr, Robert, G. and Yand, Weito "Density Functional Theory of Atoms and Molecules" (Oxford University Press 1989).
Pyykkö, P. "Relativistic Theory of Atoms and Molecules, a Bibliography 1916-1985" Springer-Verlag (Berlin 1986).
Rajagopal, A.K. (1978) *J. Phys. C.* **11** L943.
Ramana, M.V. and Rajagopal, A.K. (1983) *Adv. in Chem. Phys.* **54** 231.
Rystephanick, R.G. (1976) *Am. J. Phys.* **44** 647.
Staunton, J.B., Strange, P., Gyorffy, B.L., Matsumoto, M., Poulter, J., Ebert, H. and Archibald, N.P. (1991) (in this volume).
Strange, P., Staunton, J.B. and Gyorffy, B.L. (1984) *J. Phys. C: Solid State* **17** 3355.
Toulouse, G. and Pfeuty, P. (1975) (Presses Universitaires de Grenoble) "Introduction au Groupe de Renormalisation et a ses Applications".
Vshivtsev, A.S., Magnitskii, B.V., Maslor, I.N., Khalilov, V.R. and Perez-Fernouder, V.K. (1989) *Sov. Aston.* **33** 249.
Weinberger, P. (1990) "Electron Scattering Theory for Ordered and Disordered Matter" (Clarendon Press, Oxford 1990).
Wilson, K.G. (1975) *Rev. Mod. Phys.* **47** 773.

INFLUENCE OF RELATIVISTIC EFFECTS ON THE MAGNETIC MOMENTS AND HYPERFINE FIELDS OF 5d–IMPURITY ATOMS DISSOLVED IN FERROMAGNETIC Fe

H. Ebert[1], B. Drittler[2], R. Zeller[2] and P.H. Dederichs[2]

[1] Siemens AG, ZFE ME TPH 11, Central Research Labs.
Postfach 3220, D–8520 Erlangen, F.R. Germany

[2] Institut für Festkörperforschung, Forschungszentrum Jülich
Postfach 1913, D–5170 Jülich, F.R. Germany

INTRODUCTION

The importance of relativistic effects on the electronic structure of solids has been realized for a long time. One of the first convincing demonstration of the importance of spin–orbit coupling came from de Haas–van Alphen measurements on W (Macintosh and Andersen 1980). In these experiments it was found that the electron and hole surfaces do not touch in the ΓH–direction of the Brillouin–zone, as it was expected on the basis of non–relativistic band structure calculations. In the case of ferromagnetic Ni it was found experimentally that the de Haas–van Alphen signal vanishes for certain orientations of the external magnetic field relative to the crystal axes. This has been ascribed by Ruvalds and Falicov (1968) to a hybridisation of different spin states due to spin–orbit coupling.

The symmetry–breaking property of spin–orbit coupling (Falicov and Ruvalds 1968, Cracknell 1970), which is responsible for these and related phenomena, distinguishes spin–orbit coupling sharply from the so–called scalar–relativistic corrections, i.e. mass–velocity and Darwin term. The term "symmetry–breaking" can indeed be taken literally, as it is nicely reflected by hyperfine experiments. For example, the Au–NMR spectra of Au dissolved in Fe clearly show the presence of an electric field gradient although the Au–atoms occupy substitutionally lattice sites of (in principle) perfect cubic symmetry

(Kawakami et al. 1985). The observed field gradient is caused by an axial charge distortion which in turn comes from the uncomplete quenching of the orbital angular momentum due to spin–orbit coupling. The last example clearly shows that one can expect especially for magnetic materials a lot of interesting effects which are caused by the interplay of spin–polarization and spin–orbit coupling. In the following, this is demonstrated by a detailed discussion of the magnetic moments and hyperfine fields of 5d–atoms dissolved substitutionally in ferromagnetic Fe. Obviously, a theoretical description of the phenomena mentioned above requires a simultaneous treatment of spin–orbit coupling and spin–polarization. This is conventionally done by the use of perturbation theory. In contrast to this approach we have used the recently developed spin–polarized relativistic version of the Korringa–Kohn–Rostoker method of band structure calculations in connection with the Green's function technique. A short outline of this approach is given in the next section.

THEORETICAL FRAMEWORK

Recently Feder et al. (1983) and Strange et al. (1984) developed a spin–polarized fully relativistic version of the KKR–method of bandstructure calculations. This scheme allows to deal with the electronic structure of magnetic materials taking into account all relativistic effects in a rigorous way. Within the framework of the KKR–method or multiple scattering theory, respectively, the corresponding Green's function of the system may be written as:

$$G^+(\vec{r}, \vec{r}', E) = \sum_{\Lambda\Lambda'} Z_\Lambda(\vec{r}, E) \tau_{\Lambda\Lambda'}^{nn}(E) Z_\Lambda^x(\vec{r}', E)$$

$$- \sum_\Lambda Z_\Lambda(\vec{r}_<, E) J_\Lambda^x(\vec{r}_>, E) \ . \tag{1}$$

Here $\tau_{\Lambda\Lambda'}^{nn}(E)$ is the so–called site–diagonal scattering path operator and Z and J are the regular and irregular solutions to the single–site Dirac equation for a spin–dependent potential. In contrast to the non–spin–polarized case Z and J have no pure spin–angular character (Strange et al. 1984). For this reason the index Λ, standing for the set of quantum numbers (κ,μ) specifies only their asymptotic behaviour. For further details see Strange et al. (1989). From the knowledge of the Green's function, more or less all properties of the system can be calculated. The spin magnetic moment for example is given by the expression

$$\mu_{spin} = -\frac{\mu_B}{\pi} \text{Im Tr } \beta \underline{\underline{\Sigma}}_z \int_{-\infty}^{E_F} G^+(\vec{r}, \vec{r}, E) \, dE \tag{2}$$

where β is one of the standard Dirac matrices (Rose 1961), $\underline{\underline{\Sigma}}_z$ is the z–component of the 4 x 4 matrix vector $\underline{\underline{\vec{\Sigma}}} = \underline{I}_2 \otimes \vec{\sigma}$, with $\vec{\sigma}$ the conventional Pauli spin matrix vector.

Within a non–relativistic description of the electronic structure μ_{spin} would be the only contribution to the total magnetic moment. Spin–orbit coupling, however, causes the orbital angular moment to be only incompletely quenched, giving rise to a corresponding orbital magnetic moment. This contribution can be approximately calculated from (Singh et al. 1976):

$$\mu_{orb} = -\frac{\mu_B}{\pi} \operatorname{Im} \operatorname{Tr} \underline{\underline{\beta}} \, \underline{\underline{1}}_z \int_{-\infty}^{E_F} \underline{\underline{G}}^+(\vec{r},\vec{r},E) \, dE \tag{3}$$

where $\underline{\underline{1}}_z$ is the z–component of the 4×4 matrix vector $\underline{\underline{T}} = \underline{\underline{I}}_4 \otimes \vec{T}$ with $\vec{T}$ the conventional angular momentum operator.

Finally the correct relativistic expression for the hyperfine field is given by Breit's formula (Breit 1930) which in our notation is

$$B_{hf} = \frac{1}{\pi\mu_n} \operatorname{Im} \operatorname{Tr} \int_{-\infty}^{E_F} e \, \underline{\underline{\vec{\alpha}}} \cdot (\vec{\mu}_n \times \vec{r})/r^3 \, \underline{\underline{G}}^+(\vec{r},\vec{r},E) \, dE \tag{4}$$

where $\underline{\underline{\vec{\alpha}}}$ is the vector of the Dirac matrices and $\vec{\mu}_n$ is the nuclear moment.

In contrast to a non–relativistic description eq. (4) gives contributions to the hyperfine field which stem from non–s–electrons. As in the case of the orbital magnetic moment these contributions are again a consequence of the unquenching of orbital angular momentum due to spin–orbit coupling. On the basis of a Foldy–Wouthuysen transformation, Tterlikkis et al. (1968) have suggested a subdivision of these non–s–terms into a spin and orbital part. A more rigorous separation is possible by making use of the Gordon decomposition of the electronic current. For simplicity these non–s–contributions to the hyperfine field will be referred to as orbital contributions in the following.

RESULTS AND DISCUSSION

5d–atoms dissolved as impurities in ferromagnetic Fe, Co or Ni are interesting model cases to study the interplay of spin–polarization and spin–orbit coupling because of their high atomic number and the appreciable impurity magnetic moments induced by the host. We have calculated the electronic structure of the 5d–atoms in Fe in a charge selfconsistent way within the framework of local spin density theory using the scalar–relativistic approximation. The spin magnetic moments resulting from fully relativistic calculations using the above approach are shown in Fig. 1. These data show the typical S–like dependence on the atomic number as it is also found for 3d– and 4d– impurity atoms in Fe (see, for

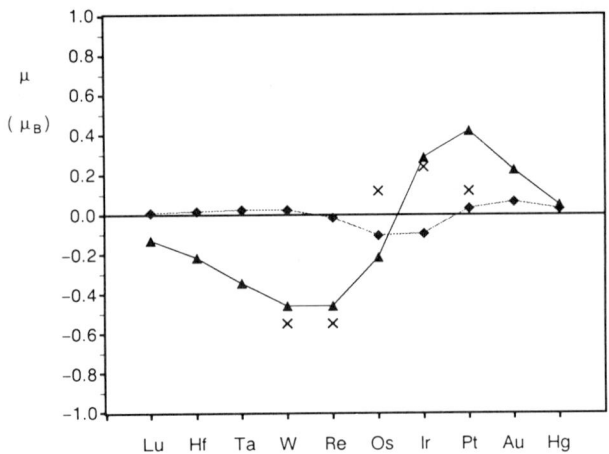

Fig. 1. Spin (▲) and orbital (♦) local moments of 5d–transition metal impurities in Fe. The crosses (x) denote experimental values (Campbell 1966) for the total moments.

example, Akai 1989). Compared to the 3d– and 4d–row atoms, the transition from negative to positive moments is however shifted to a higher atomic number. These findings are in qualitative agreement with experimental data from neutron scattering experiments (Collins and Low 1965). From these measurements it has been concluded that the Os moment should couple ferromagnetically to the host, while our calculation resulted in an antiferromagnetic coupling. This theoretical result could recently be confirmed by X–ray absorption experiments using circularly polarized radiation (Schütz et al. 1989).

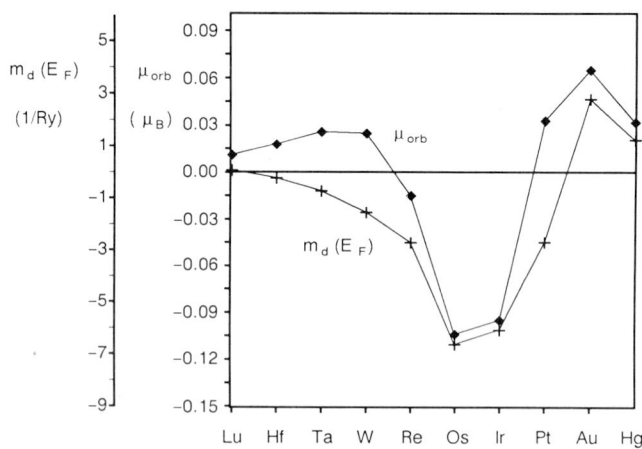

Fig. 2. Orbital moments μ_{orb} and the "magnetic" density of states $m_d(E_F) = n^{(+)}(E_F) - n^{(-)}(E_F)$ for 5d–impurities in Fe

The orbital contribution to the total magnetic moments as obtained from eq. (3) is also given in Fig. 2. Although in most cases it is relatively small compared to the spin contribution it is obviously non-negligible. For the pure elements Fe, Ni and Co it has been found that the spin and orbital moments possess the same sign (Ebert et al. 1988). However this is obviously not a general rule as it can clearly be seen from Fig. 1.

These properties of the orbital magnetic moments can be understood in a qualitative manner by considering spin-orbit coupling as a weak perturbation. Following the ideas of Gehring and Williams (1974) and Demangeat (1974), who performed tight binding model calculations for some heavy impurity atoms in pure ferromagnets, one finds for the expectation value l_z, the component of $\vec{l}$ along the magnetization, the approximate expression:

$$<l_z> \stackrel{\cdot}{=} -\frac{1}{\pi} \xi \, \text{Im} \int_{-\infty}^{E_F} dE \, [(4g_e^{(+)}(E) + g_t^{(+)}(E)) g_t^{(+)}(E) \quad (5)$$
$$- (4g_e^{(-)}(E) + g_t^{(-)}(E)) g_t^{(-)}(E)].$$

Here $g_e^{(\pm)}(E)$ and $g_t^{(\pm)}(E)$ are the tight binding matrix elements of the non-relativistic impurity Green's function for e_g- and $t_{2g}-$symmetry and for (+) and (−) spin direction and ξ is the spin orbit coupling parameter. If we assume, as in the Anderson model, that the e_g- and $t_{2g}-$Green's functions are degenerate and that the local densities of states are Lorentzians, then the energy integration can be carried out analytically yielding

$$<l_z> \approx -\xi \left(n^{(+)}(E_F) - n^{(-)}(E_F) \right) \quad (6)$$

where $n^{(+)}(E)$ and $n^{(-)}(E)$ are the local d-densities of states of the impurity for spin-up and spin-down direction. In Fig. 2 we show the orbital moments of the impurities (left scale) as compared to the "magnetic" local density of states $m_d(E_F) = n^{(+)}(E_F) - n^{(-)}(E_F)$. One clearly finds a strong correlation, even if some details as e.g. the positive orbital moments of the early 5d-impurities, are not reflected by the density of states.

Because of the short range of the hyperfine interactions the hyperfine properties are mainly determined by the electrons near the nucleus. This is just where relativistic effects on the electronic structure are most important and for that reason one must expect a strong impact on the hyperfine properties. This is convincingly demonstrated in table I by showing the hyperfine fields of Ag and Pt in the alloy Ag_6Pt_{94}, which enter a calculation of the nuclear spin lattice relaxation rate (Ebert et al. 1985). As one can see, there is a strong relativistic enhancement of the hyperfine fields which decreases for increasing l-quantum number but strongly increases with the atomic number.

TABLE I

Hyperfine fields of Ag and Pt in Ag_6Pt_{94} (in kG) calculated in a nonrelativistic (nr) and in a relativistic (rel) procedure

l	Ag nr	Ag rel	Pt nr	Pt rel	j	κ
0	8.39	10.5	14.8	34.3	1/2	−1
1	1.56	1.79	3.49	5.59	1/2	1
		1.55		3.47	3/2	−2
2		1.46		2.24	3/2	2
	1.40		1.52			
		1.43		1.80	5/2	−3

In the past several authors tried to take this influence on the hyperfine fields into account by performing scalar–relativistic bandstructure calculations in connection with the use of the conventional expression for the Fermi–contact interaction. This is however not consistent because it strongly overestimates the relativistic enhancement of the s–hyperfine field. A correct expression for the hyperfine fields within the framework of a scalar–relativistic description has been derived by Blügel et al. (1987), which results in a relativistic enhancement in complete accordance with fully relativistic calculations. Because this approach gives only access to the hyperfine fields due to s–electrons, we have adapted here the fully relativistic approach outlined above. The calculated total hyperfine fields of the 5d–atoms in Fe are shown in Fig. 3. The agreement with the corresponding experimental data is rather satisfying although some deviations occur especially at the end of the row.

From the data in table I it is clear that a non–relativistic calculation would result in hyperfine fields much too low compared with experiment. As it is normally assumed, the core part of the hyperfine field turned out to scale rather well with the impurity moment, reflecting the S–shaped variation of μ_{spin} with atomic number (Fig. 1). This holds for the s–electron contribution as well as the much smaller one due to the non–s–core electrons which has been taken into account for the first time.

As for the core part of the hyperfine field, the conduction band part is dominated by its s–contribution, which essentially determines the variation of the total field with atomic number (Fig. 3). Nevertheless the contributions coming from p– and d– electrons are appreciable. Throughout the 5d–row the p–fields are nearly constant and positive (~ 50 kG).

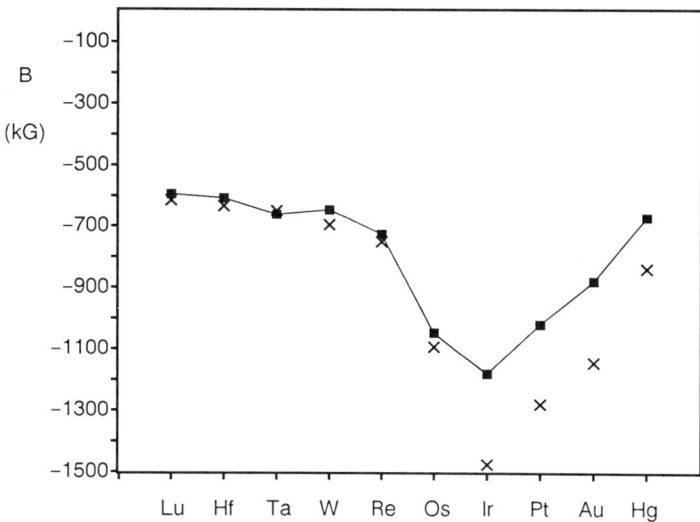

Fig. 3. Calculated hyperfine fields and experimental values (x, Rao 1979) for 5d–impurities in Fe

The d–contribution on the other hand is strongly changing with atomic number (see Fig. 4), and most important in the region where the switch–over from anti– to ferromagnetic coupling of μ_{spin} to the host magnetization occurs, in this way causing the pronounced minimum of the total field for Ir. As stated above, the source of the orbital contribution to the hyperfine field is as for the magnetic moment the uncomplete quenching of the orbital angular momentum. For this reason both quantities should be closely connected to one another. In fact the approximate expression (Abragam and Pryce 1951)

$$B_{orb} = 2\mu_B <\frac{1}{r^3}> \mu_{orb} \tag{7}$$

is often used in a discussion of experimental data. Our results for the orbital contribution to the hyperfine field and magnetic moment, shown in Fig. 4, essentially confirm the expected relation. However it is found that eq. (7) does not strictly hold. This is primarily due to the different weight of the p– and d–contributions to the hyperfine field and the magnetic moment, respectively.

In summary, it has been shown that relativistic effects have a strong impact on the properties of magnetic materials, causing phenomena that cannot be understood in a non–relativistic description. Furthermore, it has been demonstrated that especially for hyperfine interactions a consistent relativistic description is indispensible.

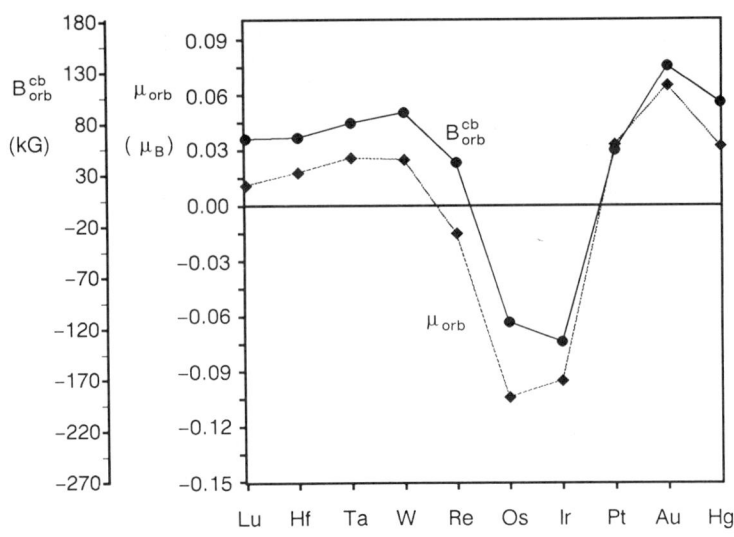

Fig. 4. Orbital contributions to the hyperfine field B_{orb}^{cb} (●, only conduction band) and to the magnetic moment μ_{orb} (♦) for 5d–impurities in Fe.

REFERENCES

Abragam, A. and Pryce, M.H.L. 1951 Proc. Roy. Soc. (London) A<u>205</u> 135
Akai, H. 1988 Hyperfine Interactions <u>43</u> 255
Blügel, S. Akai, H., Zeller, R. and Dederichs, P.H. 1987 Phys. Rev. B<u>35</u> 3271
Breit, G. 1930, Phys. Rev. <u>35</u> 1447
Campbell, I.A. 1966 Proc. Phys. Soc. (London) <u>89</u> 71
Collins, M.F. and Low, G.G. 1965 Proc. Phys. Soc. <u>86</u> 535
Cracknell, A.P. 1970 Phys. Rev. B<u>1</u> 1261
Demangeat, C. 1974 J. Phys. F: Met. Phys. <u>4</u> 169
Ebert, H., Strange, P. and Gyorffy, B.L. 1988 J. Phys. F: Met. Phys. <u>18</u> L 135
Ebert, H., Weinberger, P. and Voitländer, J. 1985 Phys. Rev. B<u>31</u> 7566
Falicov, L.M. and Ruvalds, J. 1968 Phys. Rev. <u>172</u> 498
Feder, R., Rosicky, F. and Ackermann, B. 1983 Z. Phys. B<u>52</u> 31
Gehring, G.A. and Williams, H.C.W.L. 1974 J. Phys. F: Met. Phys. <u>4</u> 291
Kawakami, M., Enokiya, H. and Okamoto, T. 1985 J. Phys. F.: Met. Phys. <u>15</u> 1613
Macintosh, A.R. and Andersen, O.K. 1980 Electrons at the Fermi Surface ed.: M. Springford (Cambridge, Cambridge University Press)
Rao, G.N. 1979 Hyperfine Interactions <u>7</u> 141
Rose, M.E. 1961 Relativistic Electron Theory (New York: Wiley)

Ruvalds, J. and Falicov, L.M. 1968 Phys. Rev. 172 508

Schütz, G., Wienke, R., Wilhelm, W., Wagner, W., Kienle, P., Zeller, R. and Frahm, R. 1989 Z. Phys. B75 495

Singh, M., Callaway, J. and Wang, C.S. 1976 Phys. Rev. B14 1214

Strange, P., Ebert, H., Staunton, J.B. and Gyorffy, B.L. 1989 J. Phys.: Condens Matter 1 2959

Strange, P., Staunton, J. and Gyorffy, B.L. 1984 J. Phys. C: Solid State Phys. 17 3355

Tterlikkis, L., Mahanti, S.D. and Das, T.P. 1968 Phys. Rev. 176 10

RELATIVISTIC SPIN-POLARIZED DENSITY-FUNCTIONAL THEORY: SIMPLIFIED METHOD FOR FULLY RELATIVISTIC CALCULATIONS

Pietro Cortona

Dipartimento di Fisica, Università degli Studi di Genova
I-16146 Genova, Italy
and CISM/MPI - GNSM/CNR, Unità di Genova, Italy

ABSTRACT

In the relativistic spin-polarized density-functional theory the one-electron wave functions are solutions of a Dirac equation containing a symmetry-breaking term. As a consequence, the wave functions cannot be separated in an angular and a radial part. No attempt to solve exactly this equation has been done till now. In general, one looks for approximate solutions which can be determined by solving a system of two coupled radial Dirac equations. The solution of this system is not easy, and only a few non-self-consistent solid-state calculations based on it have been performed. In this paper we will discuss a method which permits to find very accurate approximate solutions of the coupled equations by solving two *uncoupled* Dirac equations. The new method is easy to apply, numerically more stable and less computing time consuming than the original one. Furthermore, the tests we have performed seem to indicate that the loss of accuracy introduced by the new approximation is negligible.

INTRODUCTION: THE RELATIVISTIC SPIN-POLARIZED THEORY

Density-functional theory (DFT), both in the relativistic and in the nonrelativistic version, becomes useful for applications when approximations to the unknown exchange-correlation energy functional E_{xc} are available. Our knowledge of these approximations is less advanced in the relativistic than in the nonrelativistic theory. For example, no nonlocal approximation of E_{xc} has been elaborated till now; the local expressions of the correlation energy (Jancovici,1962; Ramana and Rajagopal, 1981) are limited to the nonmagnetic case and are not so accurate as the nonrelativistic ones derived from Monte-Carlo calculations (Vosko *et al*, 1980; Perdew and Zunger, 1981). Furthermore, applications to magnetic systems are particularly difficult. The one-electron equation, in the exact formal theory (Rajagopal,1978; MacDonald and Vosko,1979), contains the functional derivative of E_{xc} with respect to the four-current density, and it is unpractical to use. So, in general, one simplifies it by neglecting the diamagnetic effects (MacDonald and Vosko,1979). In such a way, the following equation is obtained:

$$\left(c\,\alpha\cdot\mathbf{p} + \beta\,c^2 + V_{eff}\,I_4 + B_{eff}\,\beta\,\sigma_z\right)\varphi^{(i)}(\mathbf{r}) = \varepsilon_i\,\varphi^{(i)}(\mathbf{r}) \qquad 1)$$

where α, β and σ_z are Dirac matrices, I_4 is the 4×4 identity matrix, and i is the set of quantum numbers which classify the orbitals. Eq.1 is the relativistic generalization of the nonrelativistic spin-polarized one-electron equation. The effective potential and the effective magnetic field contain now the functional derivative of E_{xc} with respect to the charge and to the spin densities while the polarization direction has been assumed to coincide with the z-axis.

The solution of Eq.1 presents two kinds of difficulties. First, one has to choice an approximate expression for E_{xc} : no expression for the correlation energy is available, while for the exchange contribution there are two different approximations (MacDonald,1983; Xu et al, 1984) corresponding to different interpretations of "local approximation" in the relativistic spin-polarized case. These expressions coincide at the first order in $1/c^2$. So, their discrepancies are probably not significant, at least for properties which are evaluated as differences of total energies. For other properties, as, for example, the transverse contribution to the exchange energy, the terms of order greater than $1/c^2$ can be important (Cortona and Sommers, 1987), and the discrepancies between the two approximations could be significant.

Second, the magnetic term $B_{eff}\beta\sigma_z$ breaks the symmetry of the system. This gives rise to technical difficulties because: i) the wave functions cannot be separated in a radial and an angular component (even if the external fields are spherically symmetric); ii) apart from m, the z component of the total angular momentum, there are no other good quantum numbers to classify the orbitals.

The usual procedure for by-passing the latter difficulties consists in using the semi-relativistic theory of Koelling and Harmon (1977), where the spin-orbit contribution is, at a first time, neglected, and then it is introduced according to the perturbation theory once the self-consistent solutions have been found. In the last years, however, attempts have been made to go beyond this approximation, and to treat spin-orbit and spin-polarization on the same footing. The initial step was done by Doniach and Sommers (1981) (DS) and then their idea was further elaborated by several authors.

In this paper we will review the DS theory and will discuss some new developments (Cortona,1989) which considerably simplify the calculations involved in the original DS method. The whole discussion will be limited to the case of spherically symmetric external fields.

THE DONIACH-SOMMERS APPROXIMATION

In order to find approximate solutions of Eq. 1, DS suggested to use wave functions which are the sum of two Dirac orbitals:

$$\varphi^{(i)} = \Psi_{nkm} + \Psi_{nk'm} = \frac{1}{r}\begin{pmatrix} i P_{nk}(r) \chi_{km}(\vartheta,\varphi) \\ Q_{nk}(r) \chi_{-km}(\vartheta,\varphi) \end{pmatrix} + \begin{pmatrix} i P_{nk'}(r) \chi_{k'm}(\vartheta,\varphi) \\ Q_{nk'}(r) \chi_{-k'm}(\vartheta,\varphi) \end{pmatrix} \qquad 2)$$

In this equation P(r) and Q(r) are the large and the small component respectively and $\chi(\vartheta,\varphi)$ are the two-dimensional Dirac spinors. The two orbitals which enter in Eq. 2 are characterized by the same values of n and m and by quantum numbers k and k' such that $\chi_{km}(\vartheta,\varphi)$ and $\chi_{k'm}(\vartheta,\varphi)$ are eigenfunctions of the orbital angular momentum L with the same eigenvalue l. We note that, for a given value of k, there is one and just one value of k' which satisfies such a condition. We also note that the s states are an exception to this rule. In this approach their wave function is given by an ordinary Dirac orbital.

In such an approximation the orbitals can be classified in terms of the quantum numbers nlm, plus an additional quantum number a which distinguishes the two states corresponding to the same values of nlm. Introducing Eq. 2 in Eq. 1, one obtains the following system of two coupled Dirac equations:

$$\frac{dP}{dr} + \frac{k}{r}P - \left(\frac{\varepsilon - V(r)}{c} + 2c\right)Q - \frac{1}{c}B(r)\,C(m,-k-k)\,Q = 0$$

$$\frac{dQ}{dr} - \frac{k}{r}Q + \frac{\varepsilon - V(r)}{c}P - \frac{1}{c}B(r)\,C(m,kk)\,P - \frac{1}{c}B(r)\,C(m,kk')\,P' = 0$$

$$\frac{dP'}{dr} + \frac{k'}{r}P' - \left(\frac{\varepsilon - V(r)}{c} + 2c\right)Q' - \frac{1}{c}B(r)\,C(m,-k'-k')\,Q' = 0$$

$$\frac{dQ'}{dr} - \frac{k'}{r}Q' + \frac{\varepsilon - V(r)}{c}P' - \frac{1}{c}B(r)\,C(m,k'k')\,P' - \frac{1}{c}B(r)\,C(m,kk')\,P = 0 \qquad 3)$$

where we have used P to indicate P_{nkm}, P' for $P_{nk'm}$ and so on. We have also introduced the coefficients $C(m,kk')$ defined by:

$$C(m,kk') = \int \langle \chi_{km} | \sigma_z^P | \chi_{k'm} \rangle \, d\Omega \qquad 4)$$

where σ_z^P is the 2×2 Pauli matrix. System 3 should be solved self-consistently using the appropriate boundary conditions.

DISCUSSION OF THE DONIACH-SOMMERS APPROXIMATION

In order to understand what the DS approximation exactly means, let us derive Eq. 2 step by step starting from the exact solutions of Eq. 1. Let ε_{nk} and ψ_{nkm} be the solutions of Eq. 1 *without* the symmetry-breaking term $B_{eff}\sigma_z$. The wave functions ψ_{nkm} are then Dirac orbitals and can be used to develop in series the wave functions $\varphi^{(i)}$:

$$\varphi^{(i)} = \sum_{nk} C_{nk}^{(i)} \psi_{nkm} \qquad 5)$$

where the coefficients $C_{nk}^{(i)}$ satisfy the following infinite system of linear equations:

$$(\varepsilon_i - \varepsilon_{nk})\,C_{nk}^{(i)} = \sum_{n'k'} C_{n'k'}^{(i)} \int d^3r\, B_{eff}(r) \langle \psi_{nkm} | \beta \sigma_z | \psi_{n'k'm} \rangle \qquad 6)$$

Using the coefficients defined by Eq. 4 and the fact that $C(m,kk')=0$ except if χ_{km} and $\chi_{k'm}$ correspond to the same value of l (σ_z^P does not change the orbital angular momentum), one can transform Eq. 6 as follows:

$$(\varepsilon_i - \varepsilon_{nk})\,C_{nk}^{(i)} = \sum_{n'} \left[C_{n'k}^{(i)}\,C(m,kk) \int dr\, B(r)\,P_{nk}(r)\,P_{n'k}(r) \right.$$

$$- C_{n'k}^{(i)}\,C(m,-k-k) \int dr\, B(r)\,Q_{nk}(r)\,Q_{n'k}(r) +$$

$$+ C_{n'k'}^{(i)}\,C(m,kk') \int dr\, B(r)\,P_{nk}(r)\,P_{n'k'}(r) -$$

$$\left. - C_{n'k''}^{(i)}\,C(m,-k-k'') \int dr\, B(r)\,Q_{nk}(r)\,Q_{n'k''}(r) \right] \qquad 7)$$

where k, k' and k'' are related to each other through the two equations: $k+k'=-1$ and

$k+k''=1$. These two conditions mean that the terms involving k and k' come from states having the same value of l relative to the large component, while the term involving k'' comes from two orbitals having that value different by ± 2.

The last term in the right member of Eq. 7 is of order $1/c^2$ with respect to the first and to the third one. By neglecting it, Eq. 7 splits in a set of subsystems, one for each pair of values k and k' for which $C(m,kk')\neq 0$. This is the first (and crucial) hypothesis introduced by DS. A discussion of it was given by Feder et al (1983), who were able to show that this assumption is equivalent to neglect terms dependent on the gradient of the effective magnetic field. From another standpoint, Eq. 7 makes explicit the fact that this approximation is not completely coherent, because some contributions which are taken into account (the second term in Eq. 7) are of the same order as those which are neglected.

Nevertheless, using this assumption, l becomes a good quantum number, and Eq. 5 can be written as follows:

$$\varphi^{(i)} = \sum_n C^{(i)}_{nk} \Psi_{nkm} + \sum_n C^{(i)}_{nk'} \Psi_{nk'm} \qquad 8)$$

Factorizing the angular parts of the wave functions and performing the two sums, one obtains two Dirac orbitals with the principal quantum numbers determined by the number of nodes of the resulting radial parts. Indicating these orbitals with Ψ_{nkm} and with $\Psi_{n'k'm}$, one can write:

$$\varphi^{(i)} = \Psi_{nkm} + \Psi_{n'k'm} \qquad 9)$$

We note that Ψ_{nkm} and $\Psi_{n'k'm}$ are not solutions of the nonmagnetic equation. We note also that, in principle, n could be different from n'. If one supposes $n=n'$, one gets to the DS wave function (Eq. 2). This is the second assumption of DS, assumption that should work very well, at least for effective magnetic fields which are not too large.

SIMPLIFICATION OF THE DONIACH-SOMMERS APPROXIMATION

The coupled DS equations (Eq. 3) have been solved self-consistently by Cortona et al (1985) for atomic systems, while several authors have performed non-self-consistent solid-state calculations (Schadler et al,1986 and 1987; Ebert, 1988; Ebert et al,1988). The methods to solve them are not easy. In the atomic case one has to determine, at each iteration, three parameters for each orbital, instead of only the one-electron energy. This requires considerably more computing time, and the whole iterative process is less stable. Furthermore, for deep core states, one of the two terms of the wave function (Eq. 2) is of greater order of magnitude than the other and system 3 becomes difficult to solve by techniques which treat the two coupled Dirac equations on the same footing. In this case the two equations tend to decouple and the wave functions tend to the one-term solutions of the following Dirac equation:

$$\frac{dP}{dr} + \frac{k}{r}P - \left(\frac{\varepsilon - V(r)}{c} + 2c\right)Q - \frac{1}{c}B(r)\,C(m,-k\text{-}k)\,Q = 0$$

$$\frac{dQ}{dr} - \frac{k}{r}Q + \frac{\varepsilon - V(r)}{c}P - \frac{1}{c}B(r)\,C(m,kk)\,P = 0 \qquad 10)$$

The results of Cortona et al (1985), and the DS calculations reported in this paper, have been obtained by assuming that Eq. 3 can be replaced by Eq. 10 for all the electronic states with $n \leq 3$ (for the rare-earth ions).

It is possible to obtain a method easier to implement by introducing a new approximation on the wave function. Let us start by looking for an approximate solution of the coupled equations of the form:

$$\varphi^{(i)} = \varphi_{anlm} = A_a \psi_{nkm} + \left(1 - A_a^2\right)^{1/2} \psi_{nk'm} \qquad 11)$$

where ψ_{nkm} and $\psi_{nk'm}$ are normalized solutions of Eq. 10 with eigenvalues ε and ε' respectively. The energy ε_a, corresponding to φ_{anlm}, is related to the parameter A_a as follows:

$$\varepsilon_a = \varepsilon + C(m,kk') \frac{\left(1 - A_a^2\right)^{1/2}}{A_a} \int dr\ B(r)\ P(r)\ P'(r)$$

$$\varepsilon_a = \varepsilon' + C(m,kk') \frac{A_a}{\left(1 - A_a^2\right)^{1/2}} \int dr\ B(r)\ P(r)\ P'(r) \qquad 12)$$

Indicating $A_a / (1 - A_a^2)^{1/2}$ with x_a and $\int B(r)P(r)P'(r)dr$ with I, we get the following equation for x_a:

$$x_a - \frac{1}{x_a} = \frac{\varepsilon - \varepsilon'}{C(m,kk')\ I} \qquad 13)$$

This equation has a positive and a negative solution which fix the two wave functions and the two eigenvalues corresponding to a given choice of quantum numbers nlm. Using the wave functions so determined one can derive the charge and the spin densities and the new potential and magnetic field to use in Eq. 10, and then to iterate the overall process to self-consistency. This gives rise to a method that, in the following, will be referred to as Method 1 (M1).

In M1, the coupling terms affect only indirectly the two basis functions which enter in Eq. 11. An approximate way to take into account more effectively the coupling is the following: We still look for approximate solutions of Eq. 3 of the form given in Eq. 11, but now we choose the orbitals ψ_{nkm} and $\psi_{nk'm}$ to be normalized solutions of the two uncoupled Dirac equations:

$$\frac{dP}{dr} + \frac{k}{r} P - \left(\frac{\varepsilon - V(r)}{c} + 2c\right) Q - \frac{1}{c} B(r)\ C(m,-k-k)\ Q = 0$$

$$\frac{dQ}{dr} - \frac{k}{r} Q + \frac{\varepsilon - V(r)}{c} P - \frac{1}{c} B(r)\ C(m,kk)\ P - \frac{1}{c} B(r)\ C(m,kk') \frac{1}{x_a} P = 0 \qquad 14)$$

and

$$\frac{dP'}{dr} + \frac{k'}{r} P' - \left(\frac{\varepsilon' - V(r)}{c} + 2c\right) Q' - \frac{1}{c} B(r)\ C(m,-k'-k')\ Q' = 0$$

$$\frac{dQ'}{dr} - \frac{k'}{r} Q' + \frac{\varepsilon' - V(r)}{c} P' - \frac{1}{c} B(r)\ C(m,k'k')\ P' - \frac{1}{c} B(r)\ C(m,kk')\ x_a\ P' = 0 \qquad 15)$$

where x_a has an arbitrary value at the beginning of the iterations and is then determined self-consistently as the effective potential and the effective magnetic field. At each iteration the one-electron energy ε_a and the new value of x_a are still given by Eqs. 12 and 13 but after substituting ε and ε' as follows:

$$\varepsilon \to \varepsilon - C(m,kk') \frac{1}{x_a} \int B(r) \, P(r) \, P(r) \, dr$$

$$\varepsilon' \to \varepsilon' - C(m,kk') \, x_a \int B(r) \, P'(r) \, P'(r) \, dr \qquad \qquad 16)$$

We shall refer to the method so modified as Method 2 (M2). Note that one is assuming, for the coupling term, $P \propto P'$. This should be quite a good approximation, and, in any case, certainly a better one than to assume $P=P'=0$ as we have done in M1.

We have tested the methods by performing calculations of the electronic properties of all the trivalent rare-earth ions. The results obtained by both the methods for Ce^{3+} are reported in Tables I to V as well as the corresponding DS results. The analogous tables for the other rare-earth ions show completely similar features. The only difference is that all the effects are magnified owing to the strongest magnetic interactions [see Cortona (1989) for the results relative to Gd^{3+}].

Table I   Total energy (in Hartrees) and the integral of the spin density of the Ce^{3+}. DS indicates results obtained from the self-consistent solution of the coupled Dirac equations of the fully relativistic spin-polarized theory. M1 and M2 are the corresponding values calculated by using the two simpler methods discussed in this work.

	M1	M2	DS
Total energy	-8834.014	-8834.014	-8834.014
Total spin	0.97	0.97	0.97

Table II   One-electron energies (in Hartrees) of the 4f states of the Ce^{3+}. Only the lowest level is occupied. M1, M2 and DS as in Table I.

m	M1	M2	DS
2.5	-1.0241	-1.0239	-1.0239
1.5	-1.0227	-1.0225	-1.0225
0.5	-1.0212	-1.0210	-1.0210
-0.5	-1.0196	-1.0194	-1.0194
-1.5	-1.0177	-1.0175	-1.0175
-2.5	-1.0155	-1.0153	-1.0153
-3.5	-1.0127	-1.0125	-1.0125
-2.5	-0.9958	-0.9954	-0.9954
-1.5	-0.9936	-0.9932	-0.9932
-0.5	-0.9918	-0.9914	-0.9914
0.5	-0.9901	-0.9898	-0.9898
1.5	-0.9887	-0.9883	-0.9883
2.5	-0.9873	-0.9869	-0.9869
3.5	-0.9861	-0.9857	-0.9857

Table III Mean values of r^6, r and r^{-3} for the 4f states of the Ce^{3+}. The first level is occupied, the other ones are empty. M1, M2 and DS as in Table I.

	$<r^6>$			$<r>$			$<r^{-3}>$		
m	M1	M2	DS	M1	M2	DS	M1	M2	DS
2.5	35.74	35.19	35.19	1.056	1.055	1.055	4.440	4.445	4.445
1.5	36.36	35.39	35.39	1.058	1.056	1.056	4.426	4.436	4.436
0.5	36.89	35.60	35.60	1.060	1.057	1.057	4.412	4.425	4.425
-0.5	37.28	35.85	35.85	1.061	1.058	1.058	4.398	4.412	4.412
-1.5	37.52	36.14	36.14	1.062	1.059	1.059	4.383	4.396	4.396
-2.5	37.52	36.50	36.50	1.063	1.061	1.061	4.365	4.374	4.374
-3.5	37.04	37.03	37.03	1.063	1.063	1.063	4.337	4.337	4.337
-2.5	37.56	38.72	38.72	1.062	1.065	1.065	4.396	4.384	4.384
-1.5	37.56	39.12	39.12	1.063	1.067	1.067	4.378	4.363	4.363
-0.5	37.82	39.44	39.44	1.064	1.068	1.068	4.363	4.347	4.347
0.5	38.25	39.72	39.72	1.065	1.069	1.069	4.348	4.334	4.334
1.5	38.83	39.98	39.98	1.067	1.070	1.070	4.334	4.323	4.323
2.5	39.52	40.21	40.21	1.069	1.071	1.071	4.320	4.313	4.313
3.5	40.32	40.42	40.42	1.072	1.072	1.072	4.305	4.304	4.304

Table IV Fourier transforms of the spin density of the occupied 4f state of the Ce^{3+}. M1, M2 and DS as in Table I. $\sin\vartheta/\lambda$ is in $Å^{-1}$.

$\sin\vartheta/\lambda$	M1	M2	DS
0.05	0.9455	0.9460	0.9460
0.10	0.8755	0.8762	0.8762
0.15	0.7745	0.7755	0.7755
0.20	0.6590	0.6601	0.6601
0.25	0.5427	0.5439	0.5439
0.30	0.4350	0.4361	0.4361
0.35	0.3405	0.3414	0.3414
0.40	0.2607	0.2614	0.2614
0.45	0.1951	0.1956	0.1956
0.50	0.1423	0.1427	0.1427
0.60	0.0681	0.0683	0.0683
0.70	0.0245	0.0245	0.0245
0.80	0.0005	0.0005	0.0005
0.90	-0.0113	-0.0113	-0.0113
1.00	-0.0160	-0.0161	-0.0161
1.10	-0.0168	-0.0169	-0.0169
1.20	-0.0156	-0.0156	-0.0156

The main remark about Tables I-V is that, while M1 gives rise to some discrepancies with respect to DS, the results obtained by M2, in the limit of precision of our Tables, *coincide* with the DS ones. This is true for all the properties we have studied, with the only exception of the Fourier transforms of the core-states' spin density, for which some minor discrepancies can still be noted. In particular, we remark that M2 eliminates the shift of the M1 energy eigenvalues and that it corrects the anomalies in the trend of the mean values of the

Table V Fourier transforms of the closed shells spin density of the Ce^{3+}. M1, M2 and DS as in Table I. $\sin\vartheta/\lambda$ is in $Å^{-1}$.

$\sin\vartheta/\lambda$	M1	M2	DS
0.05	0.0008	0.0016	0.0017
0.10	0.0027	0.0055	0.0056
0.15	0.0046	0.0094	0.0095
0.20	0.0056	0.0113	0.0113
0.25	0.0052	0.0104	0.0104
0.30	0.0039	0.0075	0.0075
0.35	0.0021	0.0039	0.0039
0.40	0.0005	0.0007	0.0007
0.45	-0.0006	-0.0015	-0.0014
0.50	-0.0012	-0.0025	-0.0024
0.60	-0.0011	-0.0021	-0.0021
0.70	-0.0005	-0.0007	-0.0007
0.80	0.0000	0.0002	0.0002
0.90	0.0001	0.0004	0.0004
1.00	-0.0001	0.0003	0.0003
1.10	-0.0002	0.0001	0.0001
1.20	-0.0003	-0.0001	-0.0001

Table VI Fourier transforms of the closed shells spin density of the Ce^{3+}. The three columns are obtained by using the Method 2 discussed in this paper with a two-terms wave function for states with $n>3$, $n>2$ and without restrictions respectively. $\sin\vartheta/\lambda$ is in $Å^{-1}$.

$\sin\vartheta/\lambda$	$n>3$	$n>2$	all electrons
0.05	0.0016	0.0015	0.0015
0.10	0.0055	0.0054	0.0054
0.15	0.0094	0.0093	0.0093
0.20	0.0113	0.0112	0.0112
0.25	0.0104	0.0103	0.0103
0.30	0.0075	0.0073	0.0073
0.35	0.0039	0.0037	0.0037
0.40	0.0007	0.0005	0.0005
0.45	-0.0015	-0.0017	-0.0017
0.50	-0.0025	-0.0028	-0.0028
0.60	-0.0021	-0.0025	-0.0025
0.70	-0.0007	-0.0012	-0.0012
0.80	0.0002	-0.0003	-0.0003
0.90	0.0004	-0.0002	-0.0002
1.00	0.0003	-0.0004	-0.0004
1.10	0.0001	-0.0006	-0.0006
1.20	-0.0001	-0.0008	-0.0008

powers of r calculated by M1 (in the case of Ce^{3+} these anomalies are just sketched in the values of $<r^6>$, but they are more evident for the other ions). We emphasize the good description of the core-states' polarization obtained by M2. This is a property which is very sensitive to the relaxation effects due to the coupling terms, as it is indirectly shown by the bad agreement between the M1 and DS results reported in Table V. The good values of the Fourier transforms of the core spin densities calculated by M2, are a strong indication in favour of the method.

The ensemble of our results indicates that, to a very high degree of accuracy, the solutions of the DS equations can be identified with the wave functions calculated by M2. This is confirmed by the fact that starting a DS calculation using the M2 self-consistent potential and magnetic field, one finds that either the usual tests of self-consistency are satisfied, or that very few iterations are required.

Finally, we point out that the method can be applied without difficulties to deep core states. In Tables I to V, in order to make a consistent comparison with the DS results, we have reported values obtained by using M2 for states with $n>3$ and Eq. 10 in the other cases. However, we have also performed calculations using M2 for states with $n>2$ and for all the electrons without restrictions. We have not found appreciable differences with respect to the results we have reported, with the only exception of the Fourier transforms of the core-states' spin density. The results of the three calculations for the latter quantities are compared in Table VI.

CONCLUSIONS

We have discussed two methods of performing fully relativistic spin-polarized calculations. They are both approximate versions of the DS method, which have the main advantage of requiring the solution of *uncoupled* Dirac equations. As a consequence, they are simpler, numerical more stable, and less computing time consuming than the original method. They both give satisfactory results. In particular, one of them seems to have a loss of accuracy with respect to the DS method, which is completely negligible.

REFERENCES

Cortona, P., Doniach, S., and Sommers, C., 1985, Phys. Rev. A, 31: 2842.
Cortona, P., and Sommers, C., 1987, J. Magn. Magn. Mat., 63&64: 658.
Cortona, P., 1989, Phys. Rev. B, 40: 12105.
Doniach, S., and Sommers, C., 1981, The use of local density functional theory for spin polarized relativistic band structure calculations, in: "Proceedings of the International Conference on Valence Fluctuations of Solids" L. M. Falicov, W. Hanke and M. B. Maple eds., North-Holland, Amsterdam.
Ebert, H., 1988, Phys. Rev. B, 38: 9390.
Ebert, H., Strange P., and Gyorffy, B. L., 1988, J. Appl. Phys., 63: 3052.
Feder, R., Rosicky, F., and Ackermann, B., 1983, Z. Phys. B, 52: 31.
Jancovici, B., 1962, Nuovo Cimento, 25: 428.
Koelling, D. D., and Harmon, B. N., 1977, J. Phys. C, 10: 3107.
MacDonald, A. H., and Vosko, S. H., 1979, J. Phys. C, 12: 2977.
MacDonald, A. H., 1983, J. Phys. C, 16: 3869.
Perdew, J. P., and Zunger, A., 1981, Phys. Rev. B, 23: 5048.
Rajagopal, A. K., 1978, J. Phys. C, 11: L943.
Ramana, M. V., and Rajagopal, A. K., 1981, Phys. Rev. A, 24: 1689.
Schadler, G., Weinberger, P., Boring, A. M., and Albers, R. C., 1986, Phys. Rev. B, 34: 713.
Schadler, G., Albers, R. C., Boring, A. M., and Weinberger, P., 1987, Phys. Rev. B, 35: 4324.
Vosko, S. H., Wilk, L., and Nusair, M., 1980, Can. J. Phys., 58: 1200.
Xu, B. X., Rajagopal, A. K., and Ramana, M. V., 1984, J. Phys. C 17: 1339.

THEORY OF MAGNETOCRYSTALLINE ANISOTROPY

J. Staunton[1], P. Strange[2,3], B.L. Gyorffy[4], M. Matsumoto[1,5],
J. Poulter[1], H. Ebert[6] and N.P. Archibald[2]

[1]Physics Department, Warwick University
Conventry, West Midlands, CV4 7L, UK

[2]Physics Department, Keele University
Keele, Staffordshire, ST5 5BG, UK

[3]Neutron Division, Rutherford-Appleton Laboratory
Chilton, Didcot, Oxfordshire, OX11 0QX, UK

[4]Physics Department, Bristol University
Tyndall Avenue, Bristol, BS8 1TL, UK

[5]Permanent Address: University of Library and
Information Science, Tsukuba 305, Japan

[6]Siemens AG, ZFE TPH 11, Paul-Gossen Strasse 100
D-8520 Erlangen, Federal Republic of Germany

Introduction

One of the first things one learns in a school course in magnetism is that a magnet has a north and a south pole. This defines a direction in the magnet, the axis of magnetisation. If one cools a single crystal sample of a magnetic material from above the magnetic transition temperature it will choose a crystal axis along which the magnetisation vector will point. Obviously, the direction chosen is the one with the lowest free energy. The magnetisation direction usually coincides with one of the principal crystal directions. In iron this is the (0,0,1) direction, whereas in nickel it is the (1,1,1) direction. The free energy of the crystal thus contains a term which couples the magnetisation direction to the crystal lattice. This is known as the magnetocrystalline anisotropy energy [1].

The anisotropy energy can be determined by measuring the field needed to rotate the magnetisation vector away from the easy axis. It is traditionally expressed in terms of the direction cosines of the magnetisation vector and some anisotropy constants [2,3]. For example, if θ is the angle between the magnetisation vector and the easy axis, then the anisotropy energy in an HCP crystal is given in terms of two anisotropy constants K_1 and K_2 as

$$E_{an} = K_1 sin^2\theta + K_2 sin^4\theta \qquad (1.1)$$

Many experiments have been performed to measure the anisotropy constants as a function of temperature for many metals (see [3] for example).

When one comes to consider what determines the anisotropy constants on a microscopic level one necessarily becomes involved in relativistic quantum theory. The angular momentum of an electron in a solid is divided into a spin and an orbital contribution. This is rather like dividing the angular momentum of the earth up into a part due to its orbital motion around the sun and a part due to its rotation about its own axis. Spin and orbital angular momentum are conserved separately in classical quantum theory. In relativistic quantum theory only total angular momentum is conserved, an electron can flip its spin by exchanging angular momentum with its orbital motion. The amount of angular momentum exchanged depends on the relative orientation of the spin and orbital angular momentum vectors. One finds that magnetocrystalline anisotropy is due to this subtle interplay between spin-orbit coupling and spin-polarisation. This turns out to be a case where these esoteric relativistic and quantum effects are observable in every day life [4].

One of the aims of condensed matter physics is to be able to calculate the properties of a material from first principles and faith in quantum theory! ie with the only input being the atomic number of the constituents and (if appropriate) the stoichiometry, (This information simply identifies the material). Such an approach has given us tremendous insight into such diverse properties as lattice constants, phonon spectra, crystal structures, superconducting transition temperatures and the occurrence of magnetism. It has also enabled us to give a detailed interpretation of such experiments as de Haas-Van Alphen measurements, positron annihilation studies and various spectroscopies[5,6].

In this chapter we develop a theory to describe magnetocrystalline anisotropy energies at zero temperature from first principles. This work is based on the relativistic density functional theory described elsewhere in this volume. The single site scattering theory is also described elsewhere in this volume (and in [7]). Here we present the multiple scattering theory [8]. In the first section we discuss the scattering theory. In the second section we restrict this to two site scattering and consider the relativistic generalisation of the RKKY interaction leading to an effective two body interaction Hamiltonian [9,10]. At the end of this section we show the results of some preliminary simulations of this Hamiltonian [11]. In the third section we discuss applying the theory to bulk crystals and illustrate this by showing explicit calculations of Fermi surfaces and magnetocrystalline anisotropy energies [12,13]. Finally we briefly mention some work in progress on a finite temperature generalisation of this work [14].

Relativistic Spin-Polarised Multiple Scattering Theory

In this section we outline the derivation of the relativistic spin-polarised Green function for a system of scatterers. The formulism presented here starts from the discussion of the relativistic free electron Green function derived by Rose [15]. This section is very similar to the non-relativistic derivation of reference [16] and the relativistic work of [8]. Equivalent derivations have been presented by Schadler et al [17] and Feder et al [18].

It is necessary to start with a series of definitions. Let us start with the quantum numbers. For completeness we include a definition of them all. The well known relativistic quantum numbers are j and μ which are the quantum number associated with the total angular momentum and its z-component respectively. The quantum number κ is the eigenvalue of the operator $(\underline{\sigma}.\underline{L} + 1)$. Here $\underline{\sigma}$ is the Pauli matrices vector and $\underline{L}$ is the orbital angular momentum operator. The eigenvectors are the spin angular functions described below. If l is the usual orbital angular momentum quantum number we have

$$\kappa = l \qquad j = l - 1/2 \qquad (2.1a)$$

$$\kappa = -l - 1 \qquad j = l + 1/2 \qquad (2.1b)$$

It can be seen from here that κ takes on all integer values except zero. The sign of κ gives the parity of the wavefunction, we define

$$S_\kappa = \kappa/|\kappa| \qquad (2.2)$$

We also introduce $\bar{l}$ which is the value of l associated with $-\kappa$. So

$$\bar{l} = \kappa - 1 \qquad \kappa > 0 \qquad (2.3a)$$

$$\bar{l} = -\kappa \qquad \kappa < 0 \qquad (2.3b)$$

For a given value of j the two possible values of κ are $j + \frac{1}{2}$ and $-j - \frac{1}{2}$. Hence

$$l - \bar{l} = S_\kappa \qquad (2.4)$$

and

$$j = l - \frac{1}{2} S_\kappa \qquad (2.5)$$

The spin angular functions are defined as

$$\chi^\mu_\kappa = \sum_{m_s=\pm 1/2} C(l, \frac{1}{2}, j; \mu - m_s, m_s) Y_l^{\mu - m_s} \chi^{m_s} \qquad (2.6)$$

Here the C is a Clebsch-Gordon coefficient as defined by Rose [15], Y_l^m is a spherical harmonic and the χs are spinors

$$\chi^{1/2} = \begin{pmatrix} 1 \\ 0 \end{pmatrix}; \qquad \chi^{-1/2} = \begin{pmatrix} 0 \\ 1 \end{pmatrix} \qquad (2.7)$$

As shown by Faulkner and Stocks [16] the scattering Green function for a particle moving in the field of a single potential located at a point labelled by n can be written in the following form.

$$G^n(\mathbf{r}, \mathbf{r}'; E) = G^0(\mathbf{r}, \mathbf{r}'; E) + \int \int G^0(\mathbf{r}, \mathbf{r}''; E) t_{nn}(\mathbf{r}'', \mathbf{r}'''; E) G^0(\mathbf{r}''', \mathbf{r}'; E) d\mathbf{r}'' d\mathbf{r}''' \qquad (2.8)$$

where the $t_{nn}(\mathbf{r}, \mathbf{r}'; E)$ is a general scattering t-matrix. This is very difficult to calculate in itself, The *on the energy shell* elements of the t-matrix are what we require, these are defined below and are fairly straightforward to evaluate. The G^0 is the free electron Green function given by Rose [15]

$$G^0(\mathbf{r}, \mathbf{r}'; E) = \begin{pmatrix} G^0_{11}(\mathbf{r}, \mathbf{r}'; E) & G^0_{12}(\mathbf{r}, \mathbf{r}'; E) \\ G^0_{21}(\mathbf{r}, \mathbf{r}'; E) & G^0_{22}(\mathbf{r}, \mathbf{r}'; E) \end{pmatrix} \qquad (2.9)$$

with

$$G^0_{11}(\mathbf{r}, \mathbf{r}'; E) = \frac{p(E+1)}{\pi} \sum_{\kappa\mu} h_l(pr) j_l(pr') \chi^\mu_\kappa(\hat{r}) \chi^{\mu*}_\kappa(\hat{r}') \qquad (2.10a)$$

$$G^0_{12}(\mathbf{r},\mathbf{r}';E) = -p^2 \sum_{\kappa\mu} S_\kappa h_{\bar{l}}(pr) j_l(pr') \chi^\mu_{-\kappa}(\hat{r}) \chi^{\mu*}_\kappa(\hat{r}') \tag{2.10b}$$

$$G^0_{21} = -G^0_{12} \tag{2.10c}$$

$$G^0_{22}(\mathbf{r},\mathbf{r}';E) = \frac{p(E-1)}{\pi} \sum_{\kappa\mu} h_l(pr) j_l(pr') \chi^\mu_\kappa(\hat{r}) \chi^{\mu*}_\kappa(\hat{r}') \tag{2.10d}$$

In these equations E is the eigenvalue of the Dirac equation, and $p = (E^2-1)^{1/2}$. h_l and j_l are the usual spherical Hankel and Bessel functions respectively. Note that we are taking an outer product of the spin-angular functions so G^0_{ij} is a 2×2 matrix and hence the full Green function (as well as the t-matrix) in (2.8) are 4×4 matrices. We can rewrite this as

$$G^0(\mathbf{r},\mathbf{r}';E) = i\pi \sum_{\kappa\mu} [H^\mu_\kappa(\mathbf{r})].[J^\mu_\kappa(\mathbf{r})]^+ \tag{2.11}$$

where

$$H^\mu_\kappa(\mathbf{r}) = \left(\frac{p(E+1)}{\pi}\right)^{1/2} \begin{pmatrix} h_l(pr)\chi^\mu_\kappa(\hat{r}) \\ \frac{ip}{E+1} S_\kappa h_{\bar{l}}(pr)\chi^\mu_{-\kappa}(\hat{r}) \end{pmatrix} \tag{2.12a}$$

and

$$J^\mu_\kappa(\mathbf{r}) = \left(\frac{p(E+1)}{\pi}\right)^{1/2} \begin{pmatrix} j_l(pr)\chi^\mu_\kappa(\hat{r}) \\ \frac{ip}{E+1} S_\kappa j_{\bar{l}}(pr)\chi^\mu_{-\kappa}(\hat{r}) \end{pmatrix} \tag{2.12b}$$

Substituting these definitions into (2.8) and making the identification

$$t^{\mu\mu'}_{\kappa\kappa'}(E) = \int\int J^{\mu+}_\kappa(\mathbf{r}'')t(\mathbf{r}'',\mathbf{r}''';E)J^{\mu'}_{\kappa'}(\mathbf{r}''')d^3r''' d^3r'' \tag{2.13}$$

and using

$$\sum_{\kappa''\mu''} t^{\mu\mu''-1}_{\kappa\kappa''}(E) t^{\mu''\mu'}_{\kappa''\kappa'}(E) = \delta_{\kappa\kappa'}\delta_{\mu\mu'} \tag{2.14}$$

leads to a form for the single site scattering Green function

$$G^n(\mathbf{r},\mathbf{r}';E) = \sum_{\kappa\kappa'\mu\mu'} Z^\mu_\kappa(pr) t^{\mu\mu'}_{\kappa\kappa'}(E) H^{\mu'+}_{\kappa'}(pr') \tag{2.15}$$

Here

$$Z^\mu_\kappa(pr) = \sum_{\kappa'\mu'} J^{\mu'}_{\kappa'}(pr) t^{\mu'\mu-1}_{\kappa'\kappa}(E) + H^\mu_\kappa(pr) \tag{2.16}$$

A more useful form, computationally is

$$G^n(\mathbf{r},\mathbf{r}';E) = \sum_{\kappa\mu\kappa'\mu'} Z^\mu_\kappa(pr) t^{\mu\mu'}_{\kappa\kappa'}(E) Z^{\mu'+}_{\kappa'}(pr') - \sum_{\mu\kappa} Z^\mu_\kappa(pr) J^\mu_\kappa(pr') \tag{2.17}$$

where $r \leq r'$. This is the final expression for the single site scattering Green function. The *on the energy shell* t-matrix is the same as that described by Gyorffy et al elsewhere in this volume, and also by Strange et al [7]. Next consider the derivation of the multiple scattering Green function. This can be written in terms of the single site Green function as

$$G(\mathbf{r},\mathbf{r}';E) = G^n(\mathbf{r},\mathbf{r}';E) + \int\int G^n(\mathbf{r},\mathbf{r}'';E)T_{nn}(\mathbf{r}'',\mathbf{r}''';E)G^n(\mathbf{r}''',\mathbf{r}';E)d\mathbf{r}''d\mathbf{r}''' \qquad (2.18)$$

Where T_{nn} is the T-matrix for the system without the scatterer at the nth site. The most easy way to proceed is to substitute (2.15) for the single site Green function into this equation. Now $H_\kappa^\mu(E,\mathbf{R_n})$ is singular at the origin, but is regular at all other points. Thus it is possible to expand it in terms of $J_\kappa^\mu(\mathbf{r})$ where $\mathbf{r}$ is in the vicinity of some other site m, ie

$$H_\kappa^\mu(\mathbf{r_n}) = \sum_{\kappa\mu} g^{nm}_{\kappa\mu\kappa'\mu'} J_\kappa^\mu(\mathbf{r_m}) \qquad (2.19)$$

In fact this is a straightforward generalisation of the non-relativistic case, because each element of the 4-vector here obeys an equation identical to the non-relativistic case. The gs here are known as structure constants. they are simply a rotation of the usual non-relativistic structure constants which are written in terms of the (l,m,s) quantum numbers.

$$g^{nm}_{\kappa\mu\kappa'\mu'} = \sum_{s=\pm\frac{1}{2}} C(l,\frac{1}{2},j;\mu-s,s) g^{nm}_{l\mu-sl'\mu'-s} C(l',\frac{1}{2},j';\mu'-s,s) \qquad (2.20)$$

The non-relativistic gs are the usual real space KKR structure constants

$$g^{nm}_{l,m,\sigma;l^1,m^1,\sigma^1}(\epsilon) = \delta_{\sigma,\sigma^1} 4\pi\sqrt{\epsilon} \sum_{l_2,m_2} i^{l_2} C^{l_2,m_2}_{l,m;l^1,m^1} Y^{m_2}_{l_2}(\hat{R}_{nm}) h^+_{l_2}(\sqrt{\epsilon}R_{nm}) \qquad (2.21)$$

Here the C-coefficients are Gaunt numbers. Note that g^{nm} is independent of the atomic species under consideration and depends only upon the geometric arrangement of scatterers. Substitution of (2.16) and (2.17) into (2.18) eventually leads to an expression for the multiple scattering Green function.

$$G^n(\mathbf{r},\mathbf{r}';E) = \sum_{\kappa\mu\kappa'\mu'} Z_\kappa^\mu(p\mathbf{r}) \tau^{\mu\mu'}_{\kappa\kappa'}(E) Z^{\mu'+}_{\kappa'}(p\mathbf{r}') - \sum_{\mu\kappa} Z_\kappa^\mu(p\mathbf{r}) J_\kappa^\mu(p\mathbf{r}') \qquad (2.22)$$

Again we have $\mathbf{r} \le \mathbf{r}'$. τ is the scattering path operator first introduced by Gyorffy [19]. It is related to the total T-matrix T_{nn} by

$$T_{nn} = \sum_{i\neq n}\sum_{j\neq n} \tau_n^{ij} \qquad (2.23)$$

It is given by an integral over the Brillouin zone of the inverse of the KKR matrix (for one atom per unit cell)

$$\tau_{\kappa\mu\kappa'\mu'}(E) = \int_{BZ} [t^{-1} - g(\mathbf{q},E)]^{-1}_{\kappa\mu\kappa'\mu'} d^3\mathbf{q} \qquad (2.23)$$

If we vary E there will be certain places where the determinant $\|t^{-1} - g(\mathbf{q},E)\|$ passes through zero. At these energies $\tau_{\kappa\mu\kappa'\mu'}(E)$ and hence $G(\mathbf{r},\mathbf{r}';E)$ diverge. At these energies there will be an amplified scattered wave even with no incident electron. These energies form the eigenvalue spectrum or electronic band structure of the system. It is expression (2.22) that is usually evaluated. It has a particularly useful property that the second term is completely real at real energies. Thus it is not necessary to evaluate it if only the imaginary part of the Green function is required as is the case for many applications.

Once we have found the Green function we can use it to calculate observables such as the density of states:-

$$n(E) = -\frac{1}{\pi} Im \int Tr G(\mathbf{r},\mathbf{r}; E) d^3r \qquad (2.24)$$

the magnetic moment

$$m(r) = -\frac{1}{\pi} Im \int_0^{Ef} Tr \underline{\beta\sigma} G(\mathbf{r},\mathbf{r}; E) dE \qquad (2.25)$$

and the charge density

$$n(r) = -\frac{1}{\pi} Im \int_0^{Ef} Tr G(\mathbf{r},\mathbf{r}; E) dE \qquad (2.26)$$

In a later section we shall show some applications of these formulae, however in the next section we restrict our multiple scattering to two sites and examine the new physics introduced into this problem by the simultaneous occurrence of spin-orbit coupling and spin-polarisation.

The relativistic R.K.K.Y. interaction - magnetic impurities in a paramagnetic host

As a first application of our multiple scattering theory we shall study the interaction between magnetic impurities in a metallic host. This work is described fully in references [9] and [10] in which the cases of (i) two and three magnetic impurities and (ii) two magnetic and a single non-magnetic impurity have been investigated. The salient results will be summarised in this section. The electrons which mediate these interactions are modelled by an infinite non-interacting electron gas of appropriate density with a uniform, positively charged background (jellium). For the conventional coupling between the two identical magnetic impurities with moments pointing along the unit vectors $\hat{s}_1$ and $\hat{s}_2$ respectively at an asymptotically large separation $\mathbf{R}_{12}$, the well known Ruderman-Kittel-Kasuya-Yoshida [20,21,22] formula is found.

$$E_{12}^{RKKY} = V \hat{s}_1 \cdot \hat{s}_2 ((2k_F R_{12} cos 2k_F R_{12} - \sin 2k_F R_{12})/R_{12}^4) \qquad (2.27)$$

where k_F is the Fermi momentum and V is a parameter which describes the interaction between an impurity and the electrons. The precise nature of V depends on how the impurities are modelled.

This result is based on non-relativistic quantum mechanics. Using the Dirac equation to describe the conduction electrons, the relativistic generalisation of equation (2.27) can be derived. In addition to the usual spatially isotropic 'spin-spin' term, owing to spin-orbit coupling effects, it has also pseudo-dipolar and *squared* Dzyaloshinskii -Moriya type terms. Evidently the anisotropy is uniaxial in form. On extending the work to the case of two non-equivalent magnetic impurities, additional unidirectional terms are found. Three impurity interactions show further magnetic anisotropic effects.

In non-relativistic quantum mechanics the exchange interaction is isotropic in spin space, but in Dirac theory the spin and orbital degrees of freedom are coupled. As stated already this coupling is the main source of magneto-crystalline anisotropies in metals. The relativistic RKKY interaction is therefore of general interest because it reveals a simple example of magnetic anisotropy. It is also of particular interest because it may contribute significantly to the still mysterious anisotropies which characterise the low temperature behaviour of the spin-glass state [23].

We use spin-dependent potentials to describe the impurities:

$$\underline{v}_i(\mathbf{r}) = v_i(\mathbf{r})\underline{1} + w_i(\mathbf{r})\underline{\sigma} \cdot \hat{s}_i \quad (2.28)$$

where $\underline{1}, \underline{\sigma}^x, \underline{\sigma}^y$, and $\underline{\sigma}^z$ are the complete set of Pauli spin matrices. All orders of $\underline{v}_i (i = 1, 2, 3, \cdots)$ are treated and hence the theory is developed in terms of the 'on the energy shell' t-matrices

$$\hat{t}_i = t^{(i)}_{l,m,s;l^1,m^1,s^1}(E) = \sum_{\kappa\kappa'\mu\mu'} C(l, \frac{1}{2}, j; \mu - s, s) t^{\mu\mu'}_{\kappa\kappa'(i)} C(l^1 \frac{1}{2}, j; \mu' - s^1, s^1) \quad (2.29)$$

described earlier and in [7]. E is the kinetic energy of the incident electron, and l,m and s are the polar, azimuthal and spin magnetic quantum numbers respectively. m and s refer to a common $\hat{z}$-axis at both sites and the matrix $\hat{t}(\epsilon)$ is non-diagonal in m and s. The orbital parts of the potentials are taken to be spherically symmetric and hence the $\hat{t}^{(i)}$ are diagonal in l. In a non-relativistic theory for a local moment along $\hat{z}$, we have

$$t_{l,m,s;l^1,m^1,s^1}(E) = \delta_{l,l^1}, \delta_{m,m^1}, \delta_{s,s^1}(\exp 2i\delta^s_l(E) - 1)/(2i\sqrt{E}) \quad (2.30)$$

whereas in a fully relativistic theory $t_{l,m,\sigma;l^1,m^1,\sigma^1}$ is non-diagonal and its diagonal elements depend on both m and σ. Both of these effects trace back to spin-orbit coupling and are the root causes of magnetic anisotropy.

The quantities to calculate are the various contributions to the Grand Potential Ω which represent the interaction between $2,3, \cdots$ impurities.

$$\Omega = \Omega_0 + \sum_i \delta\Omega_i + \sum_{ij} \delta\Omega_{ij} + \sum_{ijk} \delta\Omega_{ijk} + \cdots \quad (2.31)$$

As shown later the dominant contribution is given by the *single particle* piece.

$$\Omega = -\int dE f(E - \nu) N(E) \quad (2.32)$$

where $f(E - \nu)$ is the Fermi function, ν is the electron chemical potential and N is the appropriate integrated density of states. For an arbitrary number and arrangement of non-overlapping scatterers the integrated density of states is given by [8]

$$N(\epsilon) = (1/\pi) Im \log \begin{Vmatrix} \underline{t}_1^{-1} & -g_{12} & \cdots \\ -g_{21} & \underline{t}_2^{-1} & \cdots \\ \vdots & \vdots & \ddots \end{Vmatrix} \quad (2.33)$$

where $\|\Theta\|$ denotes the operation of taking the determinant of the matrix Θ with respect to the indices l, m, s and the site indices 1,2,... $g^{12} \to g^{12}_{l,m,s;l^1,m^1,s^1}(\mathbf{R}_{12};\epsilon)$ are the usual real space KKR structure constants. They are given by equation (2.21). In (2.21) the $Y_l^m(\hat{R})$ are the complex spherical harmonics for the angles which specify the direction $\hat{R}$. Equation (4) is the straightforward relativistic generalisation of Lloyd's formula [24]. From equations (2.31), (2.32) and (2.33)

$$\delta\Omega_{12} = -(1/\pi)\int dE f(E-\nu)\log|\underline{1} - \underline{t}_1\underline{G}_{12}\underline{t}_2\underline{G}_{21}| \qquad (2.34)$$

where $|\cdots|$ means taking the determinant with respect to the indices l, m and s only, the operation with respect to the site indices having been fully worked out. A similar analysis provides a more complicated expression for $\delta\Omega_{123}$, the interaction between three sites.

Although the exact formulae of the above form are easy to evaluate numerically, the physical effects can be extracted by studying their asymptotic forms for large separations between the impurities. In that limit $g^{ij} \sim 1/R_{ij}$ equation (7) can be expanded in powers of g^{ij}. For two impurities and to lowest order in g^{ij} we find

$$\delta\Omega_{12} = (1/\pi)Im\int dE \sum_{\Lambda,\Lambda_1,\Lambda_2,\Lambda_3} (t^{(1)}_{\Lambda,\Lambda_1}(E)g_{\Lambda_1,\Lambda_2}(\mathbf{R}_{12},E)t^{(2)}_{\Lambda_2,\Lambda_3}(E)g_{\Lambda_3,\Lambda}(\mathbf{R}_{21},E)) \qquad (2.35)$$

where Λ denotes all three indices l, m and s. In a similar fashion the expression for three impurities is

$$\delta\Omega_{123} = (1/\pi)Im\int dE$$

$$\sum_{\Lambda,\Lambda_1,\cdots,\Lambda_6} (t^{(1)}_{\Lambda,\Lambda_1}(E)g_{\Lambda_1,\Lambda_2}(\mathbf{R}_{12},E)t^{(2)}_{\Lambda_2,\Lambda_3}(E)g_{\Lambda_3,\Lambda_4}(\mathbf{R}_{23},E)t^{(3)}_{\Lambda_4,\Lambda_5}(E)g_{\Lambda_5,\Lambda}(\mathbf{R}_{31},E)$$

$$+ t^{(1)}_{\Lambda,\Lambda_1}(E)g_{\Lambda_1,\Lambda_2}(\mathbf{R}_{13},E)t^{(3)}_{\Lambda_2,\Lambda_3}(E)g_{\Lambda_3,\Lambda_4}(\mathbf{R}_{32},E)t^{(2)}_{\Lambda_4,\Lambda_5}(E)g_{\Lambda_5,\Lambda}(\mathbf{R}_{21},E)) \qquad (2.36)$$

The form of (2.35) and (2.36) show that these interactions can be interpreted quite neatly in scattering theory terms. In the non-relativistic limit, after making appropriate approximations, $\delta\Omega_{12}$ can be shown to have the usual RKKY form i.e.

$$\delta\Omega_{12} \propto \{\hat{s}_1 \cdot \hat{s}_2\}(2k_F R_{12}\cos 2k_F R_{12} - \sin 2k_F R_{12})/R_{12}^4 \qquad (2.37)$$

In the non-relativistic derivation it is found that $\delta\Omega_{12}$ depends only on $|\mathbf{R}_{12}|$ because $t_{\Lambda,\Lambda^1}(\epsilon)$ is diagonal and independent of m. As a consequence of such m-dependence in the relativistic theory a number of new terms containing the 'direction' of $\mathbf{R}_{12}$, denoted by the unit vector $\hat{R}_{12}$, arise. Equation (2.35) may be rewritten after some algebraic manipulation as

$$\delta\Omega_{12} = \sum_{n_1,n_2,n_3}^{2l_{max}} b_{n_1,n_2,n_3}(\hat{s}_1 \cdot \hat{s}_2) \qquad (2.38)$$

$$\{(\hat{R}_{12} \cdot \hat{s}_1)(\hat{R}_{12} \cdot \hat{s}_2)\}^{n_1}\{\hat{R}_{12} \cdot (\hat{s}_1 \times \hat{s}_2)\}^{2n_2}\{(\hat{R}_{12} \cdot \hat{s}_1)^2 - (\hat{R}_{12} \cdot \hat{s}_2)^2\}^{n_3}$$

where the coefficient $b_{n_1,n_2,n_3}(\hat{s}_1 \cdot \hat{s}_2)$ depends only on the relative orientation of $\hat{s}_1$ and $\hat{s}_2$, R_{12}, the scattering properties of the two impurities and various numerical factors. For 2 identical magnetic impurities $n_3 = 0$ only. This is the general form of the relativistic RKKY interaction, a polynomial function of anisotropic pseudo-dipolar terms $(\hat{R}_{12} \cdot \hat{s}_1)(\hat{R}_{12} \cdot \hat{s}_2)$ and squared Dzyaloshinskii-Moriya (DM) type terms, $\hat{R}_{12} \cdot (\hat{s}_1 \times \hat{s}_2)$.

Although equation (2.38) for two identical impurities contains terms like $\hat{R}_{12} \cdot (\hat{s}_1 \times \hat{s}_2)$ it differs from the usual DM terms in that here only even powers of $\hat{R}_{12} \cdot (\hat{s}_1 \times \hat{s}_2)$ enter. This is the inevitable consequence of the symmetry in this problem and is also a very satisfactory feature of our formula. A uniaxial anisotropy is described. Once this interchange symmetry is broken, unidirectional effects are added. This occurs in (2.38) for two non-equivalent magnetic impurities $n_3 \neq 0$. The three impurity problem, also studied by Levy and Fert [23,25], similarly reveals unidirectional contributions to $\delta\Omega_{123}$. These include terms linear in $\hat{R}_{12} \cdot (\hat{s}_1 \times \hat{s}_2)$ (as shown in [9]). In [10] we found such magnetic anisotropic effects to be comparable, in magnitude, to those predicted by (2.38).

In [9] we explored the relativistic RKKY interaction of (2.38) for two Fe and Co atoms respectively. In particular we looked at the interaction energies $\delta\Omega_{12}^{\uparrow\uparrow}$ for the two moments parallel to each other and to $\hat{R}_{12}$ and perpendicular, $\delta\Omega_{12}^{\rightarrow\rightarrow}$, to $\hat{R}_{12}$. As expected the anisotropy energy $\delta\Omega_{12}^{\uparrow\uparrow} - \delta\Omega_{12}^{\rightarrow\rightarrow}$ (on the scale of 10^{-6} a.u.) is smaller than the interaction energy $\delta\Omega_{12}^{\uparrow\uparrow}$ by a factor of 10^3. Moreover it was in the range of experimental observation. Interestingly, the anisotropic contributions to the interaction energy oscillate and are long ranged like the non-relativistic RKKY interaction itself.

An anisotropic frustration in spin glasses [26] is therefore apparent. These uniaxial contributions are unable to explain the purely directional character of the magnetic anisotropy of such systems as demonstrated by several types of experiment. These include measurements of shifted hysteresis cycles, NMR and torque measurements [25]. This observation then motivated the studies described in [2] on two non-equivalent impurities and three impurities in which unidirectional effects were found.

In this section we have outlined a general formulism for the study of the interaction between impurities in a metallic host modelled by relativistic jellium. The general relativistic multi-site, multi-component RKKY interaction has a particularly transparent functional form for the two site limit. Specific calculations relevant to two and three impurities have been carried out and described in [1] and [2]. Both uniaxial and unidirectional magnetic anisotropic effects appear naturally.

Very recently there has been some preliminary computer simulation work on a simplified version of the Hamiltonian defined by (2.38) [11]. This work has been a simple classical simulation of a Hamiltonian containing the Heisenberg term, the Dzyaloshinskii-Moriya term and the pseudo-dipolar term all to order one.

$$\hat{H} = J \sum_{i,j} \hat{s}_i \cdot \hat{s}_j + K \sum_{i,j} (\mathbf{R}_{ij} \cdot \hat{s}_i)(\mathbf{R}_{ij} \cdot \hat{s}_j) + L \sum_{i,j} \mathbf{R}_{ij} \cdot (\hat{s}_i \times \hat{s}_j) - B \sum_i \hat{s}_i \qquad (2.39)$$

The simulations were based on a molecular field approach on an 8×8 two dimensional lattice with periodic boundary conditions. Two of the ground states found via these

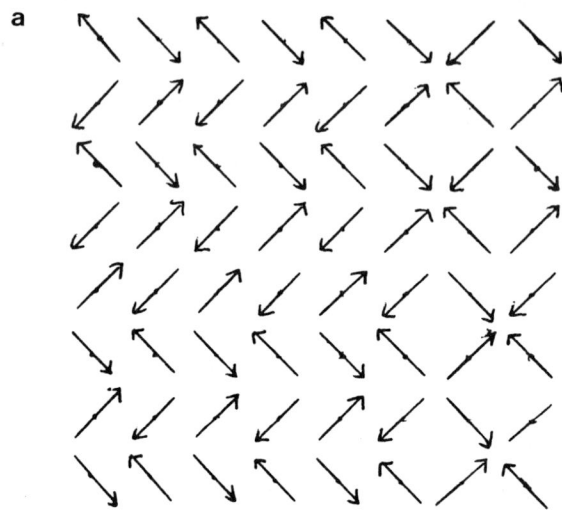

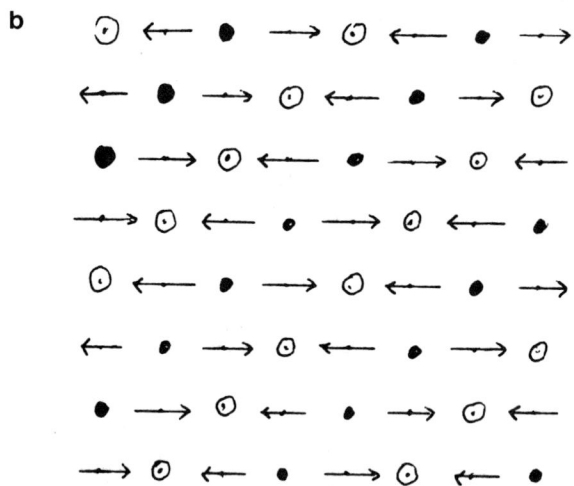

Figure (1) Ground states of the Hamiltonian (2.39) obtained by molecular field simulations on an 8×8 lattice with periodic boundaries. a) $J=B=L=0$, $K=+1$, 1b) $J=B=K=0$, $L=+1$. Open circles indicate the spin pointing up out of the page, full circles indicate the spin pointing down through the page.

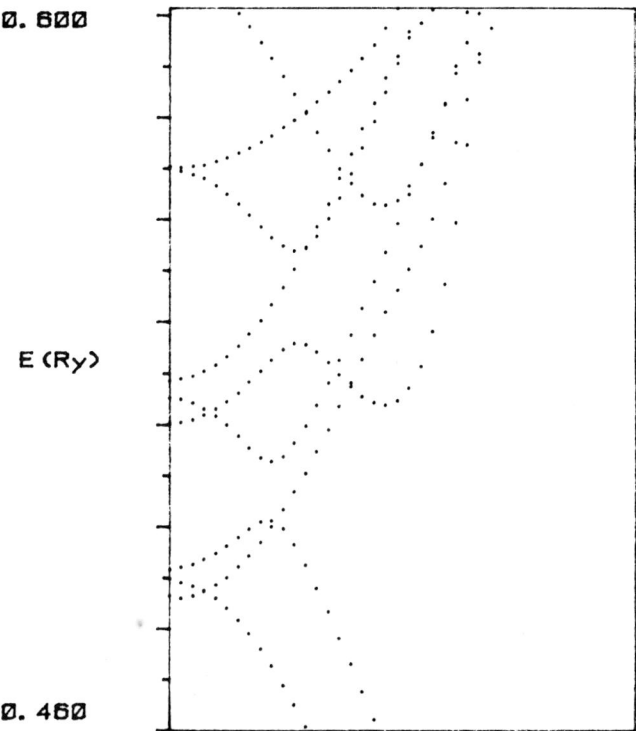

Figure (2) The relativistic spin-polarised band structure of Nickel just below the Fermi energy along $\Gamma - X(0,0,1)$ for the magnetic moment pointing along (0,0,1)

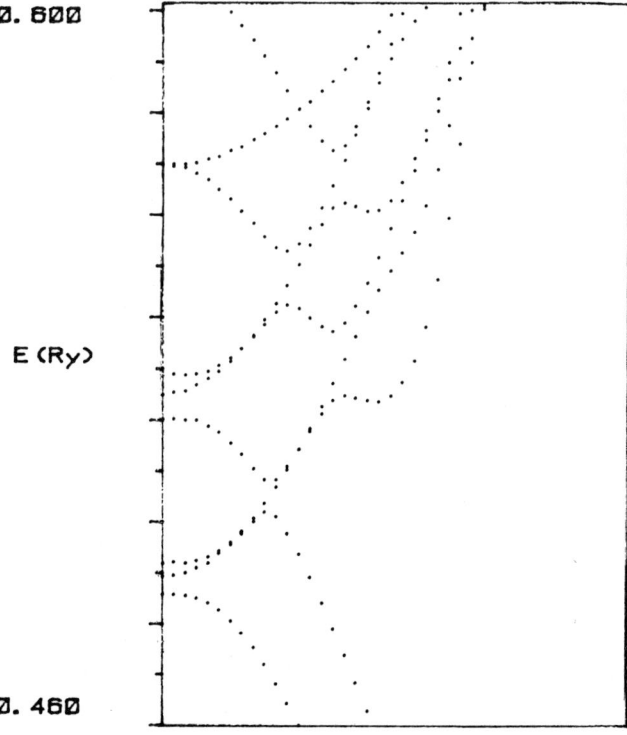

Figure (3) The relativistic spin-polarised band structure of Nickel just below the Fermi energy along $\Gamma - X(1,0,0)$ for the magnetic moment pointing along (0,0,1)

simulations are shown in figure (1). Clearly, the K-term confines the spins to a plane whereas the L-term tends to align neighbouring spins at right angles to one another. Obviously, even in such a simplified case a wealth of magnetic structures can be found. Further exploration would require three dimensional work and the full Hamiltonian. At that stage one might consider calculating spin wave spectra, so that if this interaction does determine the magnetic structure of materials this can be verified using neutron scattering.

Experimental Consequences of Magnetocrystalline Anisotropy in Metals

In the previous sections we have written down a considerable amount of scattering theory. In this section we consider the experimental consequences of implementing this theory. The principle new physics introduced by the simultaneous inclusion in the theory of spin-polarisation and spin-orbit coupling is caused by the reduction in symmetry that this brings about. The spin-orbit interaction couples the crystal lattice to the magnetisation direction. Hence in a cubic lattice, for example, some directions which were equivalent in a non-relativistic theory become inequivalent. This is illustrated in figures (2) and (3) where we plot the energy bands for nickel along the $\Gamma - X(0,0,1)$ and $\Gamma - X(1,0,0)$ directions in the Brillouin zone when the moment is along (0,0,1). Small differences and the lifting of some degeneracies at points of high symmetry can be seen between these two diagrams, which would be identical in a non-relativistic calculation. We now go on to look at how these differences can be measured experimentally. Firstly we consider how relativistic effects may be observed in the de Haas van Alphen effect. Secondly we consider the magnetocrystalline anisotropy energy itself. Of course these are not the only possibilities. For example in the magnetic X-ray dichroism theory described elsewhere in this volume the dichroism spectra will have a dependence on the direction of the magnetic moment relative to the crystal axes. This is a direct probe of the anisotropy, but will be a small effect which has not yet been observed.

The de Haas van Alphen effect

The de Haas van Alphen effect is a very sensitive probe of the Fermi surface of metals. There are several good reviews of this subject [27]. The Fermi surface is also something that is very easy to calculate from band theory. In figures (4), (5) and (6) we show cross sections through the Fermi surface of cobalt in three planes of high symmetry in the hexagonal Brillouin zone. These have been calculated in three increasingly more precise approximations. Also all these figures have been calculated within the same computer program, so they are directly comparable.

In Figure (4) we see the non-relativistic Fermi surface. This has been calculated by putting $c \to \infty$ in the program and it is in excellent agreement with previous non-relativistic calculations [28].

In figure (5) we have the Fermi surface calculated when the magnetic moment points along the c-axis of the crystal. Comparison with figure (4) shows that where the majority and minority spin bands cross in the non-relativistic case they avoid each other in the relativistic case. This leads to a rather unusual picture of the electrons behaviour. As they drift round the Fermi surface they go through regions where their spin is almost entirely in one direction, then they come to a region where the spin is undefined and

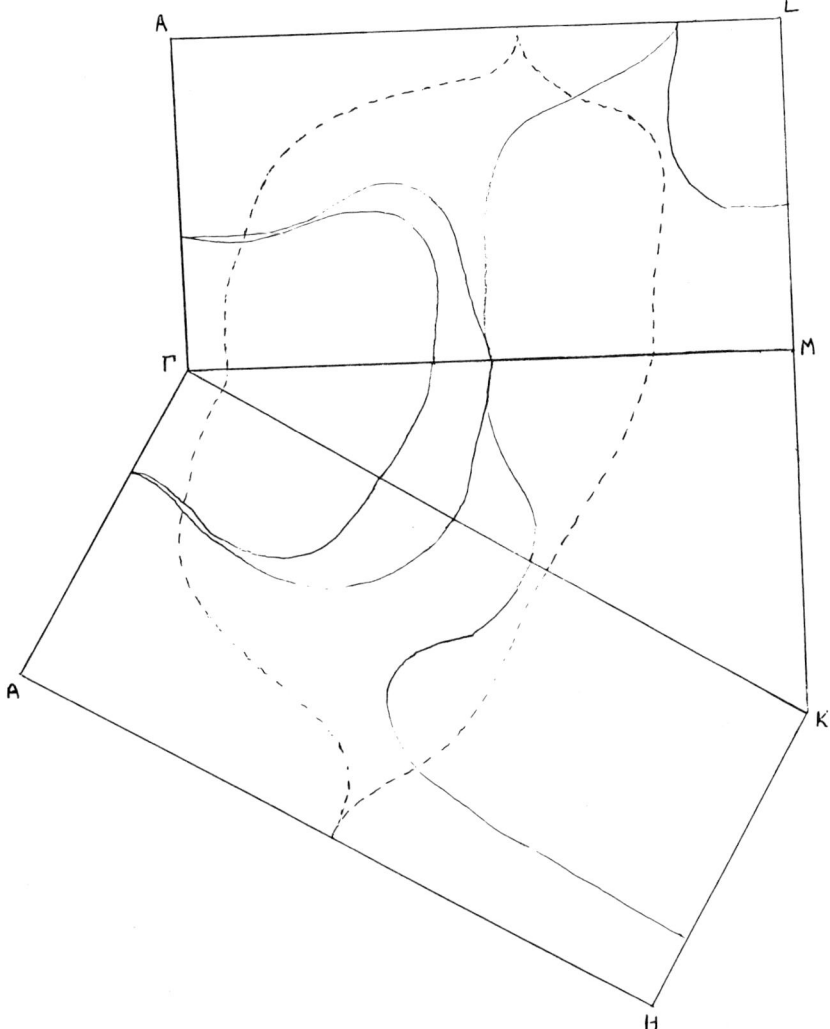

Figure (4) Cross sections through the Fermi surface of Cobalt in high symmetry planes of the H.C.P Brillouin zone, calculated non-relativistically.

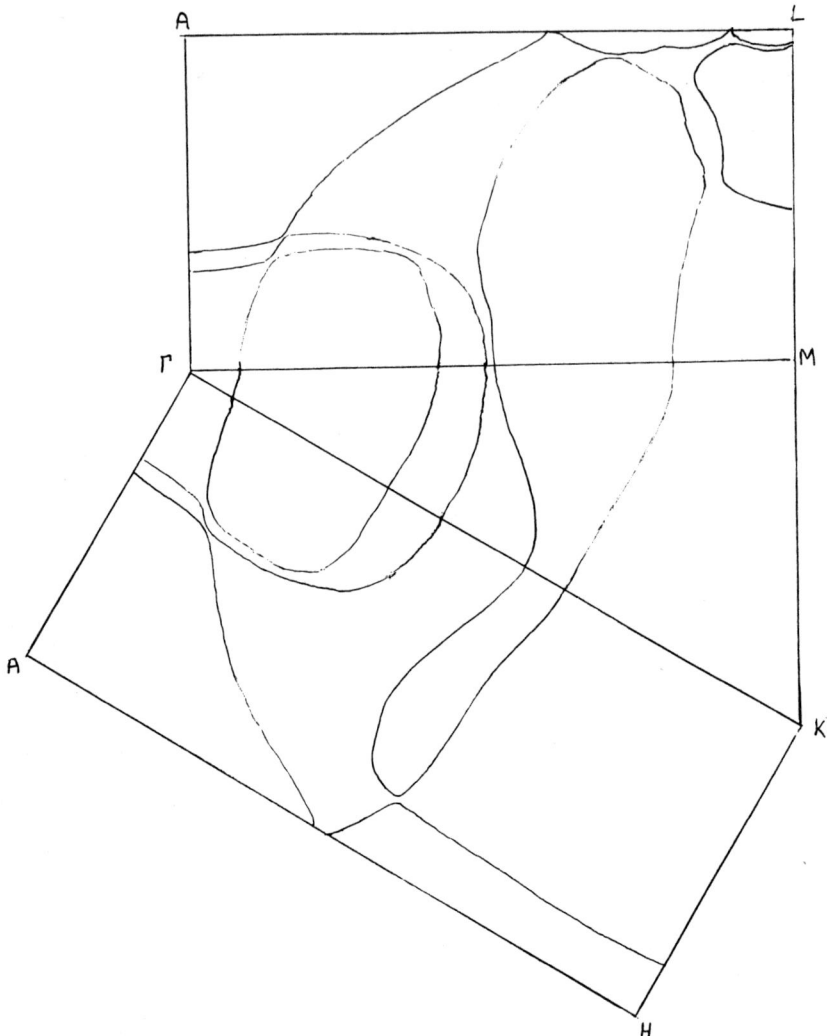

Figure (5) Cross sections through the Fermi surface of Cobalt in high symmetry planes of the H.C.P Brillouin zone, with the magnetic moment along the c-axis, calculated relativistically.

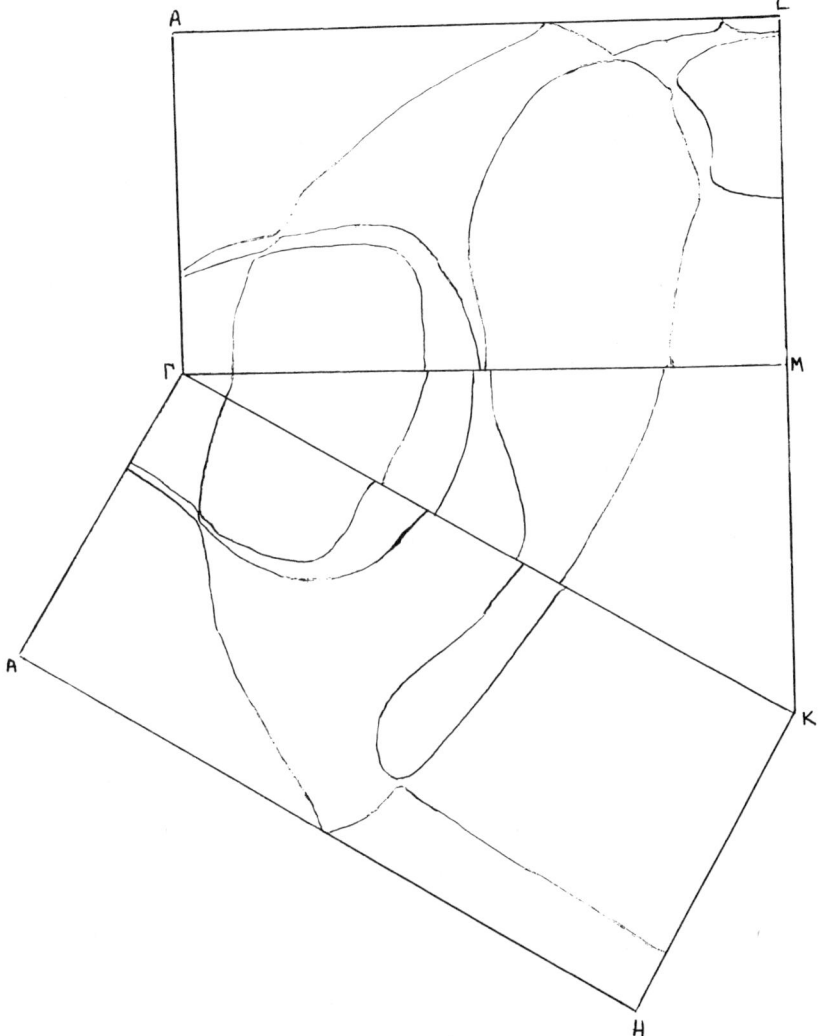

Figure (6) Cross sections through the Fermi surface of Cobalt in high symmetry planes of the H.C.P Brillouin zone, with the magnetic moment in the direction of the applied field, calculated relativistically.

the band mass is very close to zero. Later on they come to a region where the spin is almost entirely the opposite direction.

Finally in figure (6) we have the Fermi surface for the case when the magnetic moment is in the direction of the applied field. ie it is in a different direction when the experiment is looking at different planes in the Brillouin zone. This is what happens in the experiment. There is no reason why the Fermi surface plots should join up at the symmetry lines in the Brillouin zone in this case. In figure (6) we see an example of this. The two planes joined together are at right angles to one another, the bands do not join up and this effect can lead to a 2 or 3 % change in the predicted Fermi surface areas. One can imagine cases where the effect could be very large, in the case of nearly flat bands crossing very close to the Fermi energy

Some of the orbits predicted by the relativistic theory, but not by the non-relativistic theory have been confirmed in recent experiments [29].

This section shows that the usual interpretation of de Haas van Alphen experiments on magnetic materials in terms of separate spin up and spin down Fermi surfaces is fallacious [28]. Magnetic breakdown of the gaps between the Fermi surface in the figures can occur and then the experiment will measure a single spin Fermi surface. In this case this work is a warning to experimentalists that the method of measurement is changing the results and they are not measuring the ground state.

Magnetocrystalline Anisotropy Energy

The multiple scattering theory described previously has been applied to metals [8,12,13]. Before we go on to consider this work in more detail let us consider previous work on first principles calculations of magnetocrystalline anisotropy energies. Two previous attempts have been made to calculate these energies from first principles to our knowledge. Firstly Eckardt et al [30] calculated the anisotropy energy for F.C.C nickel and B.C.C iron. They found anisotropy energies of the correct order of magnitude, but the wrong sign for iron. Daalderop et al [31] also calculated these quantities but found the wrong sign for both iron and nickel. They also looked at H.C.P cobalt where the anisotropy energy is an order of magnitude larger. In this case their results were the correct sign, but their magnitude was incorrect by a factor of 2. They also looked at the anisotropy energy as a function of the number of valence electrons and found the anisotropy energy was very sensitive to this. Jansen [32] has considered the calculation of magnetocrystalline anisotropy energies from relativistic density functional theory. He has concluded that a spherical approximation for the potential means the neglect of significant contributions to the anisotropy energy. Furthermore he has stated that the Breit interaction may also play a role in determining the value of this energy. Of course, this is included, in principle, in the relativistic exchange-correlation functional. However the effect of this would only be included in the theory if we went to self-consistency within density functional theory with a suitably chosen approximation for the exchange-correlation functional.

We now go on to consider calculations of magnetocrystalline anisotropy energy using the multiple scattering theory above. In particular we look at ferromagnetic nickel on an F.C.C lattice [12]. The nickel atoms were defined by a potential and effective field from a self-consistent non-relativistic electronic structure calculation [6]. As nickel is a fairly light element this is not believed to be an important approximation.

Figures (2) and (3) are enlarged views of the energy bands for nickel in the $\Gamma - X(0,0,1)$ and $\Gamma - X(1,0,0)$ directions in the Brillouin zone with the moment pointing along $(0,0,1)$. These are very similar to previous results [6]. Comparison with these previous non-relativistic results shows that relativity introduces some small changes in band widths and lifts degeneracies at points of high symmetry. There is also some gross movement of the bands due to the mass-velocity and Darwin terms in the Hamiltonian. It can be seen that the major effect of rotating the magnetic moment is to prevent/allow band crossings and to alter some band separations. This can be understood on the basis of the $\sigma.B$ term in the Hamiltonian. When the moment rotates the scalar product varies as a cosine function.

Now, if we have different band structures for each direction of the moment we can integrate these over the Brillouin zone and we will then have different densities of states for each moment direction. This was done and a number of new mathematical techniques had to be used to ensure the numerical parts of the calculation were done to sufficient accuracy. For a discussion of these, reference should be made to the original papers [8]. The force theorem [33] states that the difference in total energy between two states with similar charge density is given by the difference in the *single electron* contributions to the energy. This was used to calculate the magnetocrystalline anisotropy energy. ie

$$E^{001} = \int_0^{Ef} En^{001}(E, n^{001}(r))dE \qquad (2.40a)$$

and

$$E^{111} = \int_0^{Ef} En^{111}(E, n^{001}(r))dE \qquad (2.40b)$$

or equivalently for any other direction of the moment in the unit cell. The difference between these two energies is the magnetocrystalline anisotropy energy. This has been calculated for nickel and Table (1) shows a full set of results. Apart from the spin-contribution to the magnetic moment all quantities shown in the table would be identically equal to zero in a non-relativistic theory. The values tabulated are those calculated when the Brillouin zone integral was performed with 36 directions per $\frac{1}{48}^{th}$ of the zone. The errors indicated are estimated on the basis of the calculation using 10 and 21 directions in the integration.

Nickel has a very small magneto-crystalline anisotropy energy. In fact any material on a cubic lattice has a much smaller anisotropy than for a non-cubic lattice [2]. Therefore many of the numerical problems associated with cubic lattices will not be present for non-cubic lattices. Unfortunately no magnetic element with one atom per unit cell exists on a non-cubic lattice. Two atoms per unit cell require four times as much cpu time and hence have not yet been examined. Therefore Strange et al [13] calculated the magnetocrystalline anisotropy energy for the fictitious compound simple tetragonal iron.

The difference in total energy for the moment pointing along (0,0,1) and (1,0,0) was calculated for a number of values of the axial ratio c/a. The potential was obtained from

Table (1)

Table showing the comparison between theory and experiment for the magnetic properties of nickel and the values of the anisotropy in energy and magnetic moment.

Quantity	Theory	Experiment
...	...	...
μ_{spin}	0.60	0.57
μ_{orb}	0.046	0.050
$\Delta E(\mu eV)$	10.5	2.7
$\Delta m(10^{-4}\mu_B)$	1.4	1.2

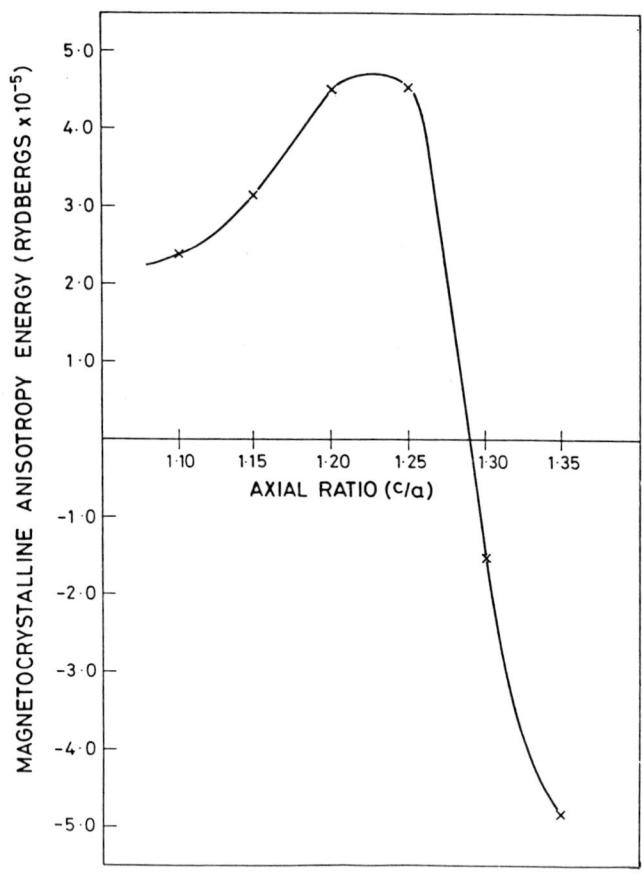

Figure (7) The magnetocrystalline anisotropy energy $E^{001} - E^{100}$ as a function of axial ratio for simple tetragonal iron.

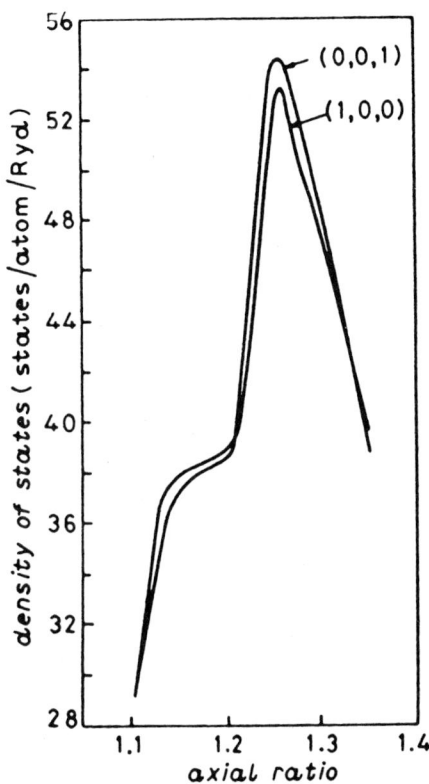

Figure (8) The density of states at the Fermi energy as a function of axial ratio for the magnetic moment (0,0,1) and (1,0,0) in simple tetragonal iron.

reference [6]. The axial ratio was varied in such a way as to keep the nearest neighbour distance constant. This was done to avoid renormalising the potential to different sphere radii for each c/a value which would lead to too large an uncertainty. Again for the numerical details of the calculation the original references should be referred to [8,13]. As expected the magnetic moment increased and the single particle energies decreased as the crystal layers got further apart. Figure (7) shows the magnetocrystalline anisotropy energy as a function of axial ratio. There are three things to note about this curve. Firstly the energy scale is one to two orders of magnitude greater than the size of the magnetocrystalline anisotropy in nickel. This is as it should be. Secondly, the curve is smoothly varying which indicates that the numerical procedures used were reliable. Thirdly the curve passes through zero at $c/a = 1.29$. This indicates that if we varied c/a through this value the moment direction would flip through ninety degrees. In fact too many approximations have gone into this work to make this number reliable. The most important of these is probably the way c/a was varied without obtaining a new self-consistent potential for each new value and hence the difference in total energies being given by the difference in single electron energies is likely to be not strictly valid. Nonetheless this work does imply the theory would be able to predict a flip in the easy axis of magnetisation as a function of some other parameter. Strange et al are able to go on to examine the density of states at E_f as a function of c/a and show a well defined peak in this quantity, figure (8). This is a d-band narrowing and rising above E_f as c/a increases and being replaced by increasing s-character below E_f, analogous to a valence transition in a rare earth metal.

Now we go on to examine the origin of the anisotropy within the formalism of relativistic spin-polarised scattering theory. If we consider equations (2.21) to (2.24) the only place where a direction with respect to the crystal lattice enters is in the spherical harmonics in the expression for $g(\mathbf{q}, E)$ Taking t^{-1} outside the square brackets of (2.23) what remains can be expanded as a binomial series. This leads to an infinite series of terms, each of which can be written $(tg)^n$ (ignoring quantum numbers), this term includes all n-site scattering processes. These can then be substituted back into (2.22). In the non-relativistic case the dependence of the t-matrices and the radial part of Z on the m-quantum numbers vanishes and they can be taken outside the sum over the m-quantum numbers. Then all that remains is a sum over Clebsch-Gordon coefficients and spherical harmonics. From simple trigonometrical relations it can be shown that the directional dependence cancels to all orders. In the relativistic case the t-matrix and Zs cannot be taken outside the m-summation. *i.e.* The symmetry which gave cancellation in the non-relativistic case has been broken and hence magnetocrystalline anisotropy is observed.

The value of the single-electron energy difference between distinct moment directions is due to different elements of the τ-matrix being dominant as the t-matrix is rotated and hence the symmetry is broken in different ways for each moment direction. Fundamentally this occurs relativistically and not non-relativistically because the relativistic theory conserves total angular momentum only, spin and orbital angular momentum are not conserved separately. In the non-relativistic case they are conserved separately.

In conclusion then, magnetocrystalline anisotropy energies can be calculated to the right order of magnitude. It is not clear why the values obtained are not in numerical agreement with experiment. For the smallest anisotropies the uncertainties in numerical procedures are nearly as large as the quantities we are trying to calculate. It may be that the potentials used just are not sufficiently precise to give the correct anisotropy energy, or that the force theorem is not a good enough approximation. However it may be that this theory is insufficient as it stands and we need to relax some of the approximations that have gone into it, for example the spherical approximation for the potential, or a more explicit inclusion of the Breit interaction. Finally as the atomic and molecular physicists have discussed elsewhere in this volume, if we try to calculate energies much smaller than these within density functional theory it will soon become necessary to examine the validity of the point nucleus approximation! Further work needs to be done to sort these problems out.

Paramagnetic spin susceptibility, Finite temperature anisotropy

In this section we report on work in progress in setting up a theory for the spin susceptibility of metals using relativistic spin polarised multiple scattering theory. The dynamical regime of this theory will be appropriate for studies of gaps in spin wave spectra and the coupling of spin wave excitations to crystallographic directions. To date, we have succeeded in carrying out explicit calculations for the static uniform paramagnetic spin susceptibility for cubic and h.c.p. systems [14]. When applied to systems which are ferromagnetic at low temperatures, the formulism will indicate the direction in the crystal lattice along which the magnetic moments initially align. Calculations for h.c.p. cobalt also show that thermally driven Stoner excitations affect this preferred direction. The work at this stage can be viewed as complementary to magnetic anisotropy energy calculations.

In short our approach is the relativistic generalisation of that of Vosko and Perdew [34], Macdonald and Vosko [35] and related work. In particular Stenzel and Winter [36] and Winter et al. [37] have set up a scheme for calculating the dynamic spin susceptibility of metals using a multiple scattering framework and our work, described fully in [14], is essentially the relativistic generalisation of the static limit of their theory. We start with an accurate relativistic treatment of the electronic structure of the paramagnetic system described by a single particle Green function. This paramagnetic system is then subjected to a small external magnetic field and the induced magnetisation determined. With the use of Euler rotation matrices, we can vary the orientation of the magnetic field with respect to the crystal axes. An expression for the paramagnetic susceptibility is obtained and has the form

$$\chi(\theta\phi;T) = \chi_0(\theta\phi;T)/(1 - I\chi_0(\theta\phi;T)) \qquad (2.41)$$

where θ and ϕ represent the orientation of the magnetic field and $\chi_0, I\chi_0$ describe the unenhanced and enhanced susceptibilities respectively. The denominator of this expression can be used to set up a relativistic Stoner condition. Full detailed expressions for these quantities are given in [14]. Once again, if the non-relativistic versions of these quantities are used the θ and ϕ dependence, i.e. the anisotropy, vanishes along with the spin-orbit coupling.

So far f.c.c. and h.c.p Co have been looked at with this theory. At low temperatures cobalt is an h.c.p. ferromagnet. As its temperature is increased it changes structure to an f.c.c. phase and at still higher temperatures becomes paramagnetic. Consequently paramagnetic h.c.p. cobalt does not exist but calculations of its spin susceptibility do provide insights into the nature of magnetic anisotropy in the presence of Stoner like excitations. In [14] they found 'Stoner' transition temperatures of 4345K and 2295K were found for h.c.p. and f.c.c. Co respectively. Owing to the neglect of spin wave fluctuations these are too high. The paramagnetic susceptibility of the cubic system in this linear response theory shows no dependence upon the orientation of the induced magnetisation but that of h.c.p. cobalt indicates that the magnetisation would initially grow in the a-b plane. On examining $\chi(\theta\phi;T)$ as a function of θ and ϕ, we found that the anisotropy varies wih temperature and ,at 4345K, it diverges when θ and ϕ describe a magnetic response in the a-b plane.

Relativistic spin-polarised scattering theory provides a new formulism for the static spin susceptibility which incorporates an anisotropic effect. Calculations show that this effect is temperature dependent. The formulism gives a generalised Stoner condition and forms the basis for future studies of dynamical effects.

References

[1] L.D. Landau, E.M. Lifshitz and L.P. Pitaevskii (1984) *Electrodynamics of Continuous Media*, Landau and Lifshitz Course on Theoretical Physics, **8** (Pergamon Press).

[2] R.F. Pearson (1979) *Magnetic Anisotropy*, in *Experimental Magnetism*, ed G.M. Kalvius and R.S. Tebble.

[3] D.M. Paige, B. Szpunar and B.K. Tanner, (1984), *Journal of Magnetism and Magnetic Materials*, **44**, 239.

[4] P. Strange (1989) *C.M.S.K. Research News*, Keele University.

[5] P. Phariseau and W.M. Temmerman (eds), (1984), *The Electronic Structure of Complex Systems*, NATO ASI Series B Vol 113.

[6] V.L. Moruzzi, J.F. Janak and A.R. Williams, (1978), *Calculated Electronic Properties of Metals*, (Pergamon Press, London).

[7] P. Strange, J.B. Staunton, and B.L. Gyorffy, (1984), *J.Phys.C* **17**, 3355.

[8] P. Strange, H. Ebert, J.B. Staunton and B.L. Gyorffy, (1989), *J. Phys.CM* **1**, 2959.

[9] J.B. Staunton, B.L. Gyorffy, J. Poulter, and P. Strange, (1987), *J.Phys.C* **21**, 1595.

[10] J.B. Staunton, B.L. Gyorffy, J. Poulter and P. Strange, (1989), *J.Phys.CM* **1**, 5157.

[11] N.P. Archibald and P. Strange, (Preprint).

[12] P. Strange, H. Ebert, J.B. Staunton and B.L. Gyorffy, (1989), *J.Phys.CM* **1**, 3947.

[13] P. Strange, J.B. Staunton and H. Ebert, (1989), *Europhys Letts* **9**, 169.

[14] M. Matsumoto, J.B. Staunton and P. Strange, submitted to J.Phys.CM.

[15] M.E. Rose, (1961), *Relativistic Electron Theory* (Wiley; New York).

[16] J.S. Faulkner and G.M. Stocks, (1980), *Phys Rev B* **21**, 3222.

[17] G. Schadler, A.M. Boring, R.C. Albers and P. Weinberger, (1987), *Phys Rev B***35**, 4324

[18] R. Feder, (1985), *Polarised Electrons in Surface Physics*, (World Scientific Publishing; Singapore) and references therein.

[19] B.L. Gyorffy and M.J. Stott, (1973), in *Band Structure Spectroscopy of Metals and Alloys*, ed D.J. Fabian and L.M. Watson, (Academic; New York)

[20] M.A Ruderman and C. Kittel, (1954), *Phys.Rev.* **96**, 99.

[21] K. Yoshida, (1957), *Phys.Rev.* **106**,893.

[22] T. Kasuya, (1956), *Progr.Theor.Phys.* **16**, 45.

[23] P.M Levy, C. Morgan-Pond, and A. Fert, (1982), *J.Appl.Phys.* **53**, 2168.

[24] P. Lloyd, (1967), *Proc.Phys.Soc.* **90**, 207.

[25] A. Fert, D. Arvanitis, and F. Hippert, (1984), *J.Appl.Phys* **55**, 1640.

[26] A.J Bray and M.A Moore, (1982), *J.Phys.C.* **15**, 3897.

[27] A.V. Gold, (1968), in *Electrons in Metals* eds J.F. Cochran and R.R. Haering, (Gordon and Breach, New York)

[28] T. Jarlborg and M.Peter, (1984). *J.Magn and Mag Mats* **42**, 89.

[29] A. Marshall, D.D. Pigram and G.G. Lonzarich, (1988), J. de Physique **49** Colloque C8, 55.

[30] H. Eckardt, L. Fritsche and J. Noffke, (1987), *J.Phys F* **17**, 943.

[31] G.H.O. Daalderop, P.J. Kelly, M.F.H Schuurmans and H.J.F Jansen, (1989), *J. de Physique* **49** Colloque C8, 93.

[32] H.J.F Jansen, (1989), *Phys Rev B* **38**, 8022.

[33] Mackintosh A.R and Andersen O.K, (1980), in *Electrons at the Fermi Surface*, Ed M. Springford, (Cambridge University Press)

[34] S.H. Vosko and J.P. Perdew, (1975) ,*Can.J.Phys***53**, 1385.

[35] A.H. Macdonald and S.H. Vosko, (1979), *J.Phys.C* **12**, 2977.

[36] E. Stenzel and H.Winter, (1985), *J.Phys.F* **15**, 1571.

[37] H. Winter, E. Stenzel, Z. Szotek and W.M. Temmerman, (1988), *J.Phys.F* **18**, 485.

THE SPIN POLARIZED PHOTOEMISSION FROM NON-MAGNETIC METALS

B. Ginatempo

Istituto di Fisica Teorica
Universita' di Messina
Messina (Italy)

B.L. Gyorffy

H.H. Wills Physics Department
University of Bristol
Bristol (UK)

INTRODUCTION

Since 1969 (1) it was well known that photoelectrons emitted from non-magnetic targets after excitation by circularly polarized radiation might be polarized. Such a theoretical prediction has its origin in the fact that, because of spin-orbit interaction, the spin-components of the electron wavefunction at a given binding energy E_i are different and the interaction with a photon in a definite helicity state selects just one spin components, making therefore the electron at energy $E_i+\hbar\omega$ polarized. Experiments on atoms, molecules and solids confirmed such predictions.

In 1984 it appeared the first photoemission experiment (2) where photoelectron extracted normally from the (1,1,1) face of Platinum, by normally incident circularly polarized radiation, were contemporarily analyzed in energy (kinetic energy), momentum (emission angle), and spin (polarization). Because these three physical quantities are represented by non-commuting quantum operators, the above sentence will be reviewed in some detail in the following section. The quoted work, actually, opened a very interesting field of research in solid state physics, showing that spin-polarized photoemission was able to gain information not only about the initial states energy dispersion but also about the spin symmetry of the electron wavefunction in non-magnetic metals. It was made feasible by the realization of a 'Mott' detector: the LEED detector, i.e. a tungsten surface onto which the photoelectron beam is shot to diffract and separate 'spin up' and 'spin down' electrons. A number of experimental papers appeared so far on this topics and Tamura, Piepke and Feder (3) have made accurate calculations and analysis about the extent of agreement between theory and experiment.

Theorists came along, in effects. In 1985, the theory and the quantitative explanations of the above experiment was provided (4,5). Later, in 1987, Feder's group was able to predict (6) that even by using linearly polarized light it was possible to extract polarized electrons, depending on the symmetry of particular surfaces (as the FCC (1,1,1) face) and experiments confirmed their prediction (7).

Another effect, still in Pt (1,1,1), was discovered in 1986 (8) in the case of off-normal emission. Opposite to the case of normal emission, in which case reversing the helicity of the light will produce a change of sign of the photoelectron spin polarization <u>but not the total photocurrent</u>, in the case of off-normal emission a sizeable asymmetry of the total photocurrent has been found, after reversing the helicity. Such phenomenon has its origin in the breakdown of the spatial and time reversal invariance introduced by the surface in addition to the above photoelectron spin-polarization effect, as it was argued by those authors. Garbe and Kirschner (9) made a more systematic study of this Photoemission Intensity Asymmetry (PIA). We have given in ref. 10 a quantitative explanation of such an effect by means of explicit calculations and symmetry analysis.

In this talk we will discuss about the fundamental points of a fully relativistic theory of photoemission, necessary to sort out the role of the electron spin and its interactions, and we will illustrate some theoretical predictions which are possible only because of the use of a fully relativistic hamiltonian.

THEORETICAL PREDICTIONS

The basic idea that made possible to elaborate a first principle theory of photoemission spectra from metals was due to Mahan (11). It consists in regarding the final state of the photoemission process as a time-reversed LEED (Low Energy Electron Diffraction) state. This is the basis of the so-called One-Step model, in which the photoemission cross section is evaluated as the rate of probability of a quantum transition for an electron in a Bloch state to such a final state, and which matches to the Three-Step model, in which the photoemission process is regarded as (i) a bulk optical excitation, followed by (ii) propagation up to the surface and (iii) diffraction. We do not discuss here about which of these models is the best, but we want just to point out that the One-Step model certainly defines the final state of the photoemission process in the most correct way. On this basis Pendry (12) was able to formulate, within the Multiple Scattering theoretical framework, a computer suitable non-relativistic theory for the One-Step model, largely using the already well established LEED theory and computer codes (see for example ref. 13). Such a non-relativistic theory af Angle Resolved Photo-Emission (ARPE) and its following improvements (14), together with its extension to the random alloy case (15), timely appeared while ARPE measurements, exploiting synchrotron radiation facilities, were coming along, and it had a remarkable success in linking the electronic structure of metals and alloys to the experimental spectra (see for example ref. 16).

Notwithstanding its merits, such a theory cannot deal with metals and alloys of high atomic numbers, because of the absence in the hamiltonian of the relativistic interactions due to the electron spin. After Pendry's theory next challenge to the theorists has been the relativistic generalization of the One-Step model. That involved the proper time reversal operation and the analysis of the photoelectron spin polarization. A very similar problem has been encountered in the literature when the relativistic electron diffraction problem has been studied (see for example refs. 17,18,19).

The existing fully relativistic theories of the One-Step model photoemission are due to Ackermann and Feder (20) and Ginatempo, Durham and Gyorffy (21). We have shown in a separate paper (9) how both theories and related codes produces very similar results. Both theories are the fully relativistic generalization of Pendry's LEED theory and differ only in few aspects: the first one can deal with both magnetic and non-magnetic metals, while the second one can deal with all non-magnetic metals, random alloys and intermetallic ordered compounds. Another big difference is about the form of the electric dipole matrix element: Ackermann and Feder use the velocity form of the matrix element, while we use a generalization of the acceleration form whose limit with $c \to \infty$ analytically recover the non-relativistic counterpart (21).

Problem set up

In the logic of time reversing the actual physical process, we start by defining the final state of the photoelectron at the Mott detector, placed far away (at Z) from the metal surface, as the following relativistic plane wave:

$$\langle r|f,\lambda\rangle = \sum_S q_\lambda^S U_\lambda^S (\varepsilon + \hbar\omega)\exp(-i k_= \cdot r)\delta(z - Z) \tag{1}$$

where λ stands for 'up' and 'down' and refers to the spin direction accepted by the Mott Detector whose internal axis of spin quantization is directed along n_d and U_λ^S is the usual four-spinor amplitude of the plane wave with $s=\pm 1$ denoting the spin state of the photoelectron respect with the surface normal. The U_λ^S are defined in terms of the Pauli spinors χ_λ and the coefficients q_λ^S are defined (19) by

$$q_{up}^+ = \cos\frac{\theta_d}{2} \ ; \ q_{up}^- = \exp(i\varphi_d)\sin\frac{\theta_d}{2}; \tag{2}$$

$$q^+_{down} = -\exp(-i\varphi_d)\sin\frac{\theta_d}{2}; \quad q^-_{down} = \cos\frac{\theta_d}{2}.$$

The final state at the detector in eq. (1) is then joined smoothly at the surface to the proper Bloch state in the bulk of energy $E_i + \hbar\omega$ Finally the amplitude of such states are calculated by the probability amplitude to have been induced by polarized light incident along the surface normal with helicity v. In terms of quantities as eqs. (2) one can write down a density matrix whose diagonal elements, for a given photon energy, photon direction, photon helicity, emission direction and Mott detector orientation the are spin resolved spectra $I^v_{up}(\varepsilon,\omega,n_q,n_e,n_d,a^v)$ and $I^v_{down}(\varepsilon,\omega,n_q,n_e,n_d,a^v)$ whose complicated expression are given in ref. 21. The full intensities when both up and down electrons are accepted by the detector (i.e. the trace of the density matrix) are

$$I^{v=+} = I^{v=+}_{up} + I^{v=+}_{down}; \quad I^{v=-} = I^{v=-}_{up} + I^{v=-}_{down} \quad (3)$$

The quantities I^v_λ allow us to calculate the component of the spin-polarization vector along the direction n_d as:

$$P^v = P \cdot n_d = \frac{I^v_{up} - I^v_{down}}{I^v_{up} + I^v_{down}} \quad (4)$$

An experiment that measures P^v for a given incidence and emission directions is somehow energy, momentum and spin resolved.

P^v is identically zero in absence of spin-orbit interaction as the following exercise will demonstrate.

An instructive exercise

Consider a single spherical muffin-tin well. In such a case the photocurrent can be written as:

$$I^v_\lambda(\varepsilon) = \sum_{\kappa\kappa'\kappa''\mu\mu'\mu''} A^\lambda_{\kappa'\mu'}(\varepsilon+\hbar\omega) D^{*v}_{\kappa'\mu'\kappa\mu} M_{\kappa\kappa'\kappa''}(\varepsilon) D^v_{\kappa\mu\kappa''\mu''} A^{*\lambda}_{\kappa'\mu'}(\varepsilon+\hbar\omega)$$

(5)

where the A's are the coefficients of the relativistic spherical waves development of the time reversed LEED state in eq. (1), the D's are the angular integrals of the electric dipole matrix element containing the selection rules (see below), the κ's are spin-angular quantum numbers, the μ's azimuthal quantum numbers and the radial integrals M's, involving the low energy Green's function, are given by:

$$M_{\kappa\kappa'\kappa''}(\varepsilon) = \int_0^{R_{mt}} r^2 dr \left\{ (g_{\kappa'}g_\kappa + f_{\kappa'}f_\kappa)\frac{\partial V}{\partial r} + \frac{i\omega V}{c}\left[g_{\kappa'}f_\kappa(\kappa'-\kappa-1) + f_{\kappa'}g_\kappa(\kappa'-\kappa+1)\right]\right\} *$$

$$\int_r^{R_{mt}} r'^2 dr' \left\{ (g^*_{\kappa''}g^+_\kappa + f^*_{\kappa''}f^+_\kappa)\frac{\partial V}{\partial r'} - \frac{i\omega V}{c}\left[g^*_{\kappa''}f^+_\kappa(\kappa''-\kappa-1) + f^*_{\kappa''}g^+_\kappa(\kappa''-\kappa+1)\right]\right\} +$$

$$+ \int_0^{R_{mt}} r^2 dr \left\{ (g_{\kappa'}g^+_\kappa + f_{\kappa'}f^+_\kappa)\frac{\partial V}{\partial r} + \frac{i\omega V}{c}\left[g_{\kappa'}f^+_\kappa(\kappa'-\kappa-1) + f_{\kappa'}g^+_\kappa(\kappa'-\kappa+1)\right]\right\} *$$

$$\int_0^r r'^2 dr' \left\{ (g^*_{\kappa''}g_\kappa + f^*_{\kappa''}f_\kappa)\frac{\partial V}{\partial r'} - \frac{i\omega V}{c}\left[g^*_{\kappa''}f_\kappa(\kappa''-\kappa-1) + f^*_{\kappa''}g_\kappa(\kappa''-\kappa+1)\right]\right\} \quad (6)$$

where $g_{\kappa'}$ and $f_{\kappa'}$ are the big and small component of the radial wavefunction for the final state (regular at the origin); g_κ and f_κ are the corresponding wavefunctions for the low energy state; g^+_κ and f^+_κ are the outgoing solution at the same energy for the radial Dirac equations (irregular at the origin). The angular integrals are:

$$D^\nu_{\kappa\mu\kappa'\mu'} = \int d\Omega \chi^*_{\kappa\mu}(\mathbf{a}\cdot\hat{\mathbf{r}})\chi^t_{\kappa'\mu'}$$

$$= \sum_s (-2s)\, C^s_{\kappa\mu} D^{\nu\,non}_{\ell\mu-s\,\ell'\mu'+s} C^{-s}_{\kappa'\mu'} \quad (7)$$

where the superscript t means time reversed and the non-relativistic angular integrals, $D^{\nu\,non}_{\ell\mu-s\,\ell'\mu'+s}$ (see ref. 12) are defined by

$$D^{\nu\,non}_{\ell\mu-s\,\ell'\mu'+s} = \frac{4\pi}{3}|\mathbf{a}|\sum_{m''=-1}^{+1} Y^*_{1m''}(\hat{\mathbf{a}}) \int d\Omega Y^*_{\ell\mu-s}(\Omega)Y_{1m''}(\Omega)Y^*_{\ell'\mu'+s}(\Omega) \quad (8)$$

In eqs. (7) and (8) $\chi_{\kappa\mu}$ are the spin angular harmonics, $Y_{\ell m}$ are the complex spherical harmonics, $C^s_{\kappa\mu}$ are Clebsh-Gordan coefficients and $\mathbf{a}$ is the amplitude of the radiation vector potential. Putting all the equations together one can achieve:

$$I^\nu_{up} - I^\nu_{down} \propto \sum_{\kappa\kappa'\kappa''} M_{\kappa\kappa'\kappa''} \sum_{\mu\mu'\mu''} \left[O^{\nu up}_{\kappa\mu\kappa'\mu'\kappa''\mu''} - O^{\nu down}_{\kappa\mu\kappa'\mu'\kappa''\mu''} \right] \quad (9)$$

where the coefficients O's at the r.h.s. are some tricky product of 3-j symbols and depend upon the polarization and the direction of the photon, the direction of the photoelectron and the direction of the Mott detector axis. If one can neglect the spin-orbit interaction, i.e. if $c \to \infty$, degeneracy is restored in the total angular momentum scattering channels, or to be more explicit

$$\lim_{c \to \infty} \left[g_{j=\ell+\frac{1}{2}} - g_{j=\ell-\frac{1}{2}} \right] = 0 \qquad (10)$$

therefore

$$\lim_{c \to \infty} \left[I^v_{up} - I^v_{down} \right] \propto \sum_{\ell\ell'\ell''} M_{\ell\ell'\ell''} \sum_{j'j''} \sum_{\mu\mu'\mu''} \left[O^{vup}_{\ell j\mu\ell'j'\mu'\ell''j''\mu''} - O^{vdown}_{\ell j\mu\ell'j'\mu'\ell''j''\mu''} \right] \qquad (11)$$

because the double sum on the j's and the μ's reduces to zero. We conclude then that the phenomenon of the spin polarization of electrons extracted from non-magnetic atoms is a direct consequence of the spin-orbit interaction.

Symmetry

The difference within the square brackets in eq. (9), however, can be zero for a number of other reasons, namely because of the polarization of the light. In fact, even if spin-orbit interaction is not neglected, in case of linearly polarized radiation ($v=0$), the above calculation yields zero again. This fact is simply due to the symmetry of the system under investigation, i.e. the spherical symmetry of the muffin-tin well. This means that for more complicates targets, as surfaces, different and sometimes curious results are to be expected, either because the formulae for the photocurrents become more complicated, involving Multiple Scattering, or because surface symmetry affects deeply the selection rules. For a deep analysis of the role of the target symmetry see for example ref. (22).

Tamura, Piepke and Feder (6) have shown that linearly polarized light extracts indeed polarized electrons from the (1,1,1) surface of platinum. Their arguments are based on the symmetry properties of such a surface. The _triangular_ symmetry of the FCC (1,1,1) face (the C_{3v} group) consists of a threefold rotation axis (the surface normal) and three mirror operation with respect to the three planes ΠLUX orthogonal to the surface (see Fig. 1) Such planes make ann angle of 120 degrees each other and their intersection is the surface normal. It follows that the right and left side, with respect to the surface normal in a mirror plane are not equivalent, because the mirror operation with respect to a plane orthogonal to a ΠLUX plan is _not_ a triangular symmetry operation. In other

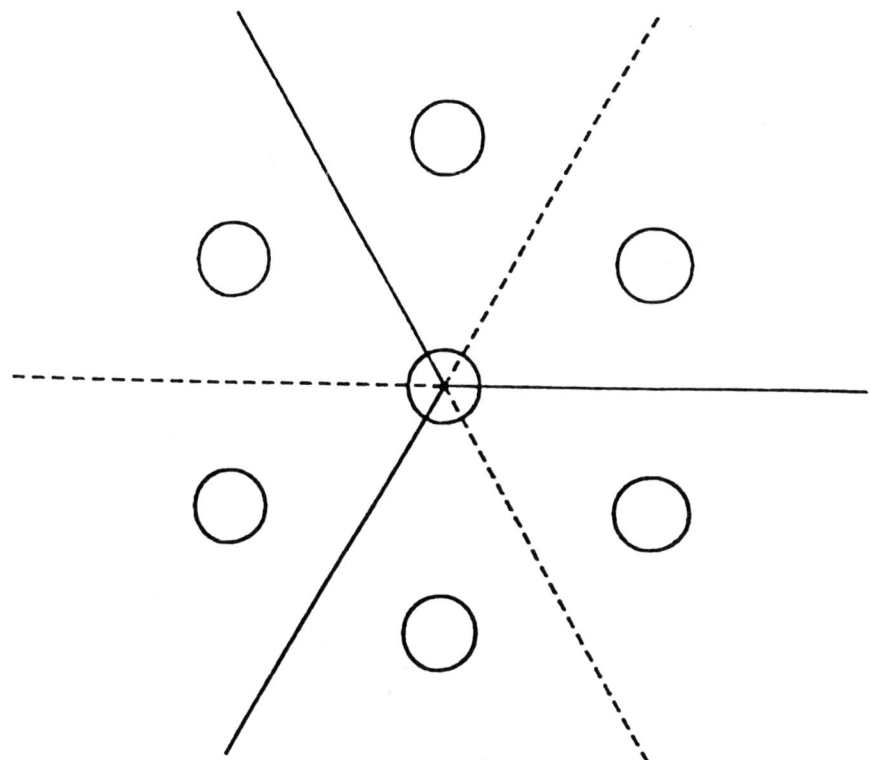

Fig. 1. The Triangular Symmetry (C_{3v}) of the FCC (1,1,1) face. Circles represent the central atom and its 6 nearest neighbours. Straight line represent the intersection of the surface with the three ΠLUX mirror planes Continue and dashed portions of the intersections represent the non equivalent sides of the mirror planes with respect to the surface normal.

words the contributions to I_λ^ν from both sides of a plane orthogonal to a mirror plane cannot be the same and therefore components of the spin-polarization vector parallel to the surface are to be expected even if linearly polarized light is used. For future reference we recall here that the symmetry of the FCC (1,1,1) surface is due to the fact that the underlying layers are shifted and rotated respect with the first.

Photoemission Intensity Asymmetry

The breakdown of translational invariance and time reversal symmetry introduced by surfaces has dramatic effects on photoemission spectra. For example, group theory arguments (23) demonstrate that a bulk optical interband transition at a given k-point along a symmetry direction must be entirely spin-polarized. The experiments never found such a results, but only partial spin polarization have been detected. The reason can be found in the fact that only few top atomic layers contribute to the spectra, because of the limited photoelectron escape depth. That means that, even if the energy position of a photocurrent peak identifies the precise band and k-point at which the transition occurs, the actual shape of a spectrum strongly depends upon the scattering history of the photoelectrons in a slab structure, before and after being excited. The reduction of translational and time reversal symmetry therefore acts broadening the photocurrent peaks and mixing the spin states. As a consequence, if photoelectrons, excited by circularly polarized radiation incident normally to the surface, are collected along an off-normal direction (the case of normal incidence and normal emission is, as we will see in the following, a highly degenerate one), one should expect for generally different spectra using photons of positive and negative helicity, because of the lack of cancellations produced by the surface. In other words, the surface acts as a spin filter, depressing or enhancing a given polarized photoelectron peak, so that affecting in different ways the shape of the spectra taken with positive and negative helicity.

Indeed, such an effect has been discovered by Oepen Hunlich and Kirschner (8) in 1986. These authors claimed also that the same kind of informations about the symmetry of electron bands one gets from the spin-polarized experiments, is obtainable from the asymmetry measurements which are in principle easier and faster than the former, because they do not need to analyze the photoelectron beam by the LEED detector. In some sense, one exploits the surface under investigation itself as a LEED detector. Such an effect is, indeed, a consequence of the fact that circularly polarized radiation excites spin-polarized electrons, however, owing to the fact that to have a sizeable PIA one has to work at off-normal emission directions and that in such a case is more difficult to identify the k-points corresponding to the initial state, this experiment should be regarded as complementary to the spin-polarized one, even if, as we will see, it brings a great deal of information about the surface properties.

Arbitrary rotation of the surface with respect to its normal must affect the spectra taken at a fixed polar angle and their asymmetry. In ref. (8) a qualitative explanation of such an effect is given in term of the Three-Step model. Garbe and Kirschner (9) presented a systematic study of the PIA for Pt (1,1,1) highlighting its behavior vs the azimuth angle. Ginatempo and Gyorffy (10) calculated the photocurrents (within the One-Step model) and their asymmetry, confirming and analyzing the experimental results, and remarked the role played by the surface and its symmetry.

Following ref. (9), one can define the PIA spectrum as

$$A = \mathbf{A} \cdot \mathbf{P} = \frac{I^+ - I^-}{I^+ + I^-} \tag{12}$$

that is the scalar product of 2 axial vectors: the photoelectron spin-polarization and the asymmetry vector, whose introduction allows us to describe the asymmetry effect independently of the polarization, and immediately evidentiate how such an effect is a relativistic one. In case of normal incidence and emission along a direction $\mathbf{n_e} \equiv (\theta_e, \varphi)$ where θ_e is measured from the surface normal and φ, the azimuth angle, is measured from a ΓLUX mirror plane of the (1,1,1) face, $A \neq 0$ is expected. Now, because $\mathbf{A}$ and $\mathbf{P}$ are axial vectors, under a mirror operation their components parallel to the mirror plane must change sign, while the normal component must not. That does not change their scalar products, but also the photon spin is an axial vector which for positive (negative) helicity is oriented parallel (antiparallel) to the surface normal (the direction of light propagation) and only has components in the mirror plane. Therefore it is inverted by mirror operation and that changes the sign of the polarization too. It follows then that the asymmetry is an antisymmetric function of the azimuth angle, i.e.:

$$A(\theta_e, +\varphi) = -A(\theta_e, -\varphi) \tag{13}$$

which in the special cases of $\varphi = \frac{n\pi}{3}$ (n =..,-2,-1,0,-1,-2,..), and in the very special case $\theta_e = 0$ gives

$$A\left(\theta_e, \frac{n\pi}{3}\right) = 0 \;;\; A(0, \varphi) = 0 \tag{14}$$

The first of equations (14) shows that the asymmetry vector must be orthogonal to the mirror planes (the polarization is parallel to them) so that in the case of normal emission it cannot be contemporarily orthogonal to the three mirror planes (which intersect at the surface normal) without being zero (second equation). The experiments of ref. (9) have demonstrated indeed (within the geometrical incertitudes) the antisymmetric behavior of A as predicted by eqs. (13) and (14). The calculations of ref. (10), obviously free from geometrical errors, systematically fulfill such

predictions. The undoubtable fact that the PIA depends upon
the surface symmetry suggests its use as a very valuable probe
in surface reconstruction analysis. Another convincing reason
is the following.

Fig.2 shows the antisymmetric behavior of the symmetry of
a feature ("band 4") in a photoemission spectrum (Pt (1,1,1)
normal incidence, $\hbar\omega = 18$ eV, $\theta_e = 62$) at about -1.5 eV below the
Fermi level as measured (9) and calculated (10). The somewhat
surprising oscillating behavior of half period 30 degrees, of
both theory and experiment (less accurate in the experiment
for negative azimuths), has a very interesting physical
interpretation. Suppose the system has the full hexagonal
symmetry (which would be true if the first layer was
isolated). In such a case eq. (13) would predict a periodic
behavior of period 60 degrees (the mirror planes in such a
case are making such an angle each other). Now, the FCC
(1,1,1) surface has only the triangular C_{3v} symmetry owing to
the displacements and rotations of the inner layers with
respect to the first. But the outer layers are the ones which
contribute more to the photocurrent, so that one could hazard
this conclusion: the 'quasi hexagonal' symmetry shown by Fig.2
arises from a combination of weak interaction between the
layers and escape depth. Because of the fact that the layers
interact and that the full symmetry is only triangular,
different peaks in the photocurrent spectrum have different
asymmetries as a function of the azimuth angle but the
tendency of such functions to oscillate around $\varphi = 30$ is a
clear surface geometry effect. The use of this function in
surface analysis is therefore promising.

Lifetimes

To make feasible such One-Step model calculations one has
to introduce two parameters the low and the high energy
inverse lifetimes, V_{0i} and V_{pi}. The LEED calculations are
carried on building a slab structure of atomic layers of
bidimensional periodicity and calculating the waves amplitude
of back and forth scattered waves in this structure, limited
by the surface and the bulk. That can be numerically achieved
introducing a negative imaginary potential to be added to the
actual energy $E_i + \hbar\omega$ of the LEED state. Photoemission
calculations consist of two LEED calculations to be carried on
for the photoelectron wavefunction and for the hole
wavefunction, so that the introduction of 2 different
parameters is required. Such parameters might be regarded just
as a numerical trick to allow the calculations to converge,
but they have interesting physical meanings. A careful
analysis of them can be found in ref. (22).

The high energy inverse lifetime V_{pi} is actually related
to the photoelectron escape depth, and, in effect, it limits
the number of layers entering the calculation. It is, in
principle a function of the photoelectron kinetic energy. For
a homogeneous gas one could take for it a constant value of 4
eV, and such choice has proved to be a successful one for most
applications (22). Looking for more sophisticated treatments,
one could write in atomic units:

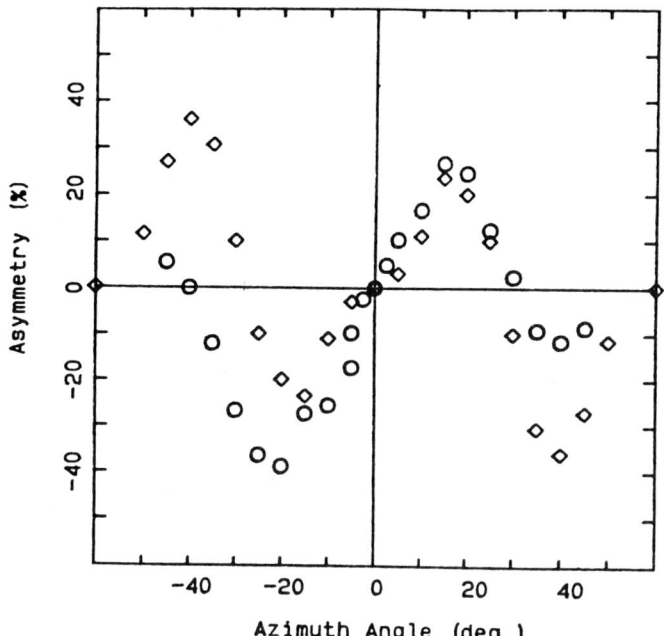

Fig. 2. The Asymmetry of one feature of off-normal photoemission spectra from Platinum (1,1,1) vs the azimuth angle. Circles: experimental data of ref.9.; rhombs: calculations of ref. 10.

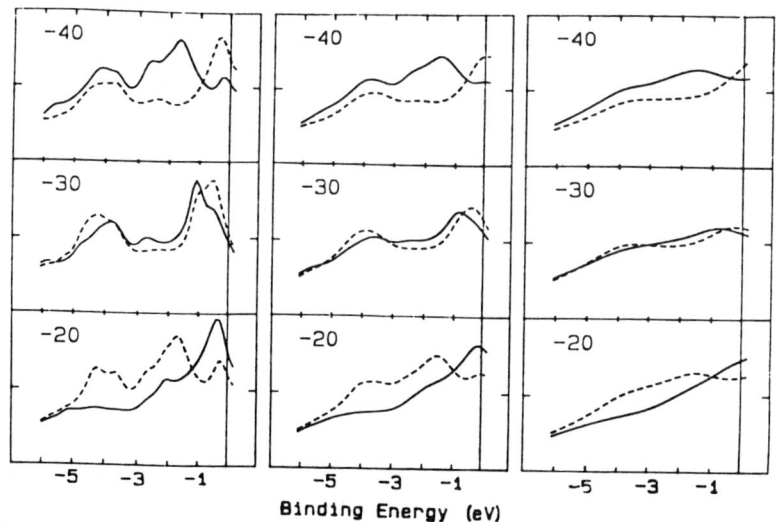

Fig. 3. The photoemission spectra vs the low energy inverse lifetime. The calculated spectra for positive (continous lines) and negative (dashed lines) helicity. Photon energy: 18 eV; emission direction 62 degrees; azimuth angle as indicated (-20,-30,-40). Left panel V_{0i} = -20 mRy,; middle panel V_{0i} =-40 mRy; right panel V_{0i} = -80 mRy.

$$\tau(\varepsilon) = \left(2V_{pi}(\varepsilon)\right)^{-1} \text{ and } \lambda = \tau\sqrt{2(\varepsilon - iV_{pi}(\varepsilon))} \tag{15}$$

and a parametric form of λ might give better results. Another possible choice could be

$$V_{pi}(\varepsilon) = \alpha(\varepsilon - \varepsilon_F)^\beta \tag{16}$$

and suggested by McRae and Caldwell(24) used by Tamura, Piepke and Feder (3).

The hole lifetime V_{0i} is a much complicated matter. It is actually the imaginary part of the self-energy to be introduced to deal with many-body effects in the photoemission process. The existing models let it varies linearly or quadratically with the distance from the Fermi level and actually they represent an important future development for the photoemission theory.

Neglecting the energy dependence of V_{0i} means in some sense to limit the calculations to the one-particle spectra. But imposing a non-zero value for it has in practice the effect of damping the low energy states, i.e. the effect of decreasing the multiple scattering of the initial states within a single layer and among the layers of the slab. Such a decrease is related to the quasi-symmetry effect illustrated in the preceding subsection to explain the oscillations of the PIA vs the azimuth angle of Fig. 2. Decreasing the interactions between the layers 'changes slowly' the symmetry from triangular to hexagonal. Fig. 3 shows such an effect. Increasing the low energy inverse lifetime absolute value has the effect to reduce the asymmetry for the angle $\varphi = 30$, and to let the spectra for $\varphi = 20$ and $\varphi = 40$ to become antisymmetric. We interpret this fact as an increase of hexagonal symmetry.

CONCLUSIONS

We like to conclude such a discussion just remarking how the spin-polarization and the asymmetry of photoelectron beams are phenomena based on the fact that the speed of the light is a universal constant. As a consequence of that, relativistic interactions affect the motion of electrons in a metal and produce, among the others, the physical effects described above. Due to their relativistic nature it is not possible to describe them out of a fully relativistic theoretical framework.

REFERENCES

1) Fano U. Phys.Rev. **178**, 131 (1969)
2) Eyers A., Schafers F., Schonhense G., Heinzmann U., Oepen H.P., Hunlich K., Kirschner J. and Borstel G. Phys.Rev.Lett. **52**, 1559, (1984)

3) Tamura E., Piepke W. and Feder R. J.Phys.:Condens. Matter $\underline{1}$, 6469 (1989)
4) Ginatempo B., Durham P.J., Gyorffy B.L. and Temmermann W.M. Phys.Rev.Lett. $\underline{58}$, 1581 (1985)
5) Ackermann B. and Feder R. Sol. State Com. $\underline{54}$, 1077 (1985)
6) Tamura E., Piepke W. and Feder R. Phys.Rev.Lett. $\underline{59}$, 934 (1987)
7) Schmiedeskamp B., Vogt B. and Heinzmann U., Phys. Rev. Lett. $\underline{60}$, 651 (1988)
8) Oepen H.P., Hunlich K. and Kirschner J. Phys. Rev. Lett. $\underline{56}$, 496 (1986)
9) Garbe J. and Kirschner J. Phys. Rev. $\underline{B39}$, 9859 (1989)
10) Ginatempo B. and Gyorffy B.L. J.Phys.: Condens. Matter, to appear (1990)
11) Mahan G.D. Phys. Rev. $\underline{B2}$, 4334 (1970)
12) Pendry J.B. Surf. Sci $\underline{57}$, 679 (1976)
13) Pendry J.B., "Low Energy Electron Diffraction" (London: Academic Press) (1974)
14) Hopkinson J.F., Pendry J.B. and Titterington D.J. Comp. Phys. Comm. $\underline{19}$ 69 (1980)
15) Durham P.J. J.Phys.F:Metal Phys. $\underline{11}$, 2475 (1981)
16) Jordan R.G. and Durham P.J. in "Alloy Phase Stability" Stocks G.M. and Gonis A. eds., NATO-ASI series E vol. 163 (Dordrecht:Kluwer Academic) (1989)
17) Jennings P.J. Surf. Sci. $\underline{20}$ 18 (1970)
18) Feder R. Phys. Stat. Sol. $\underline{46}$, K31 (1971)
19) Shen A.P. Phys. Rev. $\underline{B9}$, 1328 (1974)
20) Ackermann A. and Feder R. J.Phys.C: Sol. State Phys. $\underline{18}$, 1093 (1985)
21) Ginatempo B. Durham P.J. and Gyorffy B.L. J.Phys.: Condens. Matter $\underline{1}$ 6483 (1989)
22) Feder R. in "Polarized Electrons in Surface Physics", Feder R. ed. (Singapore: World Publishing) (1985)
23) Borstel G. Sol. State Comm. $\underline{53}$ 87 (1985)
24) McRae E.G. and Caldwell C.W. Surf. Sci. $\underline{57}$, 77 (1976)

THEORY OF MAGNETIC X-RAY DICHROISM

H. Ebert[1], B. Drittler[2], P. Strange[3], R. Zeller[2] and
B. L. Gyorffy[4]

[1]Siemens AG Research Laboratories, ZFE ME TPH 11
Postfach 3220, D-8520 Erlangen, FRG

[2]Institut fur Festkorperphysik der Kernforschungsanlage Julich,
Postfach 1913, D-5170 Julich, FRG

[3]Physics Department, Keele University
Keele, Staffordshire ST5 5BG, UK

Rutherford-Appleton Laboratory
Chilton, Oxfordshire OX11 0QX, UK

[4]Physics Department, Bristol University
Tyndall Avenue, Bristol BS8 1TL, UK

Introduction

In recent years the development of high intensity synchrotron radiation sources has enabled us to use X-ray spectroscopies as a probe of the electronic structure of materials on an extremely detailed level. In particular, the high fluxes available have made possible spectroscopies based on magnetic scattering rather than charge scattering [1,2,3]. The magnetic scattering is smaller than charge scattering by a factor of $\hbar\omega/mc^2$. Nonetheless such investigations have been carried out. Sawatzky and co-workers [4,5,6] have examined the $M_{4,5}$ absorption spectra of rare earth materials and interpreted them in terms of localised f-electrons excited from a Hunds rule ground state. Siddons and coworkers [7] have studied the Faraday rotation and ellipticity of linearly polarised radiation in transmission through a magnetic material. The Magneto-Optic Kerr Effect has been observed in many materials (see [8] for example). Schütz and co-workers [9,10,11] have used circularly polarised X-rays and have found a difference in absorption rate between left and right circularly polarised incident radiation. This effect is known as X-ray dichroism. It is a theoretical description of this that the present chapter focuses upon.

Theoretical Framework

Virtually all the first principles theoretical research into the electronic structure of solids today is based upon the density functional theory. The relativistic version of this is discussed elsewhere in this volume. By first principles we mean that the only necessary input into a calculation on a specific material is the atomic number and stoichiometry, which just identifies the material, (The crystal structure can be calculated, but usually we are not directly interested in calculating this, so we save time and use the known structure). The most important probe of electronic structures away from the Fermi energy is the large number of photon spectroscopies available. These spectroscopies excite an electron from one energy level to another, leaving a hole in the initial state, (This is the least that happens, often more complicated processes occur). Spectroscopies measure the spectrum of the excited states. Now we have reached a difficulty already. The density functional theory is only a theory of the ground-state, whereas the spectroscopies, by definition, create an excited state. It is certainly true that the density functional energy bands do not represent the quasi-particle excitation energies. However it has been argued by Durham [1] and Von-Barth [12] that the density-functional eigenvalues are probably quite close to the true quasi-particle energies. Certainly, experience has shown us that agreement between one-electron theory and experiment is surprisingly good, particularly for metals.

All the magneto-optical effects mentioned in the introduction have their origin in the subtle interplay between spin-orbit coupling and spin-polarisation. Hence we have to use relativistic density functional theory to describe them. This is implemented using relativistic band theory, in the present case this involves a multiple scattering theory approach. The important aspect of the simultaneous occurence of these two effects is the coupling of the spin to the crystal lattice and the consequent lowering of the symmetry in the material. We now go on to examine, in detail, how this symmetry breaking leads to magnetic X-ray dichroism.

Theoretical Method

The formal one-electron theory for studying spectroscopies, including X-ray absorption, using multiple scattering theory was developed by Durham [1]. It is certainly true that many electron effects also contribute to the theory, however we ignore them here, and indeed in most cases they turn out to be negligible. Ultimately though, this is a fundamental limitation of the theory presented here. This treatment assumes we know the one-electron scattering Green function. If this is known the method is applicable equally to elements, dilute alloys and ordered and disordered compounds.

In this paper we are concerned with X-ray absorption. The absorption coefficient is given by (by definition)

$$\mu(\omega) \propto \sum_i W_i^{q\lambda}(\omega) \qquad (1)$$

where $W_i^{q\lambda}(\omega)$ is the probability that a photon of frequency ω, wavevector q and polarisation state λ excites an electron in state i up into the conduction band. Note that throughout this chapter we use μ to represent two different things. Where it has an argument ω it is the absorption coefficient. Where it has no argument it is the quantum number associated with the z-component of total angular momentum operator. Durhams expression for $W_i^{q\lambda}(\omega)$ is

$$W_i^{q\lambda} = -\frac{1}{\Gamma} \int d^3r \int d^3r' \psi_i^*(\mathbf{r}) X_{\mathbf{q}\lambda}(\mathbf{r}) ImG(\mathbf{r},\mathbf{r}'; E + \hbar\omega) X_{\mathbf{q}\lambda}^*(\mathbf{r}') \psi(\mathbf{r}') \Theta(E_i + \hbar\omega - E_F) \qquad (2)$$

In this equation E_i is the energy of the core state ψ_i, which is a Dirac spinor. E_F is the Fermi energy. The Θ-function ensures that the excited electron can only occupy energetically acceptable empty conduction band states. The $X_{\mathbf{q}\lambda}$ are interaction operators.

Within a fully relatvistic and spin-polarised formalism this is given by

$$X_{q\lambda}(\mathbf{r}) = -e\underline{\alpha} \cdot \mathbf{A}(\mathbf{r}) \tag{3}$$

where $\mathbf{A}(\mathbf{r})$ is the vector potential associated with the incident radiation and $\underline{\alpha}$ are the usual Dirac matrices. It is in the interaction operator that we distinguish between left (LCP) and right (RCP) circularly polarised radiation. Of course the incident radiation may be described by a plane wave

$$E = E_o e^{i(\mathbf{q}\cdot\mathbf{r}-\omega t)}\mathbf{e}_\lambda$$

where $\mathbf{e}_\lambda$ is a vector describing the polarisation state of the wave. The convention is that the electric field associated with an incident LCP photon is $\mathbf{e}_\lambda = (1,i,0)$ and if the incident photon is RCP we have $\mathbf{e}_\lambda = (1,-i,0)$, provided $\mathbf{q}$ points along the z-axis $(0,0,1)$. Throughout this chapter we assume the incident wavevector and the magnetisation direction are parallel along the z-axis. The vector potential can easily be obtained from these via Maxwells equations. $G(\mathbf{r},\mathbf{r}';E)$ is the Green function which describes the electronic structure of the material, and in particular the final states available for the excited electron. For the present purposes we include magnetic and relativistic effects on an equal footing. The scattering theory method for calculating the Green function under such circumstances has been derived by Strange et al [13,14,15]. They show that the scattering Green function is given by

$$G(\mathbf{r},\mathbf{r}';E) = \sum_{\Lambda\Lambda'} Z_\Lambda(\mathbf{r},E)\tau^{nn}_{\Lambda\Lambda'}(E)Z^+_{\Lambda'}(\mathbf{r}',E) - \sum_\Lambda Z_\Lambda(\mathbf{r},E)J^+_\Lambda(\mathbf{r}',E) \tag{4}$$

Here $Z(\mathbf{r},E)$ and $J(\mathbf{r},E)$ are the regular and irregular solutions of the Kohn-Sham-Dirac equation respectively. Λ represents the quantum numbers. We can work in any representation in principle, but it is most convenient if Λ represents the usual relativistic quantum numbers κ and μ. $\tau^{nn}_{\Lambda\Lambda'}(E)$ is the site diagonal scattering path operator given by the integral of the inverse of the KKR matrix over the first brillouin zone.

$$\tau^{nn}_{\Lambda\Lambda'}(E) = \frac{1}{\Omega}\int_{BZ} d^3k [t^{-1}(E) - G(\mathbf{k},E)]^{-1}_{\Lambda\Lambda'} \tag{5}$$

In this equation t is the single site t-matrix and $G(\mathbf{k},E)$ are the usual KKR structure constants rotated, if necessary, into the correct representation. A full discussion of the properties of τ is given in references [16] and [17]. Details of the calculation of these quantities in the relativistic case are given by Strange et al [14]. Here we have implicitly assumed we are dealing with a single atom per unit cell. If this is not the case equation (4) can easily be generalised by allowing the superscript n to represent atomic species as well as unit cell. If we want to consider random alloys the τs can be obtained from the KKR-CPA equations [17], or for a single impurity we can find the effective site diagonal scattering path operator from (see reference [18])

$$\tau^{nn}_{imp} = \tau^{nn}_{host}[1 - (t^{-1}_{host} - t^{-1}_{imp}).\tau^{nn}_{host}]^{-1} \tag{6}$$

As shown by Durham [1], the equations (3)-(6) can be used in equation (2) to give us

$$W^{q\lambda}_i = -\frac{1}{\Gamma}\sum_{\Lambda\Lambda'} m^\Lambda_i(E_i + \hbar\omega) Im\tau^{nn}_{\Lambda\Lambda'}(E_i + \hbar\omega) m^{\Lambda'*}_i(E_i + \hbar\omega) \tag{7}$$

where the matrix elements $m^\Lambda_i(E)$ are given by:-

$$m^\Lambda_i(E) = \int \psi^*_i(\mathbf{r})(-e\underline{\alpha}a_\lambda e^{i\mathbf{q}\cdot\mathbf{r}})Z_\Lambda(r,E)d^3r \tag{8}$$

335

We can expand the exponential here as a series

$$exp(i\mathbf{q}.\mathbf{r}) = 1 + i\mathbf{q}.\mathbf{r} - (\mathbf{q}.\mathbf{r})^2 + \ldots \ldots \quad (9)$$

Retaining only the first term here is known as the electric dipole approximation. Higher terms represent magnetic dipole, electric quadrupole contributions etc. The matrix elements of equation (8) are not entirely straightforward to calculate because the functions $\psi_i(r)$ and $Z_\Lambda(r, E)$ are 4-vectors which do not have a unique spin-angular character [13,14]. This arises because the spin-polarisation and spin-orbit coupling couple states of different l-quantum number. Only μ remains a good quantum number. Henceforth we restrict ourselves to the (κ, μ) representation. In this representation the wavefunctions have the form

$$Z_{\kappa\mu}(r, E) = \sum_{\kappa'} Z^\mu_{\kappa'\kappa}(r, E) \quad (10)$$

where

$$Z^\mu_{\kappa'\kappa}(r, E) = \begin{bmatrix} g^\mu_{\kappa'\kappa}(r, E)\chi_{\kappa'\mu}(\hat{r}) \\ if^\mu_{\kappa'\kappa}(r, E)\chi_{-\kappa'\mu}(\hat{r}) \end{bmatrix} \quad (11)$$

Here the $\chi_{\kappa\mu}$s are spin angular functions defined by Rose [19]. $g^\mu_{\kappa'\kappa}(r, E)$ and $f^\mu_{\kappa'\kappa}(r, E)$ are large and small components of the Dirac wavefunction and are solutions of coupled radial Dirac equations. In the radial wavefunctions $Z^\mu_{\kappa'\kappa}$ the first quantum number represents the spin angular character and the second represents the asymptotic behaviour of $Z_{\kappa\mu}$. If we substitute equations (9)-(11) into equation (8) and work with the dipole approximation we find the matrix elements take the form

$$m_i^{\kappa\mu}(E) = ie \sum_{\kappa'\kappa'_i} \int r^2 g^\mu_{\kappa'_i\kappa_i}(r, E_i) f^{\mu'}_{\kappa'\kappa}(r, E) A^\lambda_{\kappa'_i\mu\ -\kappa'\mu'} dr$$
$$- \int r^2 f^\mu_{\kappa'_i\kappa_i}(r, E_i) g^{\mu'}_{\kappa'\kappa}(r, E) A^\lambda_{-\kappa'_i\mu\ \kappa'\mu'} dr \quad (12)$$

The $A^\lambda_{\kappa\mu\ \kappa'\mu'}$ are the angular integrals which define the selection rules.

$$A^\lambda_{\kappa\mu\ \kappa'\mu'} = \int \chi^*_{\kappa\mu}(\hat{r})\underline{\sigma}\cdot\mathbf{a}_\lambda\chi_{\kappa'\mu'}(\hat{r})d\Omega \quad (13)$$

Now a_λ is the polarisation vector of the photon field and for RCP we have $a_- = a(1, -i, 0)$ and for LCP $a_+ = a(1, i, 0)$. Taking the scalar product of these with the Pauli matrices gives

$$\underline{\sigma}\cdot a_- = 2a \begin{pmatrix} 0 & 0 \\ 1 & 0 \end{pmatrix}$$
$$\underline{\sigma}\cdot a_+ = 2a \begin{pmatrix} 0 & 1 \\ 0 & 0 \end{pmatrix} \quad (14)$$

We can substitute these matrices into (13) and perform the angular integrals straightforwardly. This leads us to:-

$$A^+_{\kappa\mu\ \kappa'\mu'} = 2aC(l, \frac{1}{2}, j; \mu - \frac{1}{2}, +\frac{1}{2})C(l', \frac{1}{2}, j; \mu' + \frac{1}{2}, -\frac{1}{2})\delta_{l,l'}\delta_{\mu-\mu',+1} \quad (15a)$$

and

$$A^-_{\kappa\mu\,\kappa'\mu'} = 2aC(l,\frac{1}{2},j;\mu+\frac{1}{2},-\frac{1}{2})C(l',\frac{1}{2},j;\mu'-\frac{1}{2},+\frac{1}{2})\delta_{l,l'}\delta_{\mu-\mu',-1} \qquad (15b)$$

Substitution of this into equation (12) gives the selection rules for the dipole approximation, $l - l' = \pm 1$ and $\mu - \mu' = \pm 1$, depending on the polarisation state. Now we have got to the source of the difference in absorption rate for LCP and RCP incident radiation. It is easier to answer the question, Why is there no difference in the absorption rate for incident LCP and RCP radiation in the non-relativistic case? In the non-relativistic case we would get these same selection rules. However the radial part of the matrix element is independent of μ in this case. Therefore, the amplitude for $\mu - \mu' = 1$ would be the same as for $\mu - \mu' = -1$, and hence the absorption rates for LCP and RCP X-rays are equal. In the relativistic case this is no longer true.

Note that (at least at real energies) both the radial and angular integrals are real, and hence the matrix element of equation (12) is completely imaginary. It is a simple generalisation to go above the dipole approximation, and include the $i\mathbf{q}.\mathbf{r}$ term from equation (9). This simply leads to an extra factor of r in the radial integrals. As these are done numerically in the computer code, this requires very little extra labour. The angular integrals in this case can still be evaluated analytically and they lead to some interesting dipole forbidden transitions.

In equation (12) we can make the simplification that we neglect the relatively small contribution from the coupling of partial waves with different l-values. Then the summations in (12) can be restricted to $\kappa'_i = \kappa_i, -\kappa_i - 1$ and $\kappa' = \kappa, -\kappa - 1$. The first application of this theory was for X-rays tuned to the K-edge of iron [20,21,22]. This was prompted by the experiments of Schütz and co-workers [9,11]. However it turns out that this was convenient from the theoretical point of view as well. This is because a core s-level has a unique spin-angular character, and hence the summation over core quantum numbers simply reduces to $\kappa'_i = \kappa_i$. This is not the case for excitations from states with a higher l-quantum number. For example let us consider initial $p_{1/2}$ and $p_{3/2}$ states. Because of the spin-dependent potential the electrons in these shells contain some admixed $p_{3/2}$ or $p_{1/2}$ character, although they can still be labelled uniquely by the μ- quantum number. A further complication arises because these states, which were degenerate in the paramagnetic case, are now slightly split by the effective magnetic field from the unpaired electrons in the outer levels. A full treatment of these effects requires a very accurate determination of core wavefunctions. Ebert [23] has developed an algorithm that provides such a determination, although he has shown that in some cases these effects can be ignored [18]. The larger the magnetic moment and the higher the principle quantum number, the more important these effects become.

Applications of the Theory

The first application of this theory was to iron [20,21]. The Green function was determined as described above using a potential obtained from a self-consistent spin-polarised non-relativistic electronic structure calculation [24]. Figure (1) shows the K-shell X-ray absorption coefficient of iron. This curve has undergone a lorentzian broadening of 0.8eV to take into account the finite lifetime of the 1s state. Further broadening of 1eV was included to take account of experimental broadening from various origins, such

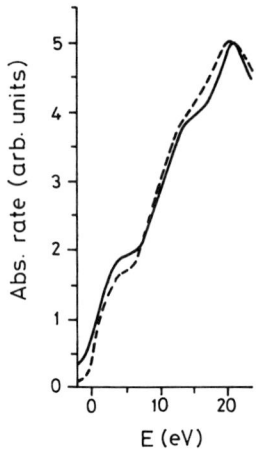

Figure (1) The average of the right and left circularly polarised X-ray absorption coefficient for the K-edge in iron. (dashed line = experiment, full line = theory)

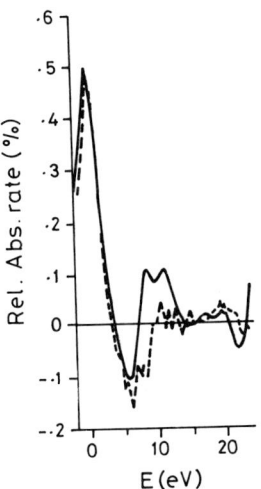

Figure (2) The relative difference between the right and left circularly polarised X-ray absorption coefficients for the K-edge in iron. (dashed line = experiment, full line = theory)

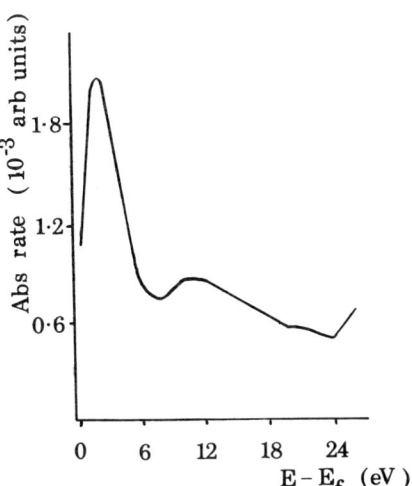

Figure (3) The additional absorption for the K-edge of iron due to the first correction to the electric dipole approximation.

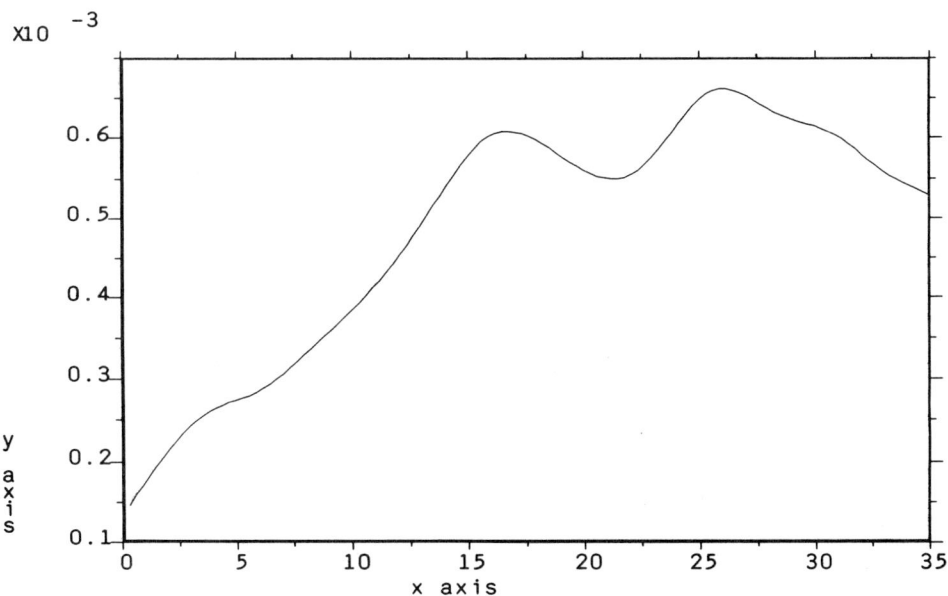

Figure (4a) The average of the right and left circularly polarised X-ray absorption coefficient for the K-edge in cobalt (from theory)

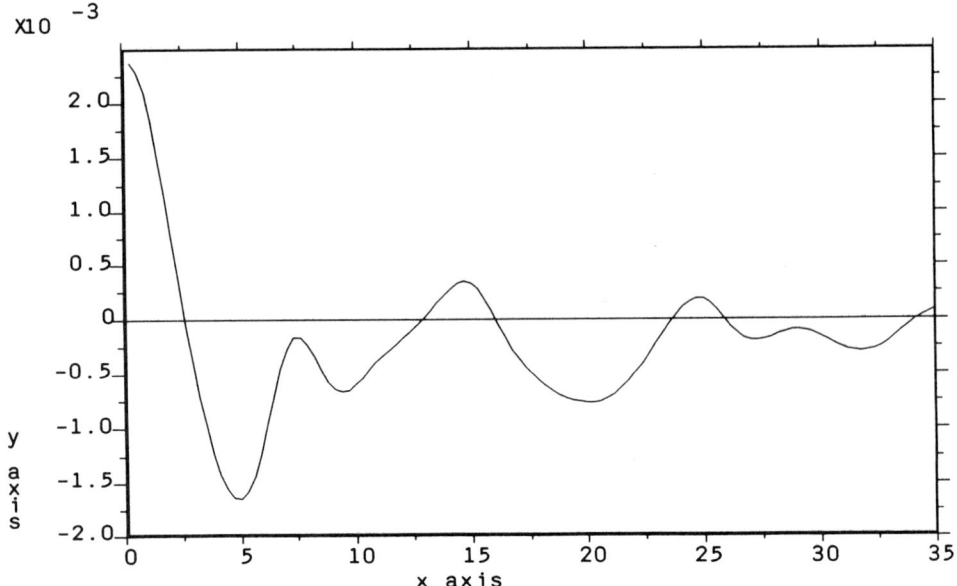

Figure (4b) The relative difference between the right and left circularly polarised X-ray absorption coefficients for the K-edge in cobalt, (from theory)

as instrumental resolution. Similar broadening has been included on all diagrams in this chapter. For iron and for all calculations the shape of the theoretical spectrum is to some extent dependent on the degree of broadening assumed. Figure (2) shows the relative difference in absorption coefficient for right and left circularly polarised incident radiation. The relative difference is defined by

$$\mu^{rel}(\omega) = \frac{\mu^{LCP}(\omega) - \mu^{RCP}(\omega)}{\mu^{LCP}(\omega) + \mu^{RCP}(\omega)} \tag{16}$$

Clearly all the main features of the experiment are well reproduced by the theory. This experiment has now been repeated by Cooper et al [25]. They have shown that the first experiment was subject to an error of a factor of 1.5. This changes quantitatively the agreement between the theory and experiment, however some of the discrepancy can be made up by decreasing the broadening factor, and anyway, the theoretical conclusions are not changed by this. The theory also gives good qualitative agreement with the corrected experiment. Figure (3) [21] shows the contribution to the absorption rate from the first correction to the electric dipole approximation. It is three orders of magnitude smaller than the leading order contribution. Nonetheless it is quite structured and certainly does contain information about the electronic structure of iron, which implies it may be worthwhile to examine it experimentally. This can be done by studying the variation of the accurately determined average intensity with changes in the crystal orientation, hence separating the magnetic and electric dipole contributions. To our knowledge this has not yet been attempted.

One further interesting feature of the relativistic theory is that the absorption depends on the orientation of the equilibrium magnetisation with respect to the crystal axes. If such an orientation dependence could be observed it would be a direct probe for studying magnetocrystalline anisotropy. However in the experiments of Schütz et al polycrystalline samples were used. For a completely valid comparison between experiment and theory averages of the theoretical results over all crystal orientations should

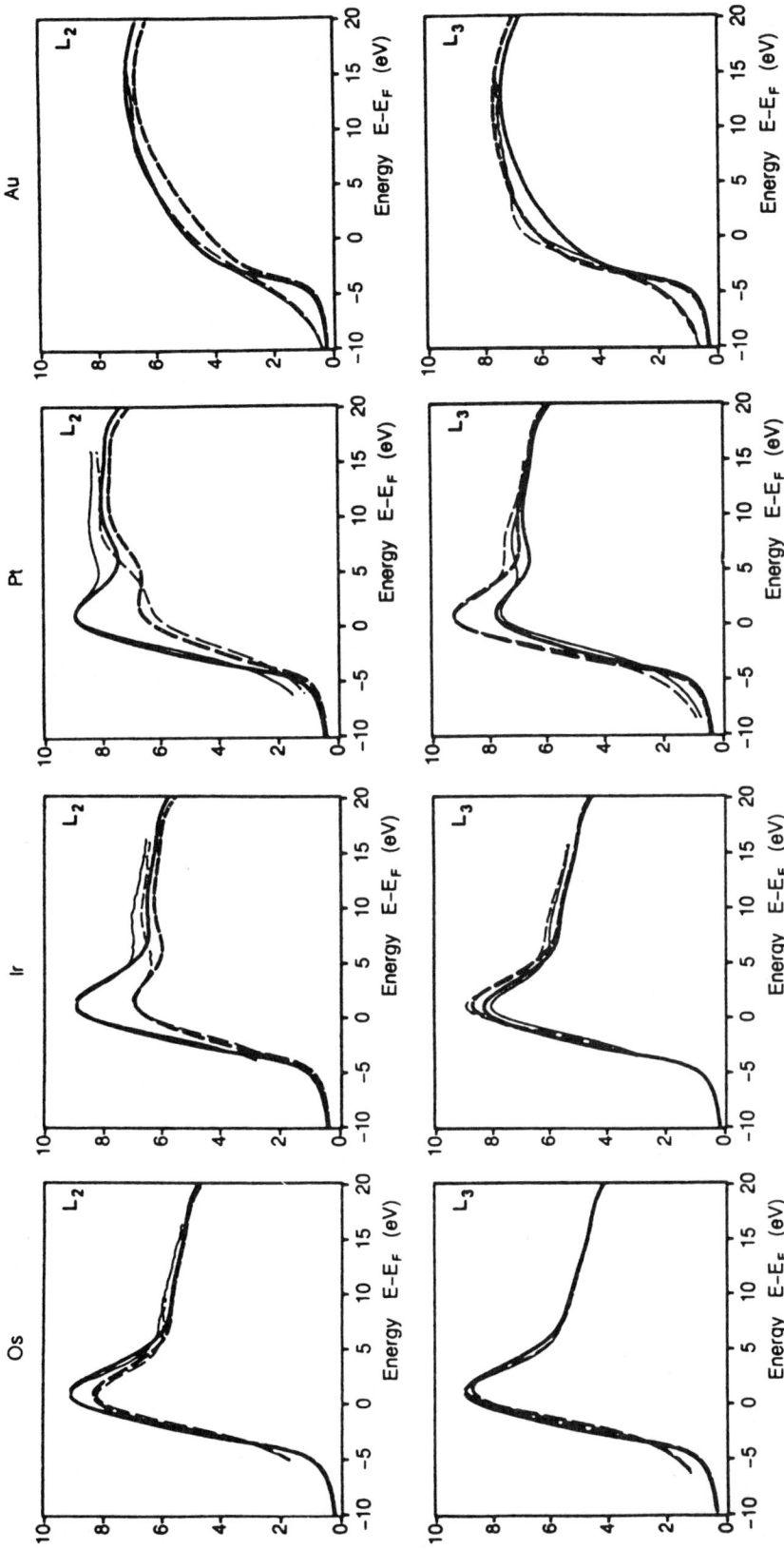

Figure (5a) The absorption coefficients μ (in arbitrary units) for left and right circularly polarised X-rays for the L_2 and L_3 edges of Os, Ir, Pt, and Au in iron. Thick (thin) lines mark theoretical (experimental) results and full (dashed) lines are for left (right) circularly polarised X-rays

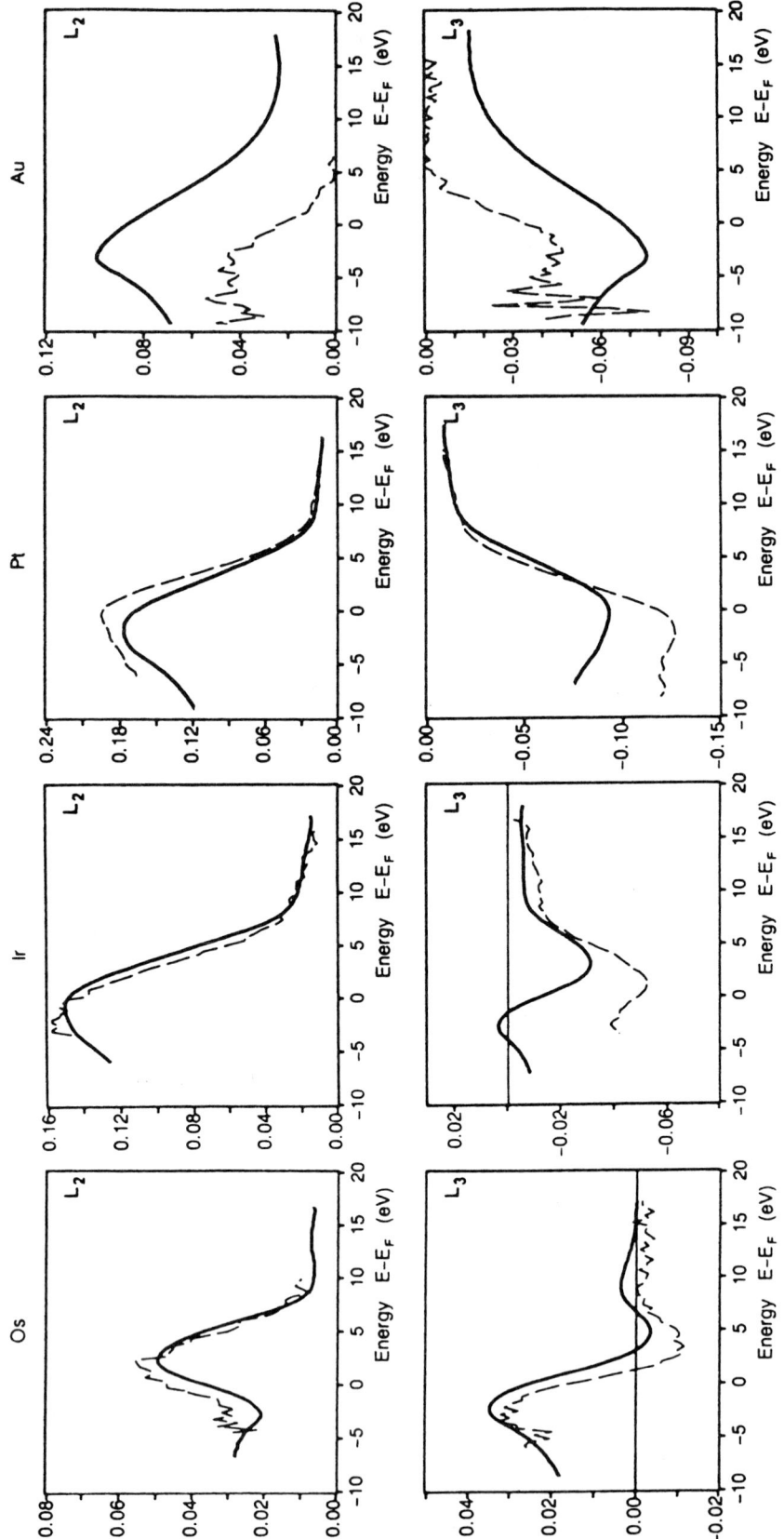

Figure (5b) Relative differences (in percent) of the absorption coefficients between left and right circularly polarised X-rays for the L_2 and L_3 edges of Os, Ir, Pt, and Au in Fe. Thick lines are for the theoretical and thin lines for the experimental results.

be made. This has not been done in the figures because of the huge numerical and computational effort that would require.

Following this success for iron other materials have been investigated, both theoretically and experimentally. Schütz et al [10,25] have performed measurements of the magnetic X-ray dichroism for the magnetic transition metals and rare earths. The spectra for different elements are all rather different, showing that this is a sensitive probe of the magnetisation.

Calculations have been performed for the K-edge absorption in nickel and cobalt. For cobalt the correspondence between experiment and theory is as good as in the case of iron, (see figure (4)). For nickel the agreement is rather poor. Nickel is a case that is notorious for not being easily describable by the local approximation to density functional theory. This is discussed in detail by von Barth [12].

The L_1, L_2 and L_3 edges of the rare earths terbium and gadolinium been investigated and they exhibit a fairly large dichroism. This has not yet been investigated theoretically for two reasons. Firstly the matrices one has to diagonalise become larger for rare earths and a large computational effort would be required. Secondly, the local approximation to density functional becomes suspect when we come to highly correlated shells like the 4f-levels in rare earths. Therefore reliable potentials for calculating the Green function are not available.

Ebert and co-workers [18,26,27,28] have developed the theory further in several directions. Implementing his algorithm for calculating core states accurately [23] has enabled them to examine dichroism from core levels other than those of s-character. The first application of this has been to the absorption of platinum atoms dissolved in ferromagnetic iron. This required the platinum atom t-matrix, which is trivial to calculate once we have a suitable potential. The potential was obtained from a self-consistent local spin density functional calculation on an embedded substitutional platinum atom in an iron matrix. A fully relativistic version of this method is not available, but for most cases the non-relativistic theory will be sufficient to generate a potential. The details of the method used to perform this calculation are described elsewhere [29]. From this they were able to find the scattering path operator of the impurity from equation (6). The absorption rate is then calculated by using equation (2) and summing over all possible core states which are defined by all possible values of μ. The authors calculated the absorption coefficient and dichroism for the L_2 and L_3 edges and their results, together with the experiments are shown in figure (5). The theory reproduces all the main features of the experiment. This is particularly so because the degree of circular polarisation in the beam is difficult to determine accurately and has an error of about 10%. A somewhat surprising feature of the curves is the difference in the magnitude of the dichroism in this case compared to the elements. Here it reaches 20% of the total whereas in elements it is only of order 0.1%.

These authors have then gone on to study dichroism in the L_2 and L_3 edges of a range of 5d impurities dissolved in iron. In all cases they have obtained excellent agreement with the experiments. In most cases they find the dichroism large compared to the 3d elements. Initially this seems rather surprising. Magnetic X-ray dichroism has its origin in the subtle interplay between spin-orbit coupling and spin-polarisation. Therefore one would naively expect it to scale with the local magnetic moment and the strength of the spin-orbit coupling. The heavy 5d elements have a rather small magnetic moment when dissolved in iron from which one would expect the dichroism to be reduced relative to iron. On the other hand these elements are substantially heavier than iron and so one expects spin-orbit coupling to be a considerably larger influence. Therefore one might expect these effects to cancel to first order. However this is not the case and the origin of the difference lies in the initial states. For the K-edge in iron the dichroism is essentially determined by the spin-orbit coupling in the 4p final states, which are the only states into which the initial 1s state can be excited, owing to the selection rules. In this case the spin splitting of the initial 1s state is negligible. For the L_2 and L_3 edges in the impurities in iron the dichroism is determined by the spin-orbit splitting of the initial 2p states as well as the splitting of the conduction band states. Therefore, contrasting these two effects is not very meaningful. A more meaningful comparison would be between the K-edge of iron and the K or L_1 edge of the impurity. According to reference [18] this comparison has been made experimentally as well as theoretically and it has been found that the smaller spin-polarisation and larger spin-orbit coupling do compensate one another.

Figure (5) shows the L_2 and L_3 dichroism for osmium, iridium, platinum and gold. One striking feature of these curves is the difference in sign of the L_2 and L_3 dichroism for Ir, Au and Pt. Schütz et al [9,18,27] have developed a simple model which explains this which is rather similar to the Fano effect [30]. Spin-orbit coupling gives rise to an effective spin-polarisation of an excited core electron which is described by a polarisation parameter P_e. This parameter relates the dichroism curves to the relative spin-polarisation $\triangle n_l/n_l$ in the component and l-projected final state bands via

$$\mu^{rel}(\omega) = P_e . \triangle n_l/n_l \qquad (17)$$

where n_l is the l-projected density of states, and the angular momentum l is fixed by the selection rules. This relationship makes it clear that the dichroism depends on the local magnetic properties of the 5d transition metals. This model artificially splits up the absorption process into two steps, Firstly the excitation and polarisation of the electron and secondly, the settling into a final state. The value of P_e is usually taken from free atom calculations [9,31]. This shows that the L_2 transitions should consist nearly exclusively of $2p_{1/2} \to 5d_{3/2}$ transitions while the L_3 spectra are primarily $2p_{3/2} \to 5d_{5/2}$ transitions. This involves an apparently rather arbitrary splitting of the final state into the κ-sub-bands. The corresponding P_e values have a ratio of about $-2:1$ and therefore explain the opposite sign of $\mu^{rel}(\omega)$ for the L_2 and L_3 spectra. According to this model (equation (16))the dichroism curves divided by P_e give the difference between spin up and spin down polarisations in the κ-projected density of states, where l is determined by the selection rules.

Ebert and Zeller [18] have performed a detailed analysis of their rigorous calculations to see if this simplified model has any validity. The example they chose to consider was platinum dissolved in iron. They found that the model does contain the essential aspects of the magnetic X-ray dichroism spectra for this system because the electron polarisation parameter deduced from atomic calculations remains more or less unchanged in the solid state environment. The transitions for the L_2 and L_3 spectra were dominated by the processes discussed above. They also found that the spin-polarisation derived from experiment cannot be interpreted as the one of κ-projected sub-bands. For the system in question it bore more resemblance to the total d-band. Since the spin-polarisation calculated scalar relativistically agrees with that calculated fully relativistically they concluded that it was probably justified in most cases to use the scalar relativistic calculations to interpret the experimental spectra.

Further interesting physics can be deduced from this work. Proceeding across the periodic table, increasing the atomic number of the 5d impurity, the coupling between

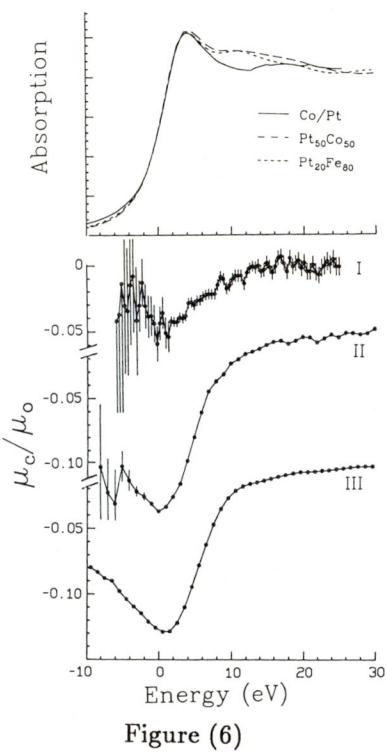

Figure (6)

Top: Pt L_3 absorption profile for Co/Pt layered structure (solid line), $Co_{50}Pt_{50}$ (dashed line) and $Pt_{20}Fe_{80}$ (dotted line).

Bottom: The corresponding dichroism curves for Co/Pt layered structure (curve I), $Co_{50}Pt_{50}$ (curve II) and $Pt_{20}Fe_{80}$ (curve III) normalised to the total saturation magnetisation and $P_c = 1$.

the host and the impurity changes from antiferromagnetic to ferromagnetic. This has been deduced from neutron scattering measurements. This can also be deduced from the dichroism results. Because the P_e parameters for all the 5d impurities are more or less equal, and from the connection with the density of states from equation (16), the change in sign of the dichroism between osmium and iridium also indicates this change of coupling. However this occurs at a different place in the row of the periodic table than indicated by neutron scattering. Neutron scattering data has been interpreted [32] as indicating ferromagnetic coupling of osmium to iron. However, because the X-ray absorption is a much more local probe than neutron scattering it is believed to provide more reliable information on the coupling.

One note of caution needs to be sounded concerning the interpretation of magnetic X-ray dichroism spectra. The information it gives about the properties of impurities will only be unambiguous if there is no clustering of impurity atoms. If the impurity atoms do cluster, there will be alot of inequivalent sites with different moments. In that case the dichroism must be viewed only as a measure of the average spin-polarisation.

One very interesting application of Magnetic X-ray dichroism has been in the investigation of magnetic multilayers. In particular, this has been applied to Co/Pt Multilayers and to $Co_{50}Pt_{50}$ [28], (see figure (6)). In these layer structures the platinum contribution to the magnetism is not known. In Co/Pd multilayer structures the saturation magnetisation is different to that for pure cobalt, suggesting a moment is induced in the Pd. Platinum is isoelectronic with palladium so one would expect the same effect to be observed. However, in Co/Pt the enhancement of the saturation magnetisation is not clearly observed [33]. Schütz and co-workers have measured the L_3 dichroism in Co/Pt multilayers and compared it with the ordered compound $Co_{50}Pt_{50}$ [28]. Their results show unambiguously that the platinum atoms in the layer compounds have a significant spin-polarisation. Their results also indicated that the average moment on Pt in the layer structure is a factor of about 3 smaller than in the ordered compound. This is in reasonable agreement with the corresponding spin moments found from standard electronic structure calculations. These were $0.37\mu_B$ for $Co_{50}Pt_{50}$ and $0.08\mu_B$ as an average value for the multilayer structure 2Co/7Pt multilayer structure [34]. These results demonstrate that the experimental investigations of magnetic X-ray dichroism are a useful step towards an understanding of the magnetism of these layer structures. It is particularly useful as a probe of the penetration of the polarisation into the normally non-magnetic platinum.

One final point to make here is a potential use of dichroism. The largest value of the relative absorption rates found so far is about 20%. It is likely that materials that have not yet been investigated will have larger values. Some Heusler alloys[8], for example, may be good candidates. If this is so we can pass an unpolarised beam through the material and one polarisation will be effectively extinguished before the other. By a suitable choice of material thickness it should be possible to have a beam exiting from the material that is purely circularly polarised. Hence we have a method of producing X-rays that are purely circularly polarised, something that is rather difficult to do using other techniques.

References

[1] Durham P.J, in *The Electronic Structure of Complex Systems*, edited by P. Phariseau and W.M. Temmerman (Plenum, New York) (1984).

[2] Blume M, *J. Appl Phys* **57** 3615, (1985).

[3] Lovesey S.W, *J. Phys C (Solid State Physics)* **20** 5625 (1987)

[4] Thole B.T, van der Laan G, and Sawatzky G.A, *Phys Rev Letts* **55** 2086, (1985).

[5] van der Laan G, Thole B.T, Sawatzky G.A, Goedkoop J.B, Fuggle J.C, Esteva J.M, Karnatak R, Remeika J.P, and Dabkowska H.A, *Phys Rev B* **34**, 6529 (1986).

[6] Goedkoop J.B, Thole B.T, van der Laan G, Sawatzky G.A, de Groot F.M.F, and Fuggle J.C, *Phys Rev B* **37** 2086 (1988)

[7] Siddons D.P, Hart M, Amemiya Y, and Hastings J.B, *Phys Rev Letts* **64**, 1967 (1990)

[8] van Engen P.G, Buschow K.H.J, Jongebreur R, and Erman M, *Appl Phys Letts* **42** 202 (1983)

[9] Schütz G, Wagner W, Wilhelm W, Kienle P, Zeller R, Frahm R, and Materlik C, *Phys Rev Letts* **58** 737 (1987)

[10] Schütz G, Knulle M, Wienke R, Wilhelm W, Wagner W, Kienle P, and Frahm R, *Z. Phys B* **73** 67 (1988)

[11] Schütz G, Wienke R, Wilhelm W, Wagner W, Kienle P, Zeller R, and Frahm R, *Z. Phys B* **75** 495 1989)

[12] von Barth U, in *The Electronic Structure of Complex Systems*, edited by P. Phariseau and W.M. Temmerman (Plenum, New York) (1984).

[13] Strange P, Staunton J.B, and Gyorffy B.L, *J. Phys C* **17** 3355 (1984)

[14] Strange P, Ebert H, Staunton J.B and Gyorffy B.L, *J. Phys: Condensed Matter* **1** 2959 (1989)

[15] Staunton J.B, Strange P, Gyorffy B.L, Matsumoto M, Ebert H, and Archibald N.P *in this volume* (1990)

[16] Gyorffy B.L and Stott M.J, in *Band Structure Spectroscopy of Metals and Alloys* edited by D.J Fabian and L.M Watson (Academic Press) (1972)

[17] G.M. Stocks and H. Winter, in *The Electronic Structure of Complex Systems*, edited by P. Phariseau and W.M. Temmerman (Plenum, New York) (1984).

[18] Ebert H and Zeller R in the press

[19] Rose M.E *Relativistic Electron Theory* (John Wiley New York) (1961)

[20] Ebert H, Strange P and Gyorffy B.L, *J. Appl Phys* **63** 3055 (1988)

[21] Ebert H, Strange P and Gyorffy B.L, *Z. Phys B* **73** 77 (1988)

[22] Ebert H, Strange P and Gyorffy B.L, *J de Physique* **49** Colloque C8, C8-31, (1988)

[23] Ebert H, *J. Phys: Condensed Matter* **1** 9111 (1989)

[24] Collins S.P, Cooper M.J, Brahmia A, Laundy D, and Pitkanen T, *J. Phys: Condensed Matter* **1** 323 (1989)

[25] Schütz G, Frahm R, Wienke R, Wilhelm W, Wagner W and Kienle P, *Rev Sci Instrum* **60** 1661 1989)

[26] Ebert H, Drittler B, Zeller R, and Schütz G, *Solid State Comms* **69** 485 (1989)

[27] Ebert H, Schütz G, and Zeller R, To be published *J. Appl Phys*

[28] Schütz G, Wienke R, Wilhelm W, Zeper W.B, Ebert H, and Sporl K, To be Published *J. Appl Phys*

[29] Zeller R, *J. Phys F* **17** 2123 (1987)

[30] Fano U, *Phys Rev* **178** 131 (1969)

[31] Schütz G, and Wienke R *Hyperfine Interactions* **50** 457 (1989)

[32] Collins M.F and Low G.G, *Proc Phys Soc* **86** 535 (1965)

[33] Zeper W.B, Greidanus F.J.A.M, Carcia P.F, and Fincher C.R, J. Appl Phys **65** 4971 (1989)

Participants

Professor E.J. Baerends, Department of Chemistry, Vrije Universiteit De Boeleaan 1083, 1081 HV Amsterdam, The Netherlands.

Dr. K.S. Baliyan, Department of Applied Mathematics and Theoretical Physics, Queen's University of Belfast, Belfast BT7 1NN, U.K.

Professor R.C. Binning, Chemistry Department, University of Puerto Rico, Rio Piedra, Puerto Rico 00931, U.S.A.

Dr. P. Cortona, Dipartimento di Fisica, Università degli Studi di Genova, I-16146 Genova, Italy.

Dr. D. Devoyhel, Universiteit te Leuven, Departement Scheikunde, Celestijnenlaan 200F, 3030 Leuven, Belgium.

Dr. K.G. Dyall, RTC 230-3, NASA Ames Research Centre, Moffett Field, CA 94035, U.S.A.

Dr. H. Ebert, Siemens AG Research Laboratories, ZFE ME TPH 11, Postfach 3220, D-8526 Erlangen, Germany.

Dr. W. Eissner, Department of Applied Mathematics and Theoretical Physics, Queen's University of Belfast, Belfast BT7 1NN, U.K.

Dr. B.C. Fawcett, Rutherford Appleton Laboratory, Chilton, Oxfordshire, OX11 0QX, U.K.

Dr. B. Ginatempo, Istituto di Fisica Teorica, Università degli Studi, Messina 98100, Italy.

Dr. I.P. Grant, Department of Theoretical Chemistry, University of Oxford, 5, South Parks Road, Oxford OX1 3UB, U.K.

Dr. O. Gropen, Institute of Mathematical and Physical Sciences, University of Tromsø, 9000 Tromsø, Norway.

Professor B.L. Gyorffy, H.H. Wills Physics Laboratory, University of Bristol, Tyndall Avenue, Bristol BS8 1TL, U.K.

Dr. A.C. Hartley, Clarendon Laboratory, University of Oxford, Parks Road, Oxford OX1 3PU, U.K.

Professor Y. Ishikawa, Department of Chemistry, University of Puerto Rico, Rio Piedras, Puerto Rico 00931, U.S.A.

Dr. T.J. Jaakko, Laboratory of Physics, Helsinki University of Technology, 02150 Espoo, Finland.

Mde. F. Keller, C.N.R.S. Laboratoire de Photophysique Moleculaire, Bâtiment 213, Université de Paris-Sud, 91405 Orsay Cedex, France.

Professor J. Kübler, Institut für Festkorperphysik, Fachgebiet Theoretische Physik, Technische Hoch schule Darmstadt, Hochschulstr. 6, D-6100 Darmstadt, Germany.

Dr. S. Lovett, H.H. Wills Physics Laboratory, University of Bristol, Tyndall Avenue, Bristol BS8 1TL, U.K.

Dr. P.E. Mijnarends, ECN, PO Box 1, 1755 ZG Petten, The Netherlands.

Dr. D.L. Moores, Department of Physics and Astronomy, University College, Gower Street, London WC1E 6BT, U.K.

Dr. K. Pierloot, Universiteit te Leuven, Departement Scheikunde, Celestijnenlaan 200F, 3030 Leuven, Belgium.

Professor P. Pyykko, Department of Chemistry, University of Helsinki, Et. Hesperiankatu 4, SF-00100 Helsinki, Finland.

Dr. H.M. Quiney, Department of Theoretical Chemistry, University of Oxford, 5, South Parks Road, Oxford OX1 3UB, U.K.

Dr. A. Renders, Universiteit te Leuven, Departement Scheikunde, Celestijnenlaan 200F, 3030 Leuven, Belgium.

Dr. A. Ron, Department of Physics and Astronomy, University College, Gower Street, London WC1E 6BT, U.K.

Professor P.G.H. Sandars, Clarendon Laboratory, University of Oxford, Parks Road, Oxford OX1 3PU, U.K.

Mr B.J. Sarpal, Department of Physics and Astronomy, University College, Gower Street, London WC1E 6BT, U.K.

Mr. T. Saue, 8A 202 Kringsjå Stud. by, 0864 Oslo 8, Norway.

Professor W.H.E. Schwarz, Department of Theoretical Chemistry, Universität-Gesamthochschule Siegen, POB 101240, D-5900 Siegen, Germany.

Dr. P. Strange, Rutherford Appleton Laboratory, Chilton, Oxfordshire, OX11 0QX, U.K.

Dr E.M. van Wezenbeek, Department of Chemistry, Vrije Universiteit De Boeleaan 1083, 1081 HV Amsterdam, The Netherlands.

Dr. L. Visscher, Laboratory for Chemical Physics, State University of Groningen, Nijenborgh 16, 9747 AG Groningen, The Netherlands.

Dr. O. Visser, Laboratory for Chemical Physics, State University of Groningen, Nijenborgh 16, 9747 AG Groningen, The Netherlands.

Dr. U. Wahlgren, Institute of Theoretical Physics, University of Stockholm, Vanadisvagen 9, 113 46 Stockholm, Sweden.

Dr. S. Wilson, Rutherford Appleton Laboratory, Chilton, Oxfordshire, OX11 0QX, U.K.

Index

Actinoids, 7
Algebraic approximation, 83*ff*, 217*ff*
Anisotropy, finite temperature, 315*ff*
Anisotropy, magnetocrystalline, 295*ff*
Atomic structure code, 17*ff*
Atoms, 15*ff*

Balance, kinetic 33*ff*, 150, 220, 174, 185*ff*, 198*ff*
Basis sets, 24*ff*, 31*ff*, 83*ff*, 149*ff*
Basis sets, even-tempered, 25, 38, 235*ff*
Basis sets, systematic sequences of even-tempered, 236*ff*
Basis sets, universal 236*ff*
Bond angles, effects of relativity on, 144*ff*
Bond lengths, effects of relativity on, 5*ff*
Breit interaction, 25, 40*ff*, 68, 89, 220*ff*
Breit-Pauli approximation, 55*ff*

Cholesky decomposition, 239*ff*
CIV3 code, 46
Compounds, relativistic, 5
Configuration interaction, 48
Contraction of Gaussian basis sets, 197*ff*, 207*ff*
Contraction, relativistic, 1
Coupled cluster equations, 72*ff*

Dacre-Elder method, 188*ff*
de Haas van Alphen effects, 306
Density functional theory, 255*ff*
Density functional theory, spin-polarized 285ff

Destabilization, relativistic, 1
Dipole integrals, 50
Dipole transitions, magnetic, 63*ff*
Dirac-Hartree-Fock calculations, molecular, 167*ff*, 185*ff*
Dirac-Hartree-Fock program, open-shell molecular 185*ff*
Dissociation energies, effects of relativity on, 5*ff*

Donrah-Sommer approximation, 286*ff*
Double perturbation theory, 138*ff*
Dynamic load balancing, 244*ff*

Effective core potential approach, 143*ff*
Electric dipole transitions, 61*ff*
Electron affinities, 8
Electron-atom collisions, 17*ff*
Electron correlation, 218*ff*
Electron impact ionization 125*ff*
Electron propagator, 105*ff*
Even-tempered basis sets, 25, 38, 235*ff*
Even-tempered basis sets, systematic sequences of, 236*ff*
Even-tempered basis sets, universal systematic sequences of 236*ff*

Finite basis sets 24*ff*, 31*ff*, 83*ff*
Finite difference methods, 17*ff*
Finite temperature anisotropy, 315*ff*
Furry picture, 26

Gaussian basis set, 149*ff*, 167*ff*, 187*ff*, 207*ff*
Gaussian basis sets, contraction of, 197*ff*, 207*ff*
Gold, 1
Gold maximum, 1
GRASP code, 17*ff*, 46, 84

H-like systems, 59*ff*
Helium-like systems, 18, 59*ff*
HFR code, 46
Hyperfine fields, 275*ff*

Impurities, magnetic, 300*ff*
Integral matrix, Cholesky decomposition of, 239*ff*
Integral matrix, two-electron, 235*ff*
Ionization cross section, 127

Jellium model, 255*ff*

Kinetic balance, 33*ff*, 150, 220, 174, 185*ff*, 198*ff*

Kohn-Sham-Dirac equations, 263*ff*
LEED, 319*ff*
Liquids, 7
Load balancing, dynamic, 244*ff*
Local density approximation, 259
London moment, 209*ff*

Magnetic dipole transitions, 63*ff*
Magnetic impurities, 300*ff*
Magnetic moments, 275
Magnetic X-ray dichroism, 333*ff*
Magnetocrystalline anisotropy, 295*ff*
Many-body perturbation theory, 24*ff*, 68*ff*, 85*ff*, 149*ff*, 207*ff*, 217*ff*
MBPT, relativistic, 24*ff*, 68*ff*, 85*ff*, 149*ff*, 207*ff*, 217*ff*
MCDF code, 46
MCHF code, 46
Molecular DHF calculations, 167*ff*, 185*ff*
Molecular geometries, effects of relativity on, 139*ff*
MOLECULE code, 178
Molecules, 133*ff*
Multiple scattering method, 263*ff*
Muntz-Szasz theorem, 236

Opacity project, 57
Open-shell molecular Dirac-Hartree-Fock program, 185*ff*
Organometallic compounds, 5
Oscillator strengths, 45*ff*

Parallel processing, 242*ff*
Paramagnetic spin susceptibility, 315*ff*
Parity non-conserving effects, 67*ff*
Periodic table, 1*ff*
Periodic trends, 1*ff*
Perturbation theory, double, 138*ff*
Perturbation theory, many-body, 24*ff*, 68*ff*, 85*ff*, 149*ff*, 207*ff*, 217*ff*
Perturbation theory, relativistic, 135*ff*
Photoemission, spin-polarized, 319*ff*
Propagator, electron, 105*ff*

QED, 26, 46, 82*ff*, 125, 259

Radiative transitions, 60*ff*
Random phase approximation, 69*ff*
Reviews of relativistic effects, 4
R matrix method, 19*ff*, 57
R.K.K.Y. interaction, 300*ff*
Russian bubble method, 76*ff*

Self-energy calculations, 31, 98*ff*
Solid state, 253*ff*
Spectral methods, 219
Spin-polarized density functional theory, 285ff
Spin-polarized multiple scattering theory, 296ff
Spin-polarized photoemission, 319*ff*
Stabilization, relativistic, 1
SSTRUCT code, 57
Superheavy elements, 7
SUPSTRUCTURE code, 46
SWIRLES, 22*ff*, 84*ff*
Systematic sequences of even-tempered basis sets, 236*ff*

Transitions, radiative, 60*ff*
Two-electron integral matrix, 235*ff*

Universal basis sets, 236*ff*
Universal systematic sequences of even-tempered basis sets, 236*ff*

Virial theorem, 163, 164

X-ray dichroism, magnetic 333*ff*